D0139303

Laboratory and Field Manual of Ecology

Richard Brewer
WESTERN MICHIGAN UNIVERSITY

Margaret T. McCann

Saunders College Publishing
Harcourt Brace College Publishers
Fort Worth Philadelphia San Diego
New York Orlando Austin San Antonio
Toronto Montreal London Sydney Tokyo

This book was set in Press Roman by Graphic Arts Composition.
The editors were Michael Brown, Lynne Gery, and Mary Hicks.
The art & design director was Richard L. Moore.
The text design was done by Ivy Fleck Strickler.
The cover design was done by Richard L. Moore.
The artwork was drawn by Tom Mallon.
The production manager was Tom O'Connor.
The book was printed by Malloy Printing.

Cover: LeFevre Bog, Kalamazoo County, Michigan. The central pond is surrounded by a bog mat. Next to the pond the mat consists of swamp loosestrife; away from it, sphagnum and leatherleaf dominate. The distinction between these two zones is not evident in the photograph. Tamaracks form the first tree zone. Beyond is a swamp forest zone dominated by broad-leaved trees, mostly red maple. The next zone of broad-leaved trees, with fuller darker crowns, is on the upland. Photograph by Clayton D. Alway.

Laboratory and Field Manual of Ecology

ISBN 0-03-057879-5

Printed in the United States of America

Library of Congress catalog card number 81-85582.

8 9 00 01 02 - 066 - 17 16 15 14 13

This book is for Ashton and Helen Brewer and Jon and Estelle Thompson.

You learn to study by studying, to play the lute by playing, to dance by dancing, to swim by swimming, and so you learn. . . . Who think to learn in any other way deceive themselves—press forward continually, and never divert yourself by looking back. Begin as a mere apprentice, and by the dint of loving you will become a master in the art.

St. Francis deSales

PREFACE

It is easy for the teaching of ecology to become too abstract. Students need direct involvement with real organisms and ecosystems and with the methods by which they are studied. This manual will provide such firsthand experience.

The book is designed as a set of exercises that demonstrate major ecological generalizations. A large range of standard ecological methods is presented, generally in the context of specific studies based on major concepts of the science. The methods may, of course, be used in situations other than the exercises where they are described. For ease in reference, a special methods index is provided on the back cover. The book also tries to facilitate two other important functions of the ecology lab—practicing the processes by which scientific knowledge is acquired and communicated and building a fund of systematic information about the ecology of local organisms and communities.

With Brewer's *Principles of Ecology* (Saunders 1979) or one of the other fifteen texts to which exercises are keyed, the manual can be used in either an introductory or an advanced course. In an introductory course, the descriptive exercises can be emphasized. Sixteen exercises, listed in Appendix 6, are especially good for courses with minimal mathematical content. Analysis in the more quantitative exercises can depend on the use of means and on graphical comparisons. We have taken special care to subdivide exercises in such a way that the more mathematical sections can be easily omitted.

The text and manual also form a functional unit for a senior capstone course. *Principles of Ecology* focuses on principles, omitting detailed mathematical treatment of most topics and also ecological methodology. Both aspects are best handled in the laboratory where students can grapple with the formulas and procedures firsthand. The generalizations of *Principles of Ecology* combined with the practical experience provided by these exercises make a unit suitable either for the sophomore- or senior-level course.

Flexibility is built into the manual in many ways. There are more exercises than can be performed in a semester, so that flexibility is available in the selection of exercises. Most exercises can be used throughout the year; no fewer than 20 can be readily used in the winter (details are given in Appendix 6). Most exercises are widely applicable geographically; probably all of them can be used in every state in the U.S. When we have recommended specific organisms we have chosen those of wide distribution. For example, the mallard, suggested for the time and energy budget, breeds from near the

Arctic Circle to Turkey and Virginia and winters south to Ethiopia, India, and Cuba. The manual is usable in highly urban situations no less than rural ones; at least 19 exercises (Appendix 6) are suitable for courses in which access to natural habitats is limited.

We have avoided exercises that depend on expensive specialized equipment and highly individual instructional skills. Of course, our exercises can be modified to take advantage of individual situations; persons having access to caves or serpentine or possessing an infrared gas analyzer or the ability to identify every depauperate sprout in a prairie should incorporate such features into their courses.

Balance was another of our goals. Individual, population, and community/ecosystem levels of organization are all represented. About half a dozen of the exercises are devoted to plants and the same number to animals; the rest deal with both or can be used with either. In several exercises, humans or their effects are important concerns.

We have also tried to retain a balance between old and new knowledge. Transition matrices and energy systems analysis are used, but there is also an exercise on ecological modifications of leaves. The fact that tools to study plant morphology were available earlier than tools to study biochemical pathways does not make thick cuticles or sunken stomata less important adaptations to aridity than the C_4 pathway.

Our exercises work. Nothing is foolproof, but these exercises can be generally relied on to give data that can be sensibly interpreted. We have tested all the exercises and most have been used several times in both undergraduate and graduate classes.

Most exercises have the following sections:

Key to Textbooks. Although the arrangement of the manual follows *Principles of Ecology,* exercises are also keyed to appropriate pages in several other contemporary texts. This should be helpful for classes where one of these is used and also as a guide for providing additional reading in a classroom or library reserve collection. The books keyed are as follows:

Brewer, R. 1979. *Principles of Ecology.* Saunders, Philadelphia.

Barbour, M.G., J.H. Burk, and W.D. Pitts. 1980. *Terrestrial Plant Ecology.* Benjamin/Cummings, Menlo Park, California.

Benton, A.H., and W.E. Werner, Jr. 1974. *Field Biology and Ecology.* 3rd ed. McGraw-Hill, New York.

Colinvaux, P. 1973. *Introduction to Ecology.* Wiley, New York.

Emlen, J.M. 1977. *Ecology: An Evolutionary Approach.* Addison-Wesley, Reading, Massachusetts.

Kendeigh, S.C. 1974. *Ecology with Special Reference to Animals and Man.* Prentice-Hall, Englewood Cliffs, New Jersey.

Krebs, C.J. 1978. *Ecology: The Experimental Analysis of Distribution and Abundance.* 2nd ed. Harper & Row, New York.

McNaughton, S.J., and L.L. Wolf. 1979. *General Ecology.* 2nd ed. Holt, Rinehart and Winston, New York.

Odum, E.P. 1971. *Fundamentals of Ecology.* 3rd ed. Saunders, Philadelphia.

Pianka, E.R. 1974. *Evolutionary Ecology.* 2nd ed. Harper & Row, New York.

Richardson, J.L. 1977. *Dimensions of Ecology.* Williams & Wilkins, Baltimore.

Ricklefs, R.E. 1979. *Ecology.* 2nd ed. Chiron Press, Newton, Massachusetts.

———. 1976. *The Economy of Nature: A Textbook in Basic Ecology.* Chiron Press, Portland, Oregon. (short)

Smith, R.L. 1980. *Ecology and Field Biology.* 3rd ed. Harper & Row, New York.
———. 1977. *Elements of Ecology and Field Biology.* Harper & Row, New York. (short)
Whittaker, R.H. 1975. *Communities and Ecosystems.* 2nd ed. Macmillan, New York.

Objectives. These are short statements of the major aims of the exercises.

Key Words. A list of important concepts and terms is provided for each exercise.

Introduction. Laboratory work needs to be put in context so that the student sees how counting insects or recording the activity of ducks is connected with the great questions of ecology. The introductions, combined with the assigned pages in *Principles of Ecology,* begin this process, without the need for lengthy oral introductions by the laboratory instructor.

Procedure. Similarly, the directions are complete enough that classes should need little in the way of oral instructions to carry out the exercises. We have designed special Data and Analysis Sheets that often substitute for many paragraphs of written descriptions of procedures. They are easily followed, so that even students lacking a good mathematical background should be able to carry out necessary computations.

The sophisticated pocket calculators now available at low prices allow most calculations of importance in beginning ecology to be done directly. Since this is pedagogically desirable as well as being a time-saver, we have set up computations assuming that this is the way they will be done. For the same reasons we have omitted tables of logarithms, exponential functions of *e*, and the like. For the occasional student who does not have a calculator, one or two can be provided in the laboratory.

Results and Discussion. These sections have three aims: (1) to draw the students' attention to the important features of the results and their ecological meaning; (2) to provide connections between the subject of the exercise and other aspects of ecology; and (3) to suggest broader environmental and societal applications of the results. The discussion questions can be used as general guidelines for formal written reports, as specific questions for which written responses are assigned, or as guidelines for oral discussion. Depending on which questions the instructor selects and the form in which the answers are assigned, considerable flexibility in aim and level is available for every exercise.

Bibliography. The bibliographies provide specific documentation for the exercises including citations for methods, manuals for identification of organisms, and recent research and review articles.

Teaching Notes. Included are an estimate of the time required for the exercise and miscellaneous comments and suggestions, including sources of materials when these are not available from any good scientific supply house or the local discount store. Mention of a specific company or product is not meant as an endorsement. The estimates of time assume that the students have read the exercise prior to the meeting.

We welcome comments on the manual from instructors and students. We are interested in what you liked and didn't like. What would you like to see added and what omitted in future editions?

Acknowledgments. We are indebted to the students in various ecology laboratories who used these exercises as we developed them. We thank the good students for their insights and the others for helping us sharpen our attempts to make the exercises foolproof. Individual students who made helpful contributions to particular exercises include David Allan (Island Biogeography), Diane Madsen (Life Tables), Kim Chapman (Soils), and Jack Eitniear (Time and Energy Budgets). Philip Brewer, John McCann, Sandra Planisek, and R.J. Planisek made helpful suggestions on Appendix 5 (Computer Usage and Simulation). Warren Abrahamson, Alan Covich, David Cowan, John Davey, Kent Fiala, David Mahan, and Robert Leo Smith made helpful comments on some or all of the exercises. Pat Adams provided some original data on owl pellet production. Susan Mathews gave us some comments on increment boring. Ralph Babcock suggested the quotation from St. Francis de Sales. A part of the exercise on the trophic ecology of humans was prepared in connection with a National Science Foundation sponsored short course on energy systems analysis taught by H.T. and Elizabeth Odum. Western Michigan University has provided support for Brewer over the years. To all of these persons and institutions go our thanks. Special thanks go to our spouses during the completion of the manual.

Richard Brewer
Margaret T. McCann

CONTENTS

Exercise 1
MICROCLIMATE

Key to Textbooks: Brewer 22–40, 120–121, 202–207, 266–269; Barbour et al. 28–31, 300–307, 331–342, 433–436; Benton and Werner 81–82, 113; Emlen 58–60; Kendeigh 116–119; Krebs 65–103, 416–417; McNaughton and Wolf 21–23; Odum 117–121, 137–138; Pianka 43–53, 86–89; Richardson 33, 92–97; Ricklefs 113–117; Ricklefs (short) 69–71; Smith 34–36, 48–61, 67–70, 81–96, 310–315; Smith (short) 147–153, 434–439.

OBJECTIVES

One point of this exercise is to learn some instruments and ways of using them in measuring temperature, humidity, light, and some other climatic factors of importance in ecological studies. Microclimatic differences among topographic sites are examined as clues to some of the causes of differences in habitat distribution of species. A forest and an open area are compared to see some of the major reactions of the forest community on the physical environment. The use of favorable microclimates by organisms is also considered.

INTRODUCTION

A site has a particular climate which, interacting with dispersal, habitat selection, and tolerance limits of individual species, determines what organisms will occur there. Once organisms occupy a site, however, they modify the climate (as they do the other physical features) by their activities. Such effects are termed *reactions*. Many organisms such as the herbs and the soil invertebrates in deciduous forests exist in a climate that owes more to other organisms of the community than to the regional climate.

Furthermore, organisms are not simply passive receptors of temperature or light. They show adaptive responses phenotypically and on the evolutionary level. Such responses for plants are illustrated in Exercise 3. Animals may burrow, turn dark by expanding chromatophores, and do many other physiological or behavioral things that allow them to get along in what would otherwise be an inhospitable climate.

This exercise is a simple introduction to these subjects. Depending on equipment and time, it could be expanded in such ways as using instruments to provide continuous readings instead of maximum-minimum thermometers, determining wind velocity at various heights or distances into the forest, or measuring the effect of the forest canopy on the spectral composition of light.

Key Words: climatic factors, microclimate, reactions, tolerance limits, topographic effects, greenhouse effect, temperature profile, humidity, sun flecks, cold air drainage, heat gain and loss, convection, conduction, radiation, solar heating.

MATERIALS

plastic flagging

12 maximum-minimum thermometers

1 large transparent plastic bag

1 twist tie

2 pieces of white cardboard, each about 60 cm × 60 cm

stiff wire, about 2½ m (8 feet; coat hangers will do)

meter stick(s)

pole (see Forest Reactions on Climate)

5–9 laboratory thermometers (a multiple-channel resistance thermometer can substitute for 4 of these)

metal rod, planting bar, or trowel

masking tape

quadrat pins (surveyor's arrows)

sling psychrometer(s)

psychrometric slide rules or tables

distilled water

anemometer(s)

metric tape or twine, at least 10 m

photometer

2 aluminum beverage cans, 1 painted black and 1 painted white

lead or steel shot

PROCEDURE

The instructor should make sure that the thermometers and other instruments used are accurate (or, at least, matched to give identical readings). A sunny day during the period of autumn when the forest canopy is still complete but nights are cold is best; however, most of the points can be made at other seasons.

Topographic Effects. Place five maximum-minimum thermometers in a line across a small open valley that runs east and west. Put one on the flat ground at the north and the south, one on the north-facing slope, one on the south-facing slope, and one at the bottom of the valley. These should be put out 24–48 hours prior to the laboratory meeting. The simplest way to position them is to lay them on the ground with the long axis north-south. Mark each site with flagging so it can be readily relocated.

During the laboratory period, pick up the thermometers and record on the Data Sheet the current, maximum, and minimum temperatures.

Forest Reactions on Climate. This involves taking two series of readings, one series within a forest to be compared with another series in an adjacent open area (either a bare construction site or low herbaceous vegetation will serve).

1. *Temperature Variation.* If feasible, put out a line of five maximum-minimum thermometers across a wooded valley as was done for an open area under Topographic Effects. Mark the sites with flagging. The two lines should be put out and picked up at about the same time. Record values on the Data Sheet.

2. *Temperature Profile.* Take one or more temperature profiles in the forest and in the open

Figure 1–1. Apparatus for the greenhouse effect (photograph by Philip M. Brewer).

using the following heights: 3 m (or as high as can conveniently be obtained), 1 m, and 2 cm above the surface; on the soil or litter surface; in the soil below the surface at 2 cm, 10 cm, and 20 cm. A multiple-channel resistance thermometer with leads taped to a pole is most suitable for the air temperatures; use similar leads inserted into small holes produced by a metal rod or a planting bar for soil temperatures. However, ordinary laboratory thermometers can be used. In the latter case, note that the thermometers should be read in place or immediately after they are moved away from the height or depth they are supposed to measure. Shade the thermometers so that they are not being affected by direct solar radiation.

3. *Humidity*. Using a sling psychrometer with the wick on the wet bulb soaked with distilled water, record wet bulb and dry bulb temperatures at several localities in the forest and open. Using a psychrometric slide rule or tables, determine relative humidity.

4. *Wind*. Use anemometers to compare wind velocity in the forest and open.

5. *Light*. Using a photometer, compare light intensity inside and outside the forest. Make sure that the orientation of the sensing element is the same in both situations. Express light intensity in the forest as a percent of intensity outside (percent transmission).

The forest reading will give the general level of light intensity in that habitat, but one aspect of the light environment of forests is that direct sunlight coming through holes in the canopy reaches the forest floor to produce ''sunflecks.'' Using metric tapes (or twine and meter sticks) take a series of 10 m line transects in which you measure the length of the line that is shaded and the length included in sunflecks.

The Greenhouse Effect. Near where you put out the maximum-minimum thermometer on the south-facing slope (or elsewhere in an area exposed to the sun), put out two maximum-minimum and two laboratory thermometers situated as follows (Fig. 1–1): Bend a piece of white cardboard down the middle. Lay it on the ground in such a position that the half projecting upward is to the south and thus shades one maximum-minimum and one laboratory thermometer that lie on the other half of the cardboard (which is flat on the ground) and help to weight it down. Make sure that the cardboard is big enough to shade the thermometers throughout the period. Also make sure that the cardboard will not be blown by the wind. If necessary pin the cardboard to the ground with quadrat pins and use masking tape to hold the upper part of the cardboard at the correct angle.

Put out a second piece of cardboard and the other two thermometers prepared exactly the same, but this time enclose the whole thing in a large, transparent plastic bag. Put a frame in the bag to keep it from collapsing (two wire coat hangers bent into hoops connected by two other coat hangers will do). Close the mouth of the bag with a twist tie.

Record initial temperatures immediately. Later in the laboratory period, read and record maximum and current temperatures (the laboratory thermometer is included in case temperatures are above the 120–130° F that is the upper limit of most maximum-minimum thermometers). It is best to make these readings in the warmest part of the day.

Influence of Color on Heat Gain. Use two aluminum beverage cans, one painted black and the other white. These are to mimic black and white lizards or birds. Place a laboratory thermometer in each can and stand them side by side in the sun. (Put a layer of shot in the bottom of the cans for stability.) Record the initial temperature within the cans and take readings every minute for ten minutes or until the temperature in the black can levels off. Also record the air temperature (outside the can; shade the thermometer). Then place the cans in the shade and determine whether they come to the same temperature.

RESULTS AND DISCUSSION

1. Compare the temperatures recorded under Topographic Effects. Where was the highest maximum recorded? The lowest minimum?
2. Account for these and other differences among the temperatures on the basis of differences in conductive, convectional, and radiational heat gain or loss at the sites.
3. What is *cold air drainage*? Was there any evidence of it in the data?
4. What besides air temperature determines the temperature registered by the thermometers as used in this part of the exercise? If you wanted to try to measure only air temperature, how would you locate or house the thermometer differently? Give details.
5. Were temperatures recorded in the inflated plastic bag different from those for the corresponding thermometer in the open? What is the *greenhouse effect*? On what two features of a greenhouse or a plastic bag (or an automobile with the windows rolled up) does it depend?
6. Summarize the differences you detected between the forest and the open in temperature, humidity, wind, and light.
7. What effect does the forest canopy have on temperatures at or near the ground?
8. What are the ways in which the forest canopy influences light within the forest?
9. What are *spring ephemerals*? How are their seasonal cycles related to conditions produced by the dominants of the deciduous forest community? About when would you predict that the leaves of the spring ephemerals should begin to die back?
10. Based on your samples, what percentage of the forest floor was covered by sunflecks? How does the location of a sunfleck produced by a given hole in the canopy change between 10 A.M. and noon? Between noon and 2:00 P.M.? What ecological effects of sunflecks can you suggest?
11. What is humidity? *Relative humidity* is defined as the vapor pressure of water in the air expressed as a percentage of the saturation vapor pressure at that temperature. What does that mean? How does a sling psychrometer measure relative humidity?
12. How does the capacity of air to hold water vapor (or, in other words, saturation vapor pressure) change with temperature? For a given mass of air, how does this affect relative humidity?
13. How did the temperatures of the black and white beverage cans differ in the sun and in the shade?
14. In what habitats would it be advantageous for an animal to be black? White? What are some animals that change back and forth between dark and light colored?
15. Are there factors other than raising and lowering body temperature that might enter into determining the selective advantage of black or white coloration for an animal?
16. Design a house and place it in the environment so that fossil fuel requirements for heating and cooling are minimized. Consider (at least) the shape and construction of the house, orientation, the slope and vegetation of the site, and seasonal differences.
17. Solar greenhouses have become popular both for growing plants and for passive solar heating (with warm air vented into the living quarters of the house to which the greenhouse is

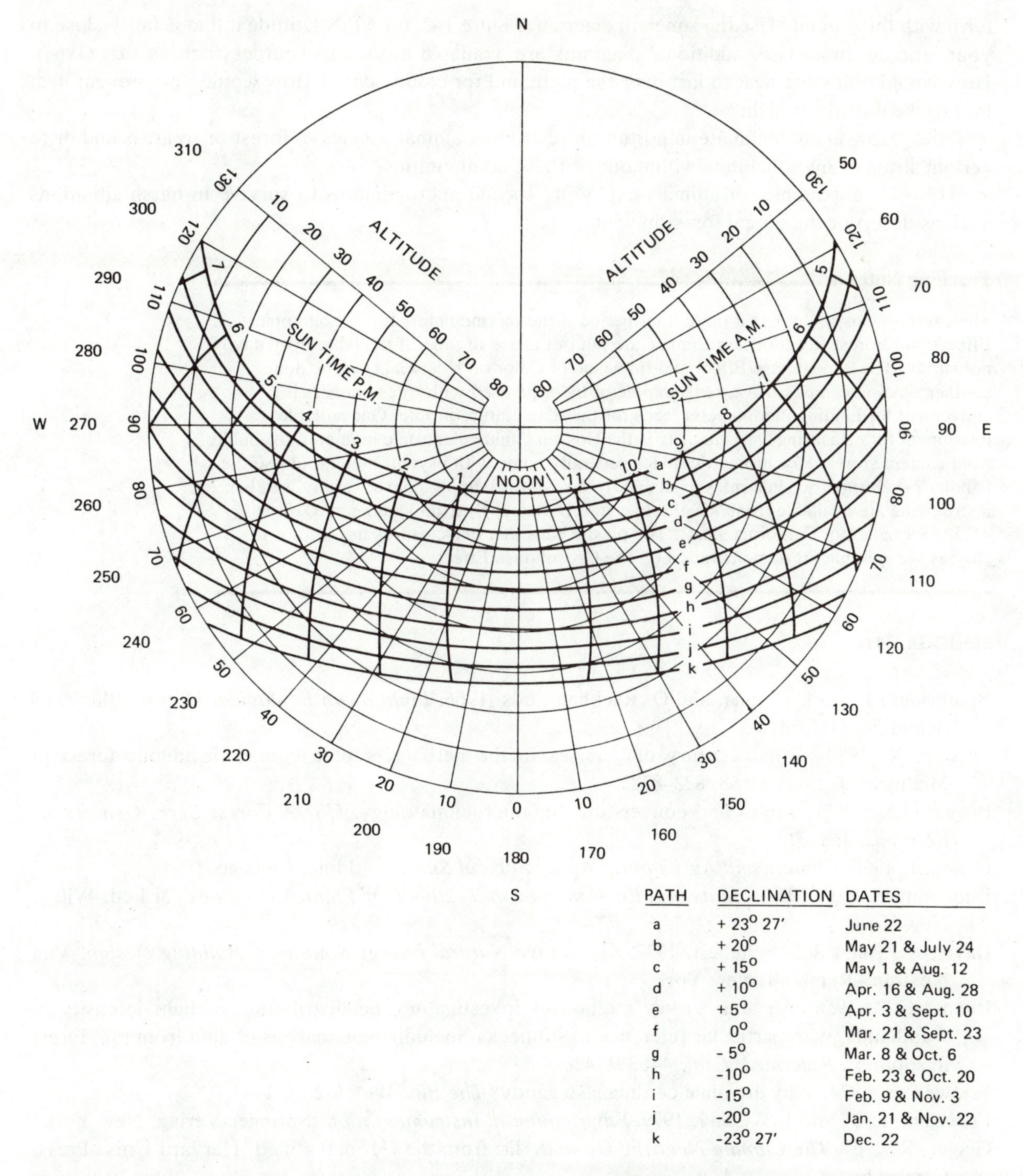

PATH	DECLINATION	DATES
a	+ 23° 27′	June 22
b	+ 20°	May 21 & July 24
c	+ 15°	May 1 & Aug. 12
d	+ 10°	Apr. 16 & Aug. 28
e	+ 5°	Apr. 3 & Sept. 10
f	0°	Mar. 21 & Sept. 23
g	- 5°	Mar. 8 & Oct. 6
h	-10°	Feb. 23 & Oct. 20
i	-15°	Feb. 9 & Nov. 3
j	-20°	Jan. 21 & Nov. 22
k	-23° 27′	Dec. 22

Figure 1–2. Sun path diagram for 40° N latitude. Times are true sun time rather than local time. The direction (or azimuth) of the sun at a given time is shown by the radiating lines. The altitude (in degrees above the horizon) of the sun is shown by the concentric circles. To find the angle of the sun above the horizon for a given time on a given date, find the horizontal curved line that corresponds to the date and follow it to the time. At this point, read up along the partial circles to the "altitude" figures. For example, on March 21 (line f) at solar noon, the sun is about 50° above the horizon. By reading down along the radiating lines, we can find the direction of the sun. At solar noon, the sun is always due south (180°), of course. At 8:00 A.M. on March 21, the sun is at about 112°, or approximately east-southeast. The sun rises on March 21 about 6:00 A.M., almost due east.
(From Brown 1973, p. 16.)

attached). Suppose your aim is to maximize heat gain during the coldest part of the year. What direction should the greenhouse face and what angle should the glass (or fiberglass or plastic) wall form with the ground? Use the sun path diagram, Figure 1–2, for 40° N latitude if that is fairly close to your latitude; otherwise, additional diagrams are available in various sources such as List (1966). How would you store heat to last over the night and for cloudy days? How would you prevent heat loss to the outside at night?

18. How is microclimate important in restricting animal species to forest or to grassland or to certain strata or microhabitats within one of these communities?

19. Give examples of animals exploiting special microclimates to survive in harsh situations such as desert or the boreal forest in winter.

Teaching Notes

The exercise can be done in a three-hour period if the thermometers for Topographic Effects and Forest Reactions on Climate are put out ahead of time. If the laboratory day is not sunny, The Greenhouse Effect and Influence of Color on Heat Gain can be done another time on campus. An extension pole of the type used with tree trimmers is convenient for holding thermometer leads for the temperature profile. One suitable instrument for measuring light intensity is the Gossen Panlux electronic luxmeter, available from camera stores. Psychrometric tables generally come with psychrometers, or use the handier psychrometric slide rules available from scientific supply houses. Many weather instruments are available from WeatherMeasure Division, Systron Donner, P.O. Box 41257, Sacramento, California 95841. The plastic bags that dry-cleaners use to return clothes are suitable for demonstration of The Greenhouse Effect.

BIBLIOGRAPHY

Bainbridge, R., G.C. Evans, and O. Rackham, eds. 1966. *Light as an Ecological Factor*. Blackwell Scientific, Oxford.

Brewer, R. 1980. A half-century of changes in the herb layer of a climax deciduous forest in Michigan. *J. Ecology* 68: 823–832.

Brown, J.M. 1973. Tables and conversions for microclimatology. *USDA Forest Serv. Gen. Tech. Rep.* NC–8: 1–31.

Chang, J. 1968. *Climate and Agriculture: An Ecological Survey*. Aldine, Chicago.

Daubenmire, R.F. 1974. *Plants and Environment; A Textbook of Plant Autecology*. 3rd ed. Wiley, New York.

Davis, A.J., and R.P. Schubert. 1977. *Alternative Natural Energy Sources in Building Design*. Van Nostrand Reinhold, New York.

Evans, G.C. 1956. An area survey method of investigating the distribution of light intensity in woodlands, with particular reference to sunflecks, including an analysis of data from rain forest in southern Nigeria. *J. Ecol.* 44: 391–428.

Fretwell, S. 1974. Why are male cardinals so gaudy? *The Bird Watch* 2(2): 1–4.

Fritschen, L.J., and L.W. Gay. 1979. *Environmental Instrumentation*. Springer-Verlag, New York.

Geiger, R. 1965. *The Climate Near the Ground*. Tr. from the German 4th ed. Harvard Univ. Press, Cambridge.

Lee, R. 1978. *Forest Microclimatology*. Columbia Univ. Press, New York.

List, R.J. 1966. *Smithsonian Meteorological Tables*. 6th rev. ed. Smithsonian Institution, Washington, D.C. (Smithsonian Misc. Collections vol. 114).

Mather, J.R. 1974. *Climatology: Fundamentals and Applications*. McGraw-Hill, New York.

Olgyay, V. 1963. *Design with Climate*. Princeton Univ. Press, Princeton, New Jersey.

Platt, R.B., and J.F. Griffiths. 1964. *Environmental Measurement and Interpretation*. Van Nostrand Reinhold, New York.

Wolfe, J.N., R.T. Wareham, and H.T. Scofield. 1949. Microclimates and macroclimate of Neotoma, a small valley in central Ohio. *Bull. Ohio Biol. Surv.* 8: 1–267.

Microclimate Data Sheet

Topographic Effects on Temperatures

Site	Open			Forest		
	Maximum	*Minimum*	*Current*	*Maximum*	*Minimum*	*Current*
Top (north)						
South-facing slope						
Valley						
North-facing slope						
Top (south)						

Greenhouse Effect

	Temperature		
	Initial	*Maximum*	*Current*
Plastic bag			
None			

Effect of Color

Air Temperature		
Temperature in	Black can	White can
Initial		
1 minute		
2 minutes		
3 minutes		
4 minutes		
5 minutes		
6 minutes		
7 minutes		
8 minutes		
9 minutes		
10 minutes		
In shade		

Forest Reactions

Temperature Profile	Open	Forest
Air 3 m		
1 m		
2 cm		
Surface		
Soil 2 cm		
10 cm		
20 cm		
Humidity		
Dry bulb temp.		
Wet bulb temp.		
Relative humidity		
Wind velocity		
Light intensity		

Sunflecks

Length	
Length shaded	
Total	
% Sunflecks	

Exercise 2

SOILS

Key to Textbooks: Brewer 36–40; Barbour et al. 398–430, 443–447; Colinvaux 43–50; Kendeigh 125–126, 140–149; Krebs 115–118, 433; McNaughton and Wolf 23–30; Odum 128–131, 368–373; Pianka 56–60; Richardson 98–101; Ricklefs 128, 696–699; Ricklefs (short) 36–49; Smith 26–30, 41–48, 68, 167, 280–286, 316–331; Smith (short) 76, 136–147; Whittaker 261–275.

OBJECTIVES

To examine several ecologically important characteristics of soil and to gain insight into some of the processes that form soils by comparing the soil found under two vegetation types.

INTRODUCTION

Soil is dynamic—it changes as time passes. Soil begins in an unstructured state, as glacial deposits or bedrock or sediments or loess. Physical weathering, chemical interactions, and organisms function in forming soil from the parent material; in time soil develops a structure more complex than its parts.

Dead plants and animals collect on the top, and roots and animals work into the soil. When they are alive, organisms break up the soil (making paths where water flows readily) and bring minerals from deep layers to the surface. When dead the organic material decomposes to add these mineral nutrients to the surface (where they are held in water and on the surface of soil particles) for new plant growth. Rain leaches minerals downward; in dry climates minerals may be moved upward by evaporation. These processes form *horizons* in the soil. The top is generally dead organic material. Under it the soil is mostly mineral; the topmost mineral horizon is called the A horizon (or A-1; the terms change over the years). Where it has some organic material deposited in it, it is called topsoil. In most forest regions, some minerals have been leached out of the A horizon into the B horizon. The B horizon is defined by these illuvial deposits. Below that is the C horizon, also called ''not-soil'' as soil is not developed there. In many instances, but not all, the C horizon is the parent material of the horizons above it.

It is not always apparent to the untrained eye where one horizon stops and the next begins, or which horizon is which, and they are not always all present. So, we will not attempt to recognize an official B or C horizon. We will use as a rough A horizon the topmost mineral soil down to where a change is readily seen.

Good study sites are two areas with different vegetation but similar terrain and parent material. A deciduous and a coniferous forest work well (plantations are suitable) but the exercise can also be

Key Words: soil horizons, soil profile, topsoil, forest floor, litter, mor, mull, moder, humus, peds, bulk density, particle size analysis, soil texture, soil moisture, field capacity, organic matter, organic gardening.

used to compare soils of an oak and a maple forest or a forest and a grassland or along a "catena" down a slope to a poorly-drained area. Visit both sites on the same day for better comparisons.

Before going into the field, find each site in the county soil survey and read the descriptions of the soils there.

MATERIALS

USDA–SCS soil survey for each site
100 cm^2 quadrat frame or ring
knife
trowel
2 quart-size Mason jars
meter stick
3-foot soil tube sampler with cutaway
flat tray
water in wash bottle
fingerbowl (culture dish)
10% HCl
LaMotte-Morgan soil pH test kit
cylinder for bulk density sample
flat piece of metal to go under cylinder
2 large jars with lids (to hold 1,000–2,000 cm^3 soil)
2 smaller jars or tennis ball cans with lids (to hold about 200 cm^3)
1–6 bulb planters
6 sturdy plastic bags and twist ties
drying oven
triple-beam balance
2 1,000 ml cylinders, preferably without pouring lips
tall jar to hold 1,000 ml, or another 1,000 ml cylinder
rubber stopper to fit 1,000 ml cylinder (optional)
no. 10 sieve (2 mm openings)
a few sheets of newspaper
wooden rolling pin
500 ml beaker or jar
3 liters sodium pyrophosphate ($Na_4P_2O_7$) solution*
milkshake-type mixer or blender
metal dispersing cup to fit mixer, preferably with baffles

---

*The commercial water softener Calgon was long used as a dispersing agent in soil particle analysis. However, as pointed out by Yaalon (1976), Calgon no longer contains sodium hexametaphosphate [$(NaPO_3)_6$], which made it an effective disperser. Sodium pyrophosphate is equally effective and, now, more easily available. For the 3 l required in this exercise, use 3 l distilled water and 4 g $Na_4P_2O_7$.

distilled water

a few ml of amyl or butyl alcohol

timer or watch with second hand, preferably with an alarm

soil hydrometer

muffle furnace

several crucibles

gloves

tongs

desiccator

analytical balance

PROCEDURE IN THE FIELD

Make notes at each site on the Field Notes Sheet.

The Forest Floor

The *forest floor* is the layer of organic material on top of the predominantly mineral soil. In the U.S. system of soil notation this is the O (or A-O) horizon or layer. The surface is *litter;* its botanical structure is readily seen. Beneath that, the organic matter is increasingly decomposed and amorphous, becoming *humus*.

Three main types of humus have been described: mor (raw humus), mull, and moder (duff mull). *Mor* generally has an accumulation of surface litter. Underneath, among the partly decomposed litter, are strands of fungal mycelia. There is little mixing of the humus with the mineral soil beneath, making a clear boundary between these horizons. Mor is usually acid. In a *mull* forest floor, there is very little litter in late summer before leaf fall—all or nearly all litter is decomposed each year. The humus is crumbly and mixed with the mineral soil beneath, so is not included in the O layer but becomes part of the A (mineral topsoil) horizon. The *moder* humus type is intermediate. There is an accumulation of litter and partly decomposed litter (as in mor) but there is mixing of the humus with the upper mineral soil.

1. Look at the litter on the soil surface. Peel it back layer by layer to the depth where its botanical structure is indistinct. Do you see mycelia (pale yellow to white threads) in it? Roots? What soil fauna do you see? Would water flow readily through the litter? Is there a distinct boundary between this (the O horizon) and the mineral soil (A horizon)?

2. Collect the forest floor—the O horizon—from 100 cm^2. If the layer is thinner than about 2 cm, collect 200 cm^2. Do not include mineral soil in this sample. (In some forests at some times of the year there will be very little or no O horizon; if that is the case, don't try to collect it.) Select a flat site, place the quadrat and cut around it, down through the litter, with a knife. Measure the thickness of the O horizon at four points at the edge of your sample. Record the average on the Field Notes Sheet. Lift the layers of litter and humus into a labeled quart Mason jar and secure the lid to prevent losing moisture.

3. Test the pH of the humus following the instructions with the pH test kit. Because of the frequency of red-green colorblindness in males, it is best for a female to judge the color. Look at the edge of the solution in indirect light.

Mineral Horizons

1. Use a soil tube sampler to bring up soil samples from various depths. The details of the procedure will depend on the sampler, but in general push it in 20–25 cm (the length of the cutaway) then pull it out, leaving a hole. Look at the soil as detailed below, then empty the sampler. Place it back in the hole and push it in an additional 20–25 cm to withdraw soil from that depth. Repeat until you are stopped by the length of the sampler or by hitting bedrock. You may have to try more than one spot to avoid rocks and large roots.

Examine the soil from the various depths. Do you see *horizons,* layers where the soil is different from the soil above or below it? For each horizon note the depth on the Field Notes Sheet and do the following:

a. Toss some soil onto a flat tray. Shake it around to break it up (this won't work well if the soil is very wet). It may break into pieces called *peds.* These are small aggregates formed by natural processes in soils, particularly where there is much clay or organic matter which aids in binding particles together. Peds are a basic unit of soil structure. Soils with a "strong" structure have distinct peds; where peds are not developed (such as in very sandy soils) the soil structure is "weak" and particles are simply packed together in no definite pattern. The size and shape of the peds affect the size and shape of the pore spaces between them, which in turn affect (among other things) root growth and water movement.

Does the soil break into peds? If so, try to classify them as (1) spherical to blocklike; (2) prismlike, having a vertical orientation; or (3) platelike, with a horizontal orientation.

b. What roots or signs of soil animals are there in this layer?

c. What color is the soil? Smear some on your Field Notes Sheet. Soil scientists (pedologists) use standard color notation from Munsell color charts to name soil colors. Here it is sufficient to compare color. Specifically, is the color different from other layers? Where there is iron in the soil, as there is in much of North America, well-drained soil has a yellow or red tone due to oxidized iron while soil that is poorly drained (though perhaps seasonally dry) has a light gray cast from the reduced iron. Organic material typically makes the soil dark brown to black.

d. Feel the soil. Is it grainy? Sticky? The sand (particles between 0.5 and 2.0 mm in diameter) feels gritty or grainy, and the clay (particles less than 0.002 mm) makes soil sticky and colors your hands. Moisten (if necessary) 10–15 cm^3 of soil, knead it, and try to mold it into a ball. Use the following table (based on Thien 1979 and Karol 1960) to make a rough classification of the soil into a texture class. (A more precise method is given later under Soil Particle Analysis.)

Field Key to Soil Texture Classes

Step	Result
1. Soil does not remain in a ball when squeezed	Sand
1′. Soil remains in a ball when squeezed	2
2. Squeeze the ball between your thumb and forefinger, attempting to make a ribbon that you push up over your finger. Soil makes no ribbon	Loamy sand
2′. Soil makes a ribbon (may be very short)	3
3. Ribbon extends less than 1 inch before breaking	4
3′. Ribbon extends an inch or more before breaking	5
4. Add excess water to small amount of soil; soil feels at least slightly gritty	Loam or sandy loam
4′. Soil feels smooth	Silt loam
5. Soil makes a ribbon that breaks when 1 to 2 inches long; cracks if bent into a ring	6
5′. Soil makes a ribbon longer than 2 inches; can be bent into a ring without cracking	7
6. Add excess water to small amount of soil; soil feels at least slightly gritty	Clay loam or sandy clay loam
6′. Soil feels smooth	Silty clay loam or silt
7. Add excess water to small amount of soil; soil feels at least slightly gritty	Clay or sandy clay
7′. Soil feels smooth	Silty clay

e. Put a small handful of soil in a fingerbowl and pour excess 10% HCl on it. Watch and listen to the effervescence using the following table as a rough measure of the amount of calcium carbonate in the soil.

% $CaCO_3$	Audible Effect	Visible Effect
<0.1	none	none
0.5	faint	none
1.0	faint-moderate	barely visible
2.0	distinct, heard away from ear	visible from very close
5.0	easily heard	bubbles up to 3 mm, easily seen
10	easily heard	strong effervescence with bubbles as large as 7 mm

(After Clarke 1957)

$CaCO_3$ is a buffer in the soil, tending to maintain the pH at 7–8.3. As the soil develops it is leached from the upper soil and deposited lower in the profile.

f. As you did with the humus, test the pH of the soil following the instructions with the pH test kit. Have a female judge the color of the edge of the solution (without colored soil particles interfering) in indirect light.

Vegetation has a strong effect on pH. Deciduous leaves, for example, contain bases which buffer and soil becomes acid as they are leached out. The importance of pH is mostly indirect; it affects what minerals are dissolved in the soil water (and thus available to plants) and what soil organisms are present.

This concludes work with the soil sampler. Be sure you have noted your observations about each soil layer, from each site, on the Field Notes Sheet.

2. Now, take a sample for determination of bulk density and soil moisture as described below. The *bulk density* of soil is the mass per unit volume of soil in its natural state, that is, with its soil structure and pore spaces intact. Take care to maintain this natural structure when taking the sample.

Select a site away from paths (where the soil may be compacted). Clear away litter, exposing mineral soil. Carefully push the cylinder into the soil until the top of it is level with the soil surface. (If the soil is so dry and hard that this is difficult, select another site and moisten the area before trying another sample.) Use the trowel to dig around the outside of the cylinder, then slip a flat piece of metal underneath the cylinder to lift it and the soil intact. The volume of the soil collected is then the volume of the cylinder.

Dump the soil from the cylinder into a labeled large jar and secure the lid to retain moisture. No special care is needed to keep the structure intact.

3. Collect about 200 cm^3 (half a tennis ball can is ample) of topsoil for particle size analysis and for measuring the amount of organic matter mixed in with the mineral soil. This does not need to be kept moist.

Soil Fauna

For a comparison of soil fauna use the bulb planter to collect three soil samples for Berlese-Tullgren funnels. Press the bulb planter in about 10 cm including litter and mineral soil. Try to keep the cores intact and moist until you get back to the laboratory. If there are enough bulb planters, just keep the soil in them in separate plastic bags. Otherwise, put the cores in plastic bags and carry gently.

PROCEDURE IN THE LABORATORY

Forest Floor

To determine the moisture content, first weigh the moist material in the jar with the lid on. Then, remove the lid, weigh it, and record. Place the jar with the moist material into a drying oven set at

Field Notes Sheet for Soils Date:________________

	Deciduous forest: location ________ Common trees: ________								Coniferous forest: location ________ Common trees: ________							
Forest floor	Mycelia	Roots	Animals	Water	Boundary	Depth	pH		Mycelia	Roots	Animals	Water	Boundary	Depth	pH	
Mineral soil, surface → **(note depth)** → **Deeper** →	Peds	Roots	Animals	Color	Feel	Texture	$CaCO_3$	pH	Peds	Roots	Animals	Color	Feel	Texture	$CaCO_3$	pH

105° C (100–110°). Weigh after one to two days and then at intervals of several hours to daily until a constant weight is reached. The drying time depends on the amount of moisture in the oven; if one oven is used for all the samples in this exercise it may take a week. Save electricity by air-drying the samples before using the oven and by drying no longer than necessary.

After oven-drying, the sample should be weighed without the addition of atmospheric moisture, so use a desiccator or weigh it quickly, immediately after taking it from the oven.

Remove the soil from the jar and weigh the jar.

Enter data in the Forest Floor Work Sheet and do the calculations.

Forest Floor Work Sheet

	Deciduous	Coniferous
A. Jar + lid + moist soil (weight) B. Jar (weight) C. Lid (weight)	g g g	g g g
Wet weight of sample (A − B − C)	g	g
D. Jar + dry sample (weight)	g	g
Dry weight of sample (D − B)	g	g
Water content of forest floor (Wet weight − dry weight)	g	g
Calculate *percent moisture by weight* $\frac{\text{Water content}}{\text{Dry weight}} \times 100$	%	%
Calculate *moisture fraction by volume* $\frac{\text{Water content}}{\text{Area of sample} \times \text{depth of sample}}$		
Calculate the *dry weight of the forest floor in g per cm²* $\frac{\text{Dry weight (g)}}{\text{Area (cm}^2\text{)}}$ Express *dry weight of the forest floor in kg per ha* by multiplying by 100,000	g/cm² kg/ha	g/cm² kg/ha
Calculate the *bulk density* of the forest floor $\frac{\text{Dry weight}}{\text{Area of sample} \times \text{depth of sample}}$	g/cm³	g/cm³

Mineral Topsoil

Soil Moisture, Bulk Density, and Pore Space. Weigh the soil sample in the closed jar. Remove the lid, weigh it; record on next page. As with the forest floor sample, put the jar and moist soil into a drying oven at 105° C and dry to a constant weight. Weigh quickly to minimize the amount of atmospheric moisture added to the sample, or use a desiccator.

Remove the soil from the jar and weigh the jar.

Make the calculations shown in the following table.

	Deciduous	Coniferous
A. Jar + lid + moist topsoil	g	g
B. Jar	g	g
C. Lid	g	g
Wet weight of topsoil (A − B − C)	g	g
D. Jar + dry topsoil	g	g
Dry weight of topsoil (D − B)	g	g
Water content of topsoil (Wet weight − dry weight)	g	g

Calculate the volume of the cylinder, which is the volume of the sample, in cm^3:

$$\text{volume} = \text{height} \times \left(\frac{\text{diameter}}{2}\right)^2 \times 3.14 = \underline{\hspace{3cm}} \text{cm}^3.$$

The *bulk density* is the dry weight of the soil divided by the volume of the sample:

_______________g/cm^3 deciduous

_______________g/cm^3 coniferous.

The bulk density has two components, the density of the particles themselves and the density of the pore spaces between them. For most mineral soil particles the particle density is 2.65 g/cm^3. The density of organic matter is much lower, about 1.4 g/cm^3. Use these values to calculate the density of the particles in your soil samples. (This calculation must wait until you know the amount of organic matter in your soil. If you are unable to determine this, use about 2% [fraction = 0.02] organic matter for the calculation.)

Particle density = (1.4 × organic matter fraction) + (2.65 × mineral fraction)

= _______________g/cm^3 deciduous

= _______________g/cm^3 coniferous.

The density of the pore spaces in dry soil is essentially zero (they are filled with air), so figure the *percent pore space* as:

$$\left(1 - \frac{\text{bulk density}}{\text{particle density}}\right) \times 100 = \underline{\hspace{3cm}}\% \text{ deciduous}$$

$$= \underline{\hspace{3cm}}\% \text{ coniferous.}$$

In field conditions some of the pore space is occupied by water and some by air. The percentage of each is calculated from the moisture content of the soil. First figure the volume of the pore space:

volume of soil sample × (percent pore space/100) = _______________cm^3 deciduous

= _______________cm^3 coniferous.

The volume of the water = weight of water in the soil* = ______________cm³ deciduous

= ______________cm³ coniferous.

Thus, the *percent of pore space occupied by water* at the time the soil sample was collected is

$$\frac{\text{volume of water}}{\text{volume of pore space}} \times 100 = \_\_\_\_\_\_\_\_\_\_\% \text{ deciduous}$$

$$= \_\_\_\_\_\_\_\_\_\_\% \text{ coniferous.}$$

The moisture content is often expressed in other ways, including the following:

Percent moisture by weight =

$$\frac{\text{weight of water in soil}}{\text{dry weight of soil}} \times 100 = \_\_\_\_\_\_\_\_\_\_\% \text{ deciduous}$$

$$= \_\_\_\_\_\_\_\_\_\_\% \text{ coniferous;}$$

Moisture fraction by volume =

$$\frac{\text{volume of water}}{\text{volume of soil}} = \_\_\_\_\_\_\_\_\_\_ \text{ deciduous}$$

$$= \_\_\_\_\_\_\_\_\_\_ \text{ coniferous.}$$

When expressed on a volume basis the moisture content is a guide to the amount of water applied to the soil surface to make that moisture. For example, it takes 30 cm of rain or irrigation to bring 1 m of dry soil to 0.3 moisture fraction by volume.

If the sampling was done within a few days of a thorough soaking of the soil, the moisture content is the *field capacity* of the soil. If the area was saturated with water so recently that the *gravitational water* has not drained away, the moisture content of the soil is higher than field capacity. Because it moves away so quickly, gravitational water is seldom directly important to plants. At the other extreme, dry soils contain some water held so tightly (both hygroscopically and in small capillary spaces) that plants cannot get it and they may wilt, even at 10–20% soil moisture by weight.

Particle Size Analysis. The distribution of the various sizes of mineral particles is one of the most important characteristics of soil, particularly if there is little organic matter. Small particles have a larger surface to volume ratio than large particles, and many soil processes occur at the particle surfaces. For example, mineral cations are held there. The particle size distribution also affects the water-holding capacity of soil; a fine-textured soil will hold more water than coarse soil.

One can get a rough idea of the sizes of particles by feeling the soil as done in the field part of this exercise, but the various settling rates of different-sized particles allow us to assess this quantitatively. For a quick demonstration of the principle, fill a jar about one-eighth full of soil, add water until nearly full, cap it, and shake it. Then let it settle. The larger particles settle quickly to the bottom, then progressively smaller particles settle. Fine particles stay in suspension very long; leave the jar for a few days and look at it now and then. Stokes's Law formally relates the diameter of a sphere to its velocity of fall, and Bouyoucos (1927) pioneered the use of a hydrometer to measure the specific gravity of the soil suspension above the settled particles to figure the particle size distribution.

The basic procedure is to remove coarse particles (which are not measured well this way), then place the soil in a cylinder to settle. Hydrometer readings are taken from the beginning of the settling to at least two hours afterward. The Bouyoucos scale on the hydrometer was designed to read

*Because 1 cm³ of water weighs 1 g.

directly the grams of soil remaining in a suspension with water, but some refinements of the original procedure have made it such that, now, correction factors* are added to the direct reading.

With one hydrometer and two (or, preferably, three) 1,000 ml cylinders you can do both the deciduous and coniferous forest soil samples, one 15–20 minutes behind the other. Keep both soil suspensions and the $Na_4P_2O_7$ solution at the same temperature during the two or more hours of hydrometer readings. Locate them away from radiators, drafts, and sunshine to prevent convection currents.

1. Air dry about 100 g soil, enough to have 70 g left after sieving.
2. Spread the soil out on a newspaper and crush the aggregates with a wooden rolling pin. The object is not to break stones or grind up particles, just to crumble clumps.
3. Pour the soil sample onto a no. 10 sieve (2 mm openings). Catch and weigh the fine earth that shakes through; this portion will be used in the rest of the analysis. Also weigh the coarse particles:

________________ g deciduous, ________________ g coniferous; fine earth ________________ g

deciduous, ________________ g coniferous.

4. Weigh out 50 g of the fine earth and place it in a beaker or the metal dispersing cup with about 500 ml of the $Na_4P_2O_7$ solution. Stir, then let soak for 15–20 hours. This time can be shortened to an hour for soils with little clay content (those you cannot make into a ball). Note that clay particles sticking together into effectively larger particles are a source of error with this method and the $Na_4P_2O_7$ soaking and stirring are to lessen that.

Recommendations as to how much soil to use range from 25 g to 100 g for soils with more than 90% sand. For most forest soils 50 g is convenient. If you want to use a different amount, make the suspension dilute enough so that the particles do not interfere with each other as they settle, and remember to use your sample weight rather than the 50 g shown below in the calculations.

5. Weigh 10–20 g of the remaining fine earth, oven-dry it at 105° C, remove and quickly reweigh. Oven-dry weight/air-dry weight × 50 g = the oven-dry weight of the 50 g used in the hydrometer analysis. Air-dry rather than oven-dry soil should be used in the analysis because oven-drying may cause some physical changes in the soil, but the equivalent oven-dry weight should be used in the calculations.
6. Put about 1,000 ml $Na_4P_2O_7$ solution in a tall jar or spare 1,000 ml cylinder to use as a calibration. Have this at room temperature, ready to use after the soil has soaked 15–20 hours.
7. Put the metal cup with the soil and $Na_4P_2O_7$ solution on the mixer. Thoroughly disperse the mixture. The standard soil mixer should go for two minutes at 16,000 rpm; with another stirrer or blender mix for 2–15 minutes. Use the longer time for fine-textured soils, as they tend to clump, and less time for sandy soils, as the sand particles might be ground down.
8. Pour the mixture into a 1,000 ml cylinder. Rinse the dispersing cup with $Na_4P_2O_7$ solution and add to the cylinder. Add $Na_4P_2O_7$ solution to the 1,000 ml mark.
9. Place a rubber stopper in the cylinder (or use a large dry palm), hold firmly, then invert completely about 60 times. Do not shake it. Place the cylinder on a stable flat surface by the $Na_4P_2O_7$ solution cylinder (out of drafts, etc.) and begin timing immediately. Remove the stopper.
10. Add 2–3 drops of amyl or butyl alcohol to reduce foaming.
11. Gently lower the hydrometer into the solution. Read the hydrometer (to the top of the meniscus) at 40 seconds and again at 120 minutes. Record below. Between readings, remove the hydrometer, rinse it with distilled water, and put it in the $Na_4P_2O_7$ solution. Read it and record in the following table. This is a correction factor for both the room temperature and the specific gravity of the $Na_4P_2O_7$ in the soil suspension. You may leave the hydrometer there but remove it in time to wipe dry the stem (for a better reading) and lower it into the soil suspension 20–30 seconds before each reading time. You may wish to set an alarm.

*The corrections used here are sufficient for general ecological work. More detailed corrections are given by the American Society for Testing and Materials in their annual book of standards (see Bibliography).

Hydrometer Readings

Time	$Na_4P_2O_7$ Solution	Deciduous Forest Soil Sample		Coniferous Forest Soil Sample	
		Uncorrected	Corrected	Uncorrected	Corrected
40 seconds					
120 minutes					

12. Where the hydrometer reading in the $Na_4P_2O_7$ solution is >0 subtract it from the soil suspension hydrometer reading to get a corrected hydrometer reading; if it is <0, add it. For example, if the soil suspension hydrometer reads 15 and the $Na_4P_2O_7$ solution reads 1, the corrected soil suspension hydrometer reading is (15−1)=14.

13. Soil texture (particle size distribution) is usually expressed as the percentage of sand, silt, and clay. By definition, sands are particles with diameters between 2.0 mm and 0.05 mm, silts are between 0.05 mm and 0.002 mm, and clays are smaller than 0.002 mm in diameter. Readings at 40 seconds show the amount of soil particles 0.05 mm in diameter in suspension and readings at 120 minutes show the amount of soil particles 0.002 mm in diameter. Accordingly, make the following calculations:

	Deciduous	Coniferous
$\% \text{ sand} = \left(1 - \frac{\text{corrected 40-second reading}}{\text{oven-dry weight of the 50 g}}\right) \times 100$	%	%
$\% \text{ clay} = \left(\frac{\text{corrected 120-minute reading}}{\text{oven-dry weight of the 50 g}}\right) \times 100$	%	%
% silt = 100 − % clay − % sand	%	%
Total of fine earth	100%	100%

14. Also calculate the percentage of particles over 2 mm (see step 3) as follows:

$\% \text{ coarse} = 100 \times \frac{\text{air-dry weight of particles over 2 mm}}{\text{air-dry weight of total sample, over 2 mm plus fine earth}}$	%	%

Note that particles over 2 mm are not included in the percentages worked out in step 13; consequently, % coarse + % sand + % silt + % clay > 100%.

Determination of the Amount of Organic Matter. There are various approaches to the measurement of organic matter in soil, some involving titration or measuring the CO_2 evolved during a chemical process and others measuring the weight loss after somehow getting rid of organic material. The various methods do not measure quite the same entity; they may be measuring oxidizable carbon or organic carbon and then assuming that multiplying that value times 1.724 gives the amount of organic matter (though surely the conversion factor varies), or they may be measuring organic material plus some carbon from carbonates and maybe some carbon of charcoal and maybe some structural water of clay minerals. For these reasons it is necessary to state the method along with the results.

Here we present two versions of one of the simpler methods. For those who wish to determine organic matter the way a soil laboratory does, see Hesse (1971) or Black et al. (1965).

1. Place about 10 g air-dry and sieved fine earth in each of several crucibles. Dry them in a drying oven at 105° C for about a day or until they are oven-dry. Remove from the oven (use asbestos gloves or tongs) and quickly weigh on an analytical balance. (As usual, try to keep atmospheric moisture from getting into the samples.)

2. Place the crucibles in a muffle furnace. Heat to 375° C and maintain for 16 hours, *or* to 850° C and maintain for ½ hour.

3. Using gloves and tongs, carefully remove the crucibles and cool them in a desiccator. Weigh them again. The loss of weight is expressed as a percent and is called the *loss on ignition* (LOI).

$$\% \text{ LOI} = \frac{\text{weight before} - \text{weight after furnace}}{\text{oven-dry weight of soil}} \times 100 = \underline{\qquad\qquad}\% \text{ deciduous}$$

$$= \underline{\qquad\qquad}\% \text{ coniferous.}$$

At 375° C, the water in clay minerals is not driven off. At higher temperatures, some is. Hesse (1971) suggests correcting for that by assuming that clay particles lose 5% of their weight in the furnace:

$$\text{corrected } \% \text{ LOI} = \% \text{ LOI} - (0.05 \times \% \text{ clay}).$$

If the deciduous and coniferous forests soils have about the same amount of clay you can make valid comparisons without this correction.

Percent loss on ignition can be roughly converted to percent organic carbon by these relationships (from Ball 1964):

$$\text{at } 375° \text{ C}, \% \text{ organic carbon} = (0.458 \times \% \text{ LOI}) - 0.4\%$$

$$\text{or, at } 850° \text{ C}, \% \text{ organic carbon} = (0.476 \times \% \text{ LOI}) - 1.87\%.$$

Then, to convert to organic *matter*, multiply the organic carbon by 1.724.

Soil Fauna. Place each soil core in a Berlese-Tullgren funnel. After two to seven days count each kind of organism that has been extracted. See Exercise 14 for complete instructions.

RESULTS AND DISCUSSION

1. Which humus type was there in the deciduous forest? The coniferous forest? Which was more acid? In which were there more earthworms?
2. Forest floor weights given in Armson (1977) range from 9,000 to 175,000 kg/ha. How do your data compare? Which humus type has more forest floor? Does this change seasonally? How?
3. You assessed the standing crop of litter. How, if at all, is that related to the rate of litter production? How are the rate and the standing crop important to plants?
4. Which site had more organic matter in the A horizon? Which had more forest floor? Is there necessarily a correlation?
5. Compare the soil fauna (extracted by the Berlese-Tullgren funnels) from the two sites. What kinds were there? Are some of these more characteristic of coniferous forest? Of deciduous forest? Are there many species in common? Was there about the same amount at each site? Use the approximate masses of soil animals given in Exercise 14 to calculate the mass of soil invertebrates at each site in kg/ha.
6. White (1979) states that microorganism biomass generally accounts for some 3% of the organic matter in woodland soil. Using that figure, what is the microbial biomass of the study sites, in kg/ha to a depth of 10 cm? Compare this to the mass of soil invertebrates from the Berlese-Tullgren funnels (also collected to 10 cm).
7. Where was soil structure strongest, near the surface or deeper in the soil? How did the ped shape change with depth?
8. What biotic and abiotic processes form peds?
9. Compare the bulk density of the forest floors and the mineral soils. Would the bulk density of the soil of a path be higher or lower than off the path? How does this affect the flow of water into the soil? How does it affect roots?
10. Which was more moist, the forest floor or the mineral topsoil? How would this change during several days after a heavy rain?

11. It is conventional to say that at field capacity 50% of the pore space in a loam is filled with water and at the wilting point about 25% of the pore space is water-filled . Do your figures fall in that range? If they are less than 50%, how many cm of rain would be needed to bring the top 20 cm of soil to field capacity?

12. Why do fine-textured soils have a higher field capacity than coarse soils? Organic matter can overcome this difference; a sandy soil with much organic matter can hold water much like a loam. What accounts for that?

13. Based on the soil color and other observations, is either site poorly drained? What does the soil survey say about the drainage of the sites? How does poor drainage affect plant growth?

14. How does air in soil pores differ in composition from atmospheric air?

15. Where was there the most $CaCO_3$? Was this correlated with pH?

16. What pH changes were there down the soil profile?

17. Study the soil triangle, Figure 2–1. Locate each of your soil sample compositions on it (from the particle size analysis). What texture class is your deciduous forest soil? Your coniferous forest soil? Were your field determinations accurate? Based on your field observations, did the lower horizons have more sand or clay? What do you think they would be called?

18. What is a loam? What is good tilth?

19. What is the *cation exchange capacity* (CEC) of soils? Very generally, how is it related to the sizes of particles? To organic matter content?

20. Clearly the two sample sites had different vegetation and their climate was the same. Did they have the same parent material? Had the soils had about the same amount of time to form? Were the sites similar in relief? Which of these five soil-forming factors accounts for the differences in the soils at the two sites? Consider all of the soil layers. (Consult the soil survey for some information here.)

21. Find the two sites on the soil survey. According to it, what kind of soil is there at each site? What are the characteristics of the soil, as given in the survey? Where you can compare information, is there agreement with your assessment of each soil?

22. If you have a modern soil survey, use it to rate each site as to favorability for septic tank sewage systems, for paths and trails, for buildings with basements, and as a source of gravel.

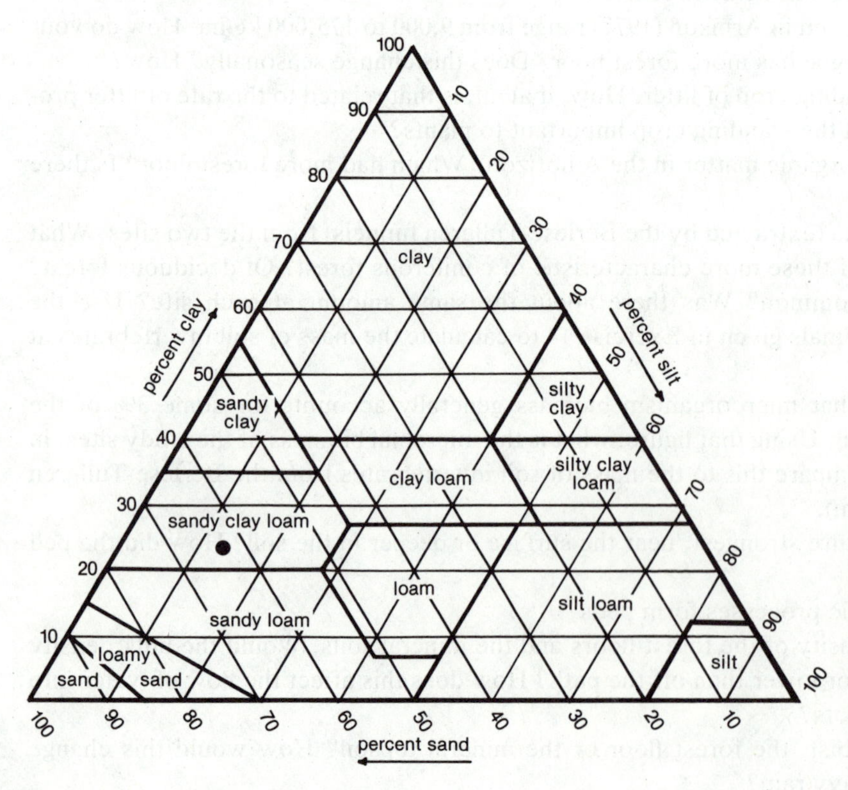

Figure 2–1. The soil triangle classifies soils into textural groups based on the percentages of sand, silt, and clay. For example, a soil with 63% sand, 14% silt, and 23% clay (indicated by the dot) is called a sandy clay loam.

23. How does repeated plowing and disking affect farm soil structure? What is a plow pan? What is minimum-tillage farming?

24. Organic gardeners advise adding organic matter to soil. How does this help plant growth? Include effects on bulk density, pH, CEC, and soil organisms.

Teaching Notes

This exercise cannot be completed in one period. On one day, make field observations and collect samples. Some samples must be dried, which takes at least a day, then these samples weighed and so on. The particle size analysis takes air-dry soil which then soaks for 15-20 hours before the analysis, which includes a reading after two hours.

Complete counts of the Berlese-Tullgren funnel samples could take most of a period. Simply checking to detect gross differences between the two sites can be done in about half an hour.

If the weather has been very dry, it might be difficult to core into the soil. If this is the case, it is best to water the sample areas thoroughly a day or so before the class visit (mark the sites). About a gallon of tap water for each bulk density sample site and another gallon for the soil coring should do it.

Suitable tube-style soil samplers (3 feet long with a 10 inch cutaway opening) are available from Forestry Suppliers, Inc., 205 West Rankin St., P.O. Box 8397, Jackson, Mississippi 39204. Several biological supply houses carry LaMotte-Morgan soil pH test kits, sieves, and soil hydrometers. Hydrometers may be identified as Bouyoucos scale or ASTM 152H or ASTM D422. Special soil test glass cylinders (KIMAX brand), available from most scientific equipment companies, are handy in that they have no pouring spout and have only the 1,000 ml calibration mark.

The metal cylinders for bulk density samples should be short and wide and contain 1,000–2,000 cm^3. A shortening can with both ends cut off and the rim of one end trimmed for sharpness is satisfactory; use it with a larger can lid to slip under it. A very nice sampler can be made from furnace ducting, with a square of sheet metal to slip beneath it. A 100 cm^2 quadrat frame (or approximately that size) can similarly be made from a can or ducting.

Tennis ball or baking powder cans with the ends and one rim cut off can be substituted for bulb planters.

Soil surveys are available from the local Soil Conservation Service office or may be in the library.

BIBLIOGRAPHY*

American Society for Testing and Materials. 1980. *Annual Book of ASTM Standards. Part 19. Natural Building Stones; Soil and Rock*. ASTM D 422–63. Am. Soc. Testing and Materials, Philadelphia.

Armson, K.A. 1977. *Forest Soils: Properties and Processes*. Univ. Toronto Press, Toronto.

Ball, D.F. 1964. Loss-on-ignition as an estimate of organic matter and organic carbon in non-calcareous soils. *J. Soil Science* 15: 84–92.

Black, C.A., D.D. Evans, J.L. White, L.E. Ensmiger, and F.E. Clark, eds. 1965. *Methods of Soil Analysis, Part 2*. Agronomy Monogr. 9. Amer. Soc. Agronomy, Madison, Wisconsin.

Bouyoucos, G.J. 1927. The hydrometer as a new method for the mechanical analysis of soils. *Soil Science* 23: 343–354.

______. 1962. Hydrometer method improved for making particle size analysis of soils. *Agronomy J.* 54: 464–465.

Clarke, G.R. 1957. *The Study of Soil in the Field*. 4th ed. Oxford Univ. Press, Cambridge.

Foth, H.D. 1978. *Fundamentals of Soil Science*. 6th ed. Wiley, New York.

Hesse, P.R. 1971. *A Textbook of Soil Chemical Analysis*. Chemical Publ., New York.

*References for soil fauna are listed in Exercise 14.

Jenny, H. 1941. *Factors of Soil Formation: A System of Quantitative Pedology*. McGraw-Hill, New York.

———. 1980. *The Soil Resource: Origin and Behavior*. Springer-Verlag, New York.

Karol, R.H. 1960. *Soils and Soil Engineering*. Prentice-Hall, Englewood Cliffs, New Jersey.

McFee, W.W., and E.L. Stone. 1965. Quantity, distribution and variability of organic matter and nutrients in a forest podzol in New York. *Soil Sci. Amer. Proc*. 29: 432–436.

Moir, W.H., and H. Grier. 1969. Weight and nitrogen, phosphorus, potassium, and calcium content of forest floor humus of lodgepole pine stands in Colorado. *Soil Sci. Amer. Proc*. 33: 137–140.

Nikol'skii, N.N. 1963. *Practical Soil Science*. Tr. from the Russian by R. Nadel, Israel Prog. for Sci. Trans., Jerusalem. Office of Technological Services, U.S. Dept. Commerce, Washington, D.C.

Robinson, W.O. 1927. The determination of organic matter in soils by means of hydrogen peroxide. *J. Agric. Research* 34: 339–356.

Romell, L.G., and S.O. Heiberg. 1931. Types of humus layer in the forests of northeastern United States. *Ecology* 12: 567–608.

Soil Conservation Service, U.S. Dept. Agriculture. 1972. *Soil Survey Laboratory Methods and Procedures for Collecting Soil Samples, Soil Survey Investigations Report No. 1*. U.S. Dept. Agriculture, Washington, D.C.

Soil Survey Staff. 1951. *Soil Survey Manual. U.S. Dept. Agriculture Handbook No. 18*. Gov't. Printing Office, Washington, D.C.

———. 1962. *Supplement to USDA Handbook No. 18*. Gov't. Printing Office, Washington, D.C.

Thien, S.J. 1979. A flow diagram for teaching texture-by-feel analysis. *J. Agronom. Ed*. 8: 54–55.

White, R.E. 1979. *Introduction to the Principles and Practice of Soil Science*. Wiley, New York.

Wollum, A.G. II. 1973. Characterization of the forest floor in stands along a moisture gradient in southern New Mexico. *Soil Sci. Amer. Proc*. 37: 637–644.

Yaalon, D.H. 1976. "Calgon" no longer suitable. *Soil Sci. Am. J*. 40: 333.

Exercise 3

ECOLOGICAL MODIFICATIONS OF LEAVES

Key to Textbooks: Brewer 25–30; Barbour et al. 307–326, 450–478, 538–543; Colinvaux 274–275, 280–295; Kendeigh 363; Krebs 87–89; McNaughton and Wolf 528–529; Odum 302–305, 393–394; Pianka 53–56, 89–90; Richardson 149, 193–200; Ricklefs 99–105; Ricklefs (short) 32–33, 192–195; Smith 21–23, 63–64, 305; Smith (short) 288–289, 411–412; Whittaker 168–171.

OBJECTIVES

In this exercise we examine structural adaptations of plants to an ecological factor, moisture.

INTRODUCTION

Plants require water and gases (CO_2 and O_2) for photosynthesis. However, water can escape through the openings that let the gases come in, so in arid regions conflicts arise. Where moisture is most plentiful (in a pond, for example), gases may be in short supply because water at 20° C (68° F) holds only 0.00005 times as much O_2 as air and only 0.002 as much CO_2. Plants cannot move to a more favorable environment, but features of life history, physiology, and morphology or structure have evolved that allow success even in seemingly unfavorable places. Plasticity of some traits allows organisms to take advantage of the good features of varying conditions. In this exercise we consider primarily structural modifications of leaves.

Mesophytes, plants of moist but not soggy places, are the typical examples of plants in botany textbooks.

Plants that live in water are called *hydrophytes*. Their special adaptations have less to do with water control than with aeration. Oxygen is in lower concentration in water than in air and may be altogether absent from the muck of marshes and swamps. Large air spaces called *lacunae* are found throughout water plants; they store O_2 and CO_2 and provide buoyancy. The characteristic spongy texture of roots and stems of hydrophytes is due to the lacunae, which form a continuous system such that submerged organs can exchange gases to the air.

Hydrophytes are often classified as emergent, floating-leaved, or submerged. Emergent leaves are held out of the water and are often much like mesophyte leaves. Leaves that float on the surface, such as water lilies and duckweeds (Lemnaceae) are round or oval. Leaves with lobes or teeth are more likely to tear and sink. In the Victoria waterlily the edges are turned up, but for plants growing

Key Words: mesophyte, hydrophyte, xerophyte, morphological adaptations, water relations, succulent, phenotypic response.

where waves could fill the leaves and swamp them this would be a liability. The petiole of floating-leaved plants is often attached near the center of the leaf blade, the mechanically best-suited place for floating. The leaves are sturdy and withstand waves as well as rain and hail which batter the leaves against the water surface. Submerged leaves can be much flimsier. They are typically ribbon-like (many monocots, such as *Vallisneria*) or dissected (dicots, such as *Myriophyllum, Cabomba, Ranunculus*).

Hydrophytes with some leaves under water and some leaves at the surface or emergent are usually *heterophyllous,* that is, they have more than one leaf shape on a single plant. The below-water leaves look like submerged-plant leaves; those above may be broad and floating *(Potamogeton natans)* or held out of the water *(Myriophyllum heterophyllum, Bidens beckii, Ranunculus* spp.). This is a phenotypic response to environmental conditions; plants with alternating groups of water and air leaves can be produced by repeated changes in the water level. Air leaves, even in the bud below water, are distinguishable from water leaves.

Xerophytes are adapted to grow where water is in short supply, whether in a desert or on an alpine peak. They withstand drought, but are not directly benefited by it. Several variations of life history, physiology, and structure can help a plant's water balance in arid situations. For instance, an annual life history allows a plant to pass the dry time as a seed. When water is available, it can germinate, grow, and ripen seed all within a period of four to six weeks. Perennials may slow or even stop growth during drought. Some can rapidly produce leaves with little structural material which they drop with little loss when drought hits.

Two variations from the common type of photosynthesis are found in xerophytes more than in other plants. Succulents (which can endure drought by storing water in fleshy stems or leaves) may use crassulacean acid metabolism (CAM), which allows stomata to be closed during the day. Some other xerophytes have the C_4 pathway of photosynthesis, rather than the C_3 pathway. These plants can use CO_2 at very low concentrations, thus requiring less opening of stomata.

Structural adaptations of xerophytes show wide variety. Roots may be very deep, or very shallow (to exploit light rains), or fast growing (to stay below the drying soil after a rain). Leaves may show features that reduce water loss. In this exercise you will be looking at the leaf structure of a "true" xerophyte: the plant does not stop or slow growth during drought, and the leaves are not drought-deciduous.

MATERIALS

slides, 1 for every 2 students:

- air and water leaves (on a single slide)
- hydrophyte, mesophyte, xerophyte leaves on a single slide, or slides of *Nymphaea, Ligustrum,* and *Ammophila*

microscope, 1 per student

stage and ocular micrometers (optional)

live or herbarium specimens, or pictures, of xerophytes and hydrophytes, particularly heterophyllous hydrophytes

picture of mesophyte leaf cross section (such as Fig. 2–7 in Brewer 1979)

PROCEDURE

1. Review the anatomy of mesophyte leaves. Be able to recognize the cuticle, epidermis, stomata, palisade and spongy mesophyll, and support and conductive tissue. To find stomata look for the dark-stained guard cells in the epidermal layer. Use high (430×) magnification until you can find them easily. Use lower magnification (dissecting microscope or 100×) for the rest of the exercise.

2. Compare leaf sections and fill in the boxes in the Results and Discussion section. If stage micrometers are available, calibrate the microscopes* and measure the leaves. If not, use comparative terms like ''thinner'' or ''more,'' or count numbers per microscope field.

3. Examine any demonstration material available.

RESULTS AND DISCUSSION

Leaf Sections

Character	Xerophyte	Mesophyte	Hydrophyte
Cross-sectional shape of leaf (sketch)			
Thickness of cuticle			
Thickness of epidermal cell walls			
Presence of palisade layer			
Location of stomata			
Amount of conducting tissue			
Amount and kind of supporting cells			
Presence and size of lacunae			
Presence and location of thin-walled, bulliform cells			

1. How do plants gain water? How do they lose water?
2. Discuss the selective advantage of each departure from the mesophyte ''norm'' shown by the xerophyte and the hydrophyte.
3. Is the hydrophyte leaf a floating leaf or a submerged leaf? How can you tell?
4. How may xerophytes be indirectly benefited by drought?
5. In drought conditions, the xerophyte grass blade rolls up. How is this accomplished?
6. What is crassulacean acid metabolism? How is it adaptive for plants of arid regions?

The air and water leaves are from a single, amphibious, hydrophyte. Look at the demonstration examples to see the gross differences.

---

*Directions for calibrating a microscope are in microscopy manuals.

7. Is the air leaf floating or emergent? How can you tell?

8. What might be the proximate factor in producing air or water leaves from a given point on the stem? Design an experiment to test your answer.

9. What might be some ultimate factors important in the differences between the two kinds of leaves? How might you test these ideas?

Character	Water Leaf	Air Leaf
Cross-sectional shape (sketch)		
Overall shape (dissected?)		
Thickness of epidermal cell walls		
Number of stomata		
Position of stomata		
Presence of palisade layer		
Amount of supportive tissue		
Amount of conductive tissue		

Teaching Notes

This work combined with a tour of the greenhouse to see succulents, etc., fits easily into a three-hour period. Air and water leaves on a single slide and single slides with hydrophyte, mesophyte, and xerophyte leaves are available from Triarch Incorporated, P.O. Box 98, Ripon, Wisconsin 54971. Several biological supply houses carry slides of suitable individual leaves.

BIBLIOGRAPHY

Arber, A. 1920. *Water Plants: A Study of Aquatic Angiosperms*. Cambridge Univ. Press. (Reprint 1963 by J. Cramer, Weinheim and Hafner, New York.)

Bannister, P. 1976. *Introduction to Physiological Plant Ecology*. Wiley, New York.

Coulter, J. M., C. R. Barnes, and H. C. Cowles. 1931. *A Textbook of Botany for Colleges and Universities. Vol. 3. Ecology*. Rev. and enlarged by G. D. Fuller. American Book, New York.

Daubenmire, R. F. 1974. *Plants and Environment; A Textbook of Plant Autecology*. 3rd ed. Wiley, New York.

Haberlandt, G. 1914. *Physiological Plant Anatomy*. Tr. from the German 4th ed. by M. Drummond. Macmillan, London.

Hutchinson, G. E. 1975. *A Treatise on Limnology. Vol. III. Limnological Botany*. Wiley, New York.

Jensen, W. A., and F. B. Salisbury. 1972. *Botany: An Ecological Approach*. Wadsworth, Belmont, California.

Schimper, A. F. W. 1908. *Pflanzen-geographie auf physiologischer Grundlage*. Fischer, Jena.

Schulthorpe, C. D. 1967. *The Biology of Aquatic Vascular Plants*. Edward Arnold, London.

Stowe, L. G., and J. A. Teeri. 1978. The geographic distribution of C_4 species of the Dicotyledonae in relation to climate. *Am. Nat.* 112: 609–623.

Weaver, J. E., and F. E. Clements. 1938. *Plant Ecology*. 2nd ed. McGraw-Hill, New York.

Exercise 4

TIME AND ENERGY BUDGETS

Key to Textbooks: Brewer 11–17, 122–135; Benton and Werner 101–120; Colinvaux 285–296; Emlen 163–164, 341–351; Kendeigh 173–180; Krebs 543–549; McNaughton and Wolf 38–41, 137–147; Odum 37–89; Pianka 77–82, 257–260; Richardson 185–191; Ricklefs 643–661; Ricklefs (short) 136–139, 203–207; Smith 54–61, 126–129; Smith (short) 59–61, 117–118; Whittaker 192–200.

OBJECTIVES

We will see how an animal's activity and environment determine how much energy it uses and, consequently, must take in as food. We will then relate the animal's energy relations to its seasonal cycle of activity, its allocation of time to foraging and other activities, and its place in the trophic structure of the ecosystem where it occurs.

INTRODUCTION

Plants and animals run on energy. For animals, the energy is taken in as food, digested and assimilated, packaged physiologically in ways that need not concern us here, and used within the body. The energy content of the food eaten by an animal is its *gross energy intake*. The energy that is contained in material not digested or not assimilated and, accordingly, egested as feces is a fraction termed *excretory energy*. In this fraction is placed also the small amount of energy in the organic compounds in urine, sweat, and any other excretions.

The energy left to the organism after excretory energy is subtracted is *assimilated energy*. Assimilated energy can be subdivided various ways depending on our purpose. Here we will divide it into five categories:

1. **Standard Metabolism.** This is the energy that the animal uses at rest just to stay alive. For humans, the term *basal metabolism* is generally used. It is the energy needed to pump blood, breathe, repair or replace a few cells, etc. In the laboratory standard metabolism is measured with the animal at rest, at a thermoneutral environmental temperature (where it does not have to spend any extra energy for heating or cooling itself), and long enough after eating that it is not digesting or absorbing food.

2. **Specific Dynamic Action (SDA).** Digesting and assimilating food require energy; this fraction of an individual's energy budget is given the curious name of specific dynamic action.

3. **Thermoregulation.** Below the thermoneutral zone homoiotherms must use energy to increase their heat production to keep their body temperature up. At high environmental temperatures,

Key Words: time budget, energy budget, trophic level, gross energy intake, assimilated energy, excretory energy, standard metabolism, thermoregulation, production, homoiotherm, poikilotherm, hibernation.

both homoiotherms and poikilotherms may need to use energy to lower their body temperature (for example, by panting).

4. **Activity.** Any activity an animal engages in takes energy. When sitting at a table, humans use energy at 1.2× standard or basal metabolism. For standing at rest the increase is about 1.7×, walking rapidly 6×, and running uphill 15×.

5. **Production.** This is energy used for new tissue in, for example, growth of a young animal. The energy content of the eggs a female lays is production; so, too, is the energy required to make the eggs, even though it is not incorporated in them. So too is the energy needed for milk production by a lactating female and energy stored as fat by a bird preparing to migrate or a mammal preparing to hibernate.

In this exercise we will try to estimate the energy requirements of an animal by use of a time and energy budget. Briefly what we do is watch an animal through the course of one of its days, figure out the amount of time it spends at various activities, multiply the time by the energetic cost of the activity, and then add the results up to get energy usage for the whole day.

MATERIALS

binoculars

watch

calculators with y^x key

maximum-minimum or recording thermometer

PROCEDURE IN THE FIELD

We suggest that observations be made on mallards *(Anas platyrhynchos)* or some other species of dabbling duck at and around some local pond or lake; however, some other water bird could be used, if necessary. Set up a maximum-minimum or recording thermometer to obtain an average night temperature for the day or days of observation. Calculate the average either as the mean of the various hourly temperatures or as (maximum + minimum)/2 for a maximum-minimum thermometer put out at dark and read at dawn.

Divide the class into pairs and then divide the time from sunrise to sunset into as many observation periods as there are pairs. Assign each pair one of the periods. The observer pair having the first period should locate a member of the species as close to sunrise as possible and make observations until the second pair takes over. If possible, the same individual bird should be watched through the whole day but if, as is likely, the bird being observed disappears, switch to another as rapidly as possible.

The specific procedure during observation is this: Watch the bird continually. Stay far enough away that your presence does not influence the bird's actions. At 30-second intervals note and tally on the Observation Sheet the activity of the bird at that instant. (Having one person observe and the other keep track of time and record data works best.) Occasional missed observations when the bird is briefly out of sight will not affect the results. The activity categories we will use are as follows:

Resting (including sleeping during daylight hours)

Standing

Preening and other maintenance activities except bathing

Walking

Feeding on land

Feeding in water (dabbling, up-ending, etc.)

Feeding while swimming

Alert

Slow swimming (up to 0.5 m/second, about 1.1 mph)

Medium swimming (0.55 m/second to 0.65 m/second, or about 1.2–1.5 mph)

Fast swimming (0.7 m/second, about 1.6 mph; this is the top swimming speed of a mallard)

Bathing

Calling and displaying

Wing flapping

Flying

It is desirable to add a comparative element to the study, for example by observing the same species during September and November, by comparing two species, or by gathering data on the two sexes separately.

PROCEDURE IN THE LABORATORY

Calculating a Time Budget

1. Obtain the time of sunrise and sunset from your local paper or astronomical tables. Enter below the number of hours of light (sunrise to sunset) and dark (sunset to sunrise). (Divide minutes by 60 to express as tenths of hours.)

Hours of light ____________

Hours of darkness ____________

2. Using the data on your Observation Sheet, divide the number of times a particular activity was tallied by the total number of observations; then multiply by 100 to express as a percentage. For example, if the period was one hour long you will have about 120 observations. If 10 of these were of feeding, the percentage of that hour spent feeding was approximately $10/120 \times 100$, or 8.3%.

3. Enter these figures in the Time Budget Sheet in the appropriate time period; also enter similar data from the other observers.

4. Add the figures for every observation period for each activity and divide by the number of observation periods to obtain an average. This will be the mean percentage of time spent in each activity during the daylight hours of the bird. (The sum of all these activities added across the bottom should be 100%.)

5. To produce a 24-hour time budget, multiply the percentages calculated across the bottom of the Time Budget Sheet by the number of daylight hours (from part 1). Enter the resulting figures in column A of the Energy Budget Sheet. For "Night" on that sheet, enter the hours of darkness directly from part 1. When you are done, the sum of column A should be 24.

Observation Sheet*

Activity	Tally of Observations	Percentage
Resting		
Standing		
Preening, etc.		
Walking		
Feeding on land		
Feeding in water		
Feeding while swimming		
Alert		
Slow swimming		
Medium swimming		
Fast swimming		
Bathing		
Calling		
Wing flapping		
Flying		
Total		

*Use this sheet to tally the observations made every 30 seconds during your observation period.

Time Budget Sheet

Observation Period	Activity														
	Resting	Standing	Preening	Walking	Feeding, Land	Feeding, Water	Feeding, Swimming	Alert	Slow Swimming	Medium Swimming	Fast Swimming	Bathing	Calling	Wing Flapping	Flying
Total															
Mean %															

Converting to an Energy Budget

1. **Standard Metabolism.** One of the important aspects of an animal affecting its rate of energy use is size. Larger animals use more energy than smaller but not proportionately more; in other words, the energy usage per gram is less in large animals than small. The mathematical expression of this relationship has the general form

$$SMR = a\ W^b$$

where SMR is standard metabolic rate, W is weight, and a and b are constants. The actual equations vary among different groups of organisms. For non-passerine birds, like mallards and other water birds, the equation is

$$SMR = 78\ W^{0.72}$$

SMR is in kilocalories per bird per day and weight is in kilograms.

Obtain the weight of the bird you are watching and express it in kilograms. If feasible, capture and weigh the bird you are observing. If not, obtain an approximate weight from another source such as Bellrose (1976). Male mallards weigh about 1.25 kg; females weigh about 1.11 kg.

To calculate standard metabolism, use a calculator with a y^x key and raise the bird weight (kg) to the 0.72 power. $W^{0.72}$ = ______________________________

Multiply this value by 78. SMR (kcal/bird/day) = ______________________________

2. **Specific Dynamic Action (SDA).** The energy necessary for specific dynamic action varies greatly based on the kind of food ingested. As a rough figure we will use an estimate of 30% of standard metabolism. Accordingly, multiply your estimate of standard metabolism (kcal/bird/day) by 0.3. SDA (kcal/bird/day) = ______________________________

3. **Thermoregulation.** For simplicity we will assume here that no energy is needed for thermoregulation except at night (and we will include a factor for that below). If this study is done during very cold periods of the year, this assumption will be wrong and as a result the energy usage of the animal will be underestimated.

4. **Activity.** We will express the various activities of the bird as multiples of standard metabolic rate *(SMR)*. To make the calculations obtain a standard metabolic rate per hour by dividing the daily SMR (calculated in part 1) by 24. SMR (kcal/bird/hr) = ______________________________. To calculate the amount of energy used during a day by a bird in any one activity, we will use a factor that indicates how much energy above standard metabolism the activity requires, multiply this by SMR (per hour), and multiply the result by the number of hours spent in each activity (from the Time Budget). The specific multiples relating the various activities to standard metabolic rate are given in Column C, "Cost Factor of Activity," of the Energy Budget Sheet. The multiple for the night hours has to be calculated. Because waterfowl remain active at night, metabolic rate does not drop to approximately SMR levels as it does in animals that sleep all night. We will follow this rule in assigning a cost factor: If the average night temperature is 20° C or above, use a cost factor (enter in column C) of 1.3 for the nighttime hours (20° is approximately the lower limit of the thermoneutral zone for mallards). Below 20°, omit column C for nighttime activity and calculate a value to be entered directly in column D by use of the formula below.

$$\text{kcal per hour of nighttime activity} = (1.3 \times SMR) + (0.15 \times [20 - T])$$

where T is average night temperature in °C.

Energy Budget Sheet

Activity	A Hours	B *SMR*/Bird/Hour (Enter from 4)	C Cost Factor of Activity	D Kcal per Hour of Activity (B × C)	E Amount of Energy Spent per Day on This Activity (D × A)
Resting			1.1		
Standing			1.6		
Preening, etc.			1.6		
Walking			1.7		
Feeding, land			1.7		
Feeding, water			1.7		
Feeding, swimming			2.1		
Alert			2.1		
Slow swimming			2.2		
Medium swimming			3.4		
Fast swimming			4.5		
Bathing			2.9		
Calling			2.9		
Wing flapping			3.0		
Flying			12.0		
Night					
Total	**24**				

Obtain a total for column E. Add to this the energy used in specific dynamic action (from part 2). The sum is the total energy used by the bird during one day.

Total energy used (kcal/bird/day) = ______________________.

DISCUSSION

1. At what activities did the bird spend the most time? Did there seem to be a daily pattern with more time spent feeding, for example, at one time of the day than another?
2. If the class work included a comparison, how did the two sets of observations compare as to the percentages and actual number of hours spent at various activities? Account for any differences.
3. You have calculated the approximate energy required for an individual bird to live a free existence. Are there other times of the year when the energy requirements for a single individual might be more or less? When and why?
4. Eventually every individual animal dies; for the species to persist in an area, reproduction must occur. What additional energetic demands will be associated with reproduction?
5. The bird will have to take in food that contains more energy than the amount you calculated as its total energy usage. Why is that?
6. If you assume that 65% of the food ingested (gross energy intake) is assimilated by the bird you studied, what has to be the daily intake of food energy to provide the energy actually expended?
7. Suppose that on very cold days the species you studied requires twice as much energy as it did on the day you studied it because of the additional costs of thermoregulation. Could it double its foraging time in order to meet this need?
8. If energy usage of a bird is substantially greater than energy intake for a period of several days, what is likely to happen? If energy intake is substantially greater than energy usage for a few days, what happens? Are there times of the year when the latter is a natural occurrence with some birds?
9. What are some ways in which land birds can reduce their energy usage while roosting (that is, while asleep at night)?
10. Break down foraging into as many sequential activities as you can. Would the time required for some of these change seasonally or with the type of food?
11. Construct a food chain or food web in which the species you studied is a link. What trophic level does it occupy?
12. The Golden Eagle Audubon Society of Nampa, Idaho, has counted as many as 316,061 mallards spending the winter on and around the Deer Flat National Wildlife Refuge. Using the figure you calculated in question 6, how many kilocalories per day would be required to feed 316,061 mallards?
13. What is the shape of the curve (both axes arithmetic) relating standard metabolic rate to weight of animal? Why isn't the increase a straight line (that is, why do bigger animals need less energy per gram of weight)?
14. What happens to the energy that an animal uses in its metabolism?
15. Suppose that you had studied a poikilotherm, say a lizard, over the same 24 hours you studied the bird. Suggest some ways in which its time and energy budget might have been different.
16. Why do some animals hibernate? Suggest both proximate and ultimate factors.

Teaching Notes

This exercise needs to be assigned and explained, then the work is done outside of class from sunrise to sunset. The instructor should be at the site at the beginning of the day to pick out an animal and start the observations.

BIBLIOGRAPHY

Bellrose, F. C. 1976. *Ducks, Geese and Swans of North America.* 2nd ed. Stackpole, Harrisburg, Pennsylvania.

Calder, W. A., and J. R. King. 1974. Thermal and caloric relations of birds. Pp. 259–413 in D. S. Farner and J. R. King (eds.). *Avian Biology,* vol. 4. Academic Press, New York.

Dwyer, T. J. 1975. Time budget of breeding gadwalls. *Wilson Bull.* 87: 335–343.

Ettinger, A. O., and J. R. King. 1980. Time and energy budgets of the willow flycatcher *(Empidonax traillii)* during the breeding season. *Auk* 97: 533–546.

Gessaman, J. A., ed. 1973. Ecological energetics of homeotherms: a view compatible with ecological modeling. *Utah State Univ. Press Monograph Series* 20: 1–155, Logan.

Hart, J. S., and M. Berger. 1972. Energetics, water economy and temperature regulation during flight. Pp. 189–199 in *Proc. 15th International Ornithological Congress*. E. J. Brill, Leiden.

Holmes, R. T., C. P. Black, and R. W. Sherry. 1979. Comparative population bioenergetics of three insectivorous passerines in a deciduous forest. *Condor* 81: 9–20.

Kendeigh, S. C. 1969. Tolerance of cold and Bergmann's rule. *Auk* 86: 13–25.

Pough, F. H. 1980. The advantages of ectothermy to tetrapods. *Am. Nat.* 115: 92–112.

Prange, H. D., and K. Schmidt-Nielsen. 1970. The metabolic cost of swimming in ducks. *J. Exp. Biol.* 53: 763–777.

Schwartz, R. L., and J. L. Zimmerman. 1971. The time and energy budget of the male dickcissel *(Spiza americana)*. *Condor* 73: 65–76.

Swanson, G. A., and A. B. Sargeant. 1972. Observation of nighttime feeding behavior of ducks. *J. Wildl. Manage.* 36: 959–961.

Wooley, J. B., Jr., and R. B. Owen, Jr. 1977. Energy costs of activity and daily energy expenditure in the black duck. *J. Wildl. Manage.* 42: 739–745.

Exercise 5

SAMPLING AND DENSITY ESTIMATION

Key to Textbooks: Barbour et al. 156–182; Benton and Werner 366–369; Kendeigh 30–34; Krebs 133–149, 608–609; Odum 163–166; Ricklefs 31–35; Ricklefs (short) 223; Smith 664–666, 686–690.

OBJECTIVES

We wish here to learn some principles of sampling and then apply these to study abundance, especially density. We also learn some specifics of several widely used methods of estimating density and population size.

INTRODUCTION

Scientists frequently want to be able to characterize a population or to compare populations. For example, they may want to know the density (number per unit area) of beech trees in a forest or to know whether high blood pressure is more frequent in persons living close to metropolitan airports than in the persons living away from them. In some circumstances it may be possible to deal with complete populations, to count every tree in a forest or examine every human in a population. For example, we might visit a small island and live-trap and weigh every mouse on it. We would then know, among other things, the total population size and weight of that population of mice. Traits of whole populations are called *parameters;* the weight of the island mouse population would be a parameter, as would be the average weight of all the adult female mice.

More often we cannot or do not care to spend the time or money to deal with total populations. Instead we *sample,* that is, we examine a fraction of the population. We then generalize from the traits the sample possesses (called *statistics*) to the traits (parameters) of the population. If the sample is *representative* of the whole population, we may do this and draw conclusions that (after allowing for some statistical uncertainty) will be correct. Two features determine whether a sample will be representative: the sample must be unbiased and it must be adequate in size.

These two matters are discussed in the sections on choosing samples and determining the size of samples. In the rest of this exercise sampling is directed toward the estimation of the abundance of organisms (or stand-ins for them such as poker chips or dots on paper); however, any trait of a

Key Words: population, sample, parameter, statistic, random and systematic sampling, quadrat, density, frequency, species:area curve, performance curve, confidence interval, two-step sampling, mark and recapture method (Lincoln index), catch per unit effort (removal) method.

population of organisms or of a community or ecosystem can be studied by sampling, and basically the same principles will apply.

Artificial populations are used for most sections because we can then tell how good or bad the estimates actually are; however, applying the methods in the field may be more interesting and is instructive in other ways.

MATERIALS

artificial fish population: 2 plastic buckets each with 600–1,000 poker chips of which half to two-thirds are white

2 plastic cups holding 30–60 poker chips

2 plastic cups holding 20–30 poker chips

random numbers table (inside front cover of this book)

artificial plant population (pp. 39–45, this book)

plastic quadrats (p. 161, this book)

calculator

graph paper

press-on labels or masking tape

CHOOSING SAMPLES

The usual approach to obtaining an unbiased sample is to sample randomly. A *random sample* is one in which every member of the population (every individual animal or every point of ground, for example) has an equal and independent probability of being included. Contrasted with random samples are (1) *systematic samples* and (2) any other kind, which amount to various kinds of selected samples. We might, for example, try to select only representative sites in a forest, so as to avoid wasting time on sites that were unusually open or dense, or odd in some other way. If we had a very good ecological eye such a sample might give us an excellent representation of the average conditions in the forest (although not of extremes, which may also be worth knowing); however, the results might be much different if some unperceptive clod were to try the same thing.

Systematic sampling involves taking samples that have some sort of systematic or regular arrangement; sample plots may be located every 100 meters or every third person may be interviewed. The advantage of systematic sampling is that it is usually simpler than random sampling. Bias will be present only if the pattern of the sampling is picking up some pattern in the population. As a simple-minded example, if you were sampling a cornfield using small plots spaced 40 inches apart, your estimate of corn plant density might be far too high or low, depending on whether your plots fell on the rows or between them. If systematic sampling is used, it is up to the investigator to demonstrate that no such bias is present. Because statistical theory has been developed on random samples, random rather than systematic samples are usually recommended.

Taking a random sample requires procedures to assure randomness. It is not sufficient simply to try to place samples haphazardly, avoiding conscious bias. When samples are located in this way, tests usually disclose definite biases—the samples somehow avoid poison ivy patches, include the largest trees, or turn out to be too evenly spaced. Since the aim is to give every unit of the population an equal and independent likelihood of being chosen, the basic approach to randomizing samples is to use chance to determine the samples. For example, we may wish to locate several square-meter sample plots in a field. We could establish a system of coordinates on the field with letters for the north-south coordinates and numbers for the east-west coordinates. We could then put tags with the letters in one hat, tags with the numbers in another hat and draw them out in pairs. If we pulled out A and 3, then one sample would be located in the field where line A crossed line 3. If we wanted to introduce some further randomization, we might flip a coin twice to decide whether to locate the plot NE, NW, SE, or SW from the point. Drawing numbers from hats, flipping coins, rolling dice, and the like are all appropriate ways of making random decisions; however, we can usually achieve the same effect more simply by using random numbers generated by a calculator or a computer or included in tables.

A random numbers table (see inside front cover) consists of a long series of digits that has been checked for non-randomness. To use it, simply enter the table in some way that prevents your exercising choice in the first sample (for example, pick a starting point anywhere in the table, then omit the first five digits and begin with the sixth); then simply take the numbers in some predetermined order such as left to right. The numbers in the table here are in pairs; if your procedure requires two-digit numbers from 00 to 99 you can use them as they are. If you need two-digit numbers from, say, 15 to 60, just skip the pairs that are outside that range. The table can be used for other than two-digit numbers by grouping the numbers as needed, ignoring the spaces. For example, for three-digit numbers, block off the digits in threes:

35 60 79 18
14 16 29 02 for 356, 079, 181, 416 and 290.

These are blocked off from left to right but you could go in some other direction. For three-digit numbers between 200 and 900 you would skip 079 and 181.

QUADRAT SAMPLING

The quadrat method originated with F. E. Clements (Pound and Clements 1898). The following description is for the artificial plant population. On it each symbol corresponds to a plant stem. Take random quadrats 5 cm on a side (using the transparent plastic quadrats provided). Imagine that your quadrat corresponds to a square meter quadrat in the field. Count the numbers of squares (which we will consider to represent a species of mint), triangles (a kind of sedge), suns (sunflower), and stars (starflower) in each of 12 quadrats and enter the figures in the Data Sheet.

To deal with plants that are partly in and partly out of the quadrat, adopt some reasonable convention. One possibility is to include plants that are more than half in and exclude ones less than half in. Another is to include all the border plants on the north and east sides and exclude all of those on the south and west sides.

Density. After two quadrats and again after 12, calculate an average density (number of stems per quadrat) and enter those figures in the Data Sheet. The actual densities of the four species are 3.3 mints, 3.3 starflowers, 3.3 sunflowers, and 0.33 sedges per quadrat. Calculate the percentage errors for each species using the formula:

$$\% \text{ error} = \frac{\text{estimated density} - \text{actual density}}{\text{actual density}} \times 100.$$

Discussion. (1) In general (use results from the rest of the class as well as your own), how does increasing the size of the sample affect the accuracy of results? (2) Why was there a difference in the

---

The Artificial Plant Population Map

On the following four pages are the four quarters of an artificial plant population map designed for use with Exercise 5. Including the map as a single 17 × 22 inch foldout would have necessitated a considerably higher price for the book. Shrinking the map to fit on an 8½ × 11 inch page also seemed unsatisfactory. Our solution, admittedly also imperfect, was to include the map on these pages which can be torn out and taped together. This will produce a map of about 19½ × 14½ inches. It is primarily for use with this exercise but, in a pinch, can be used for Exercises 6 and 16.

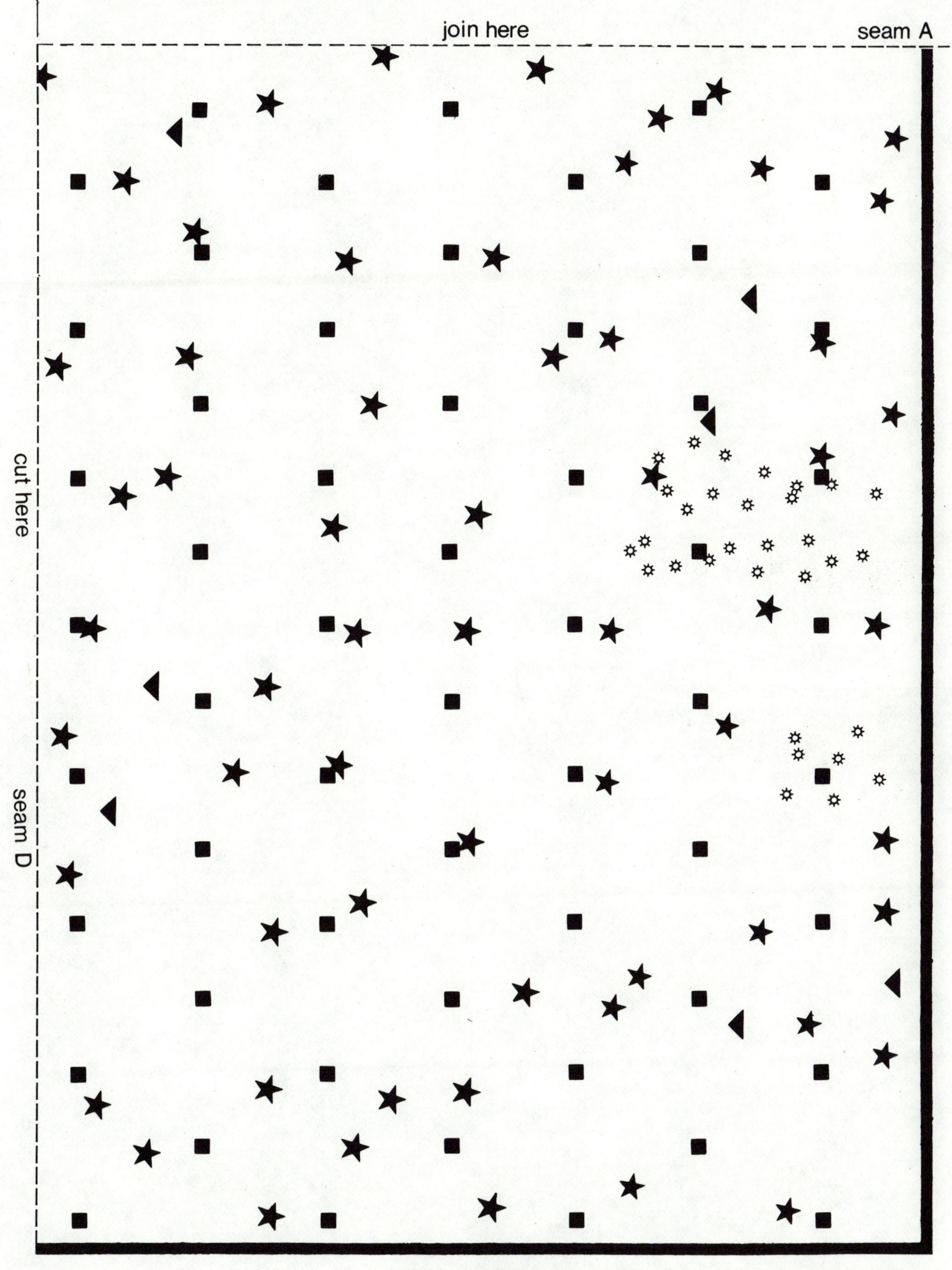
join here
seam A
cut here
seam D
Upper Right

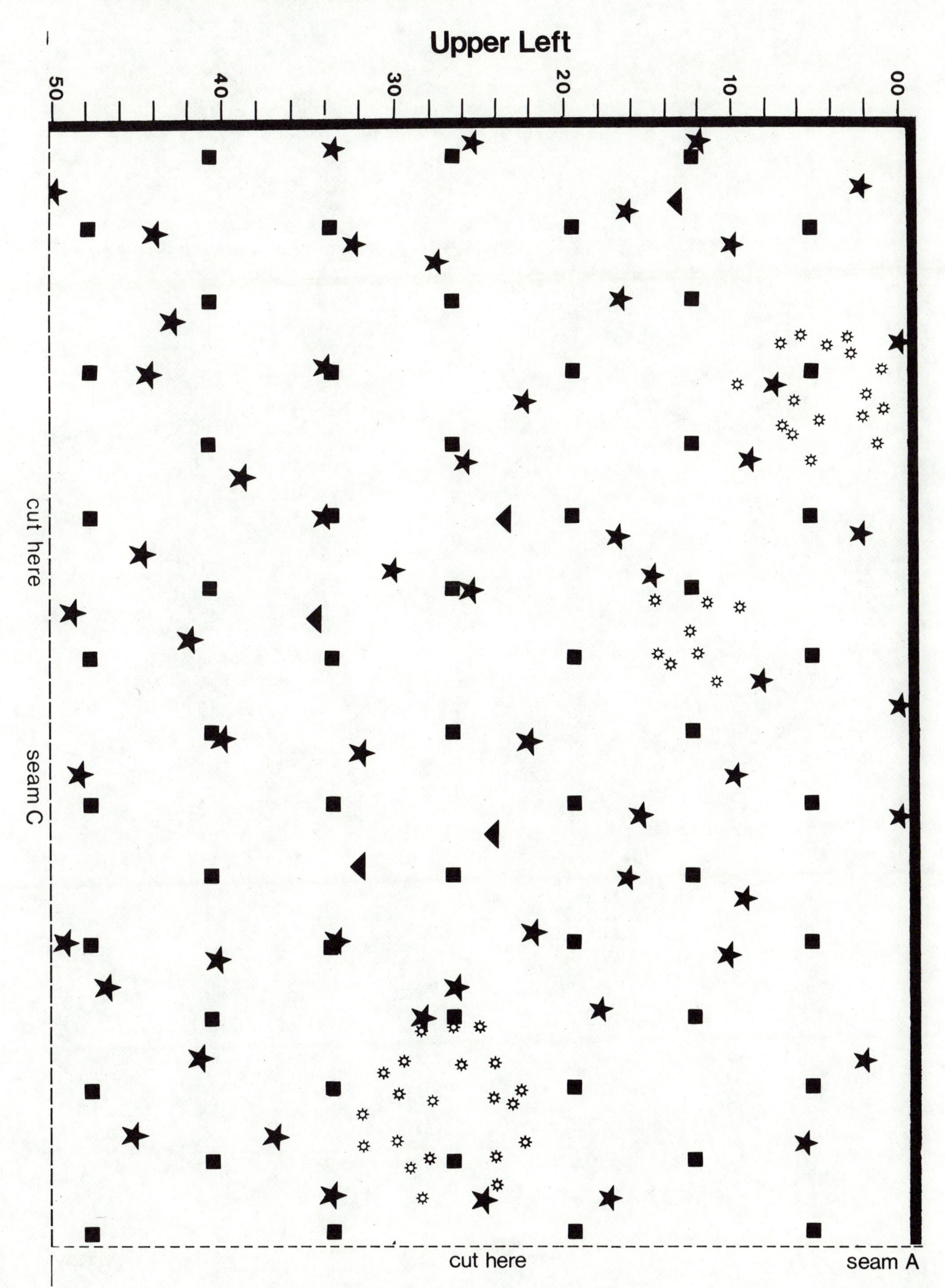
Upper Left
50
40
30
20
10
00
cut here
seam C
cut here
seam A

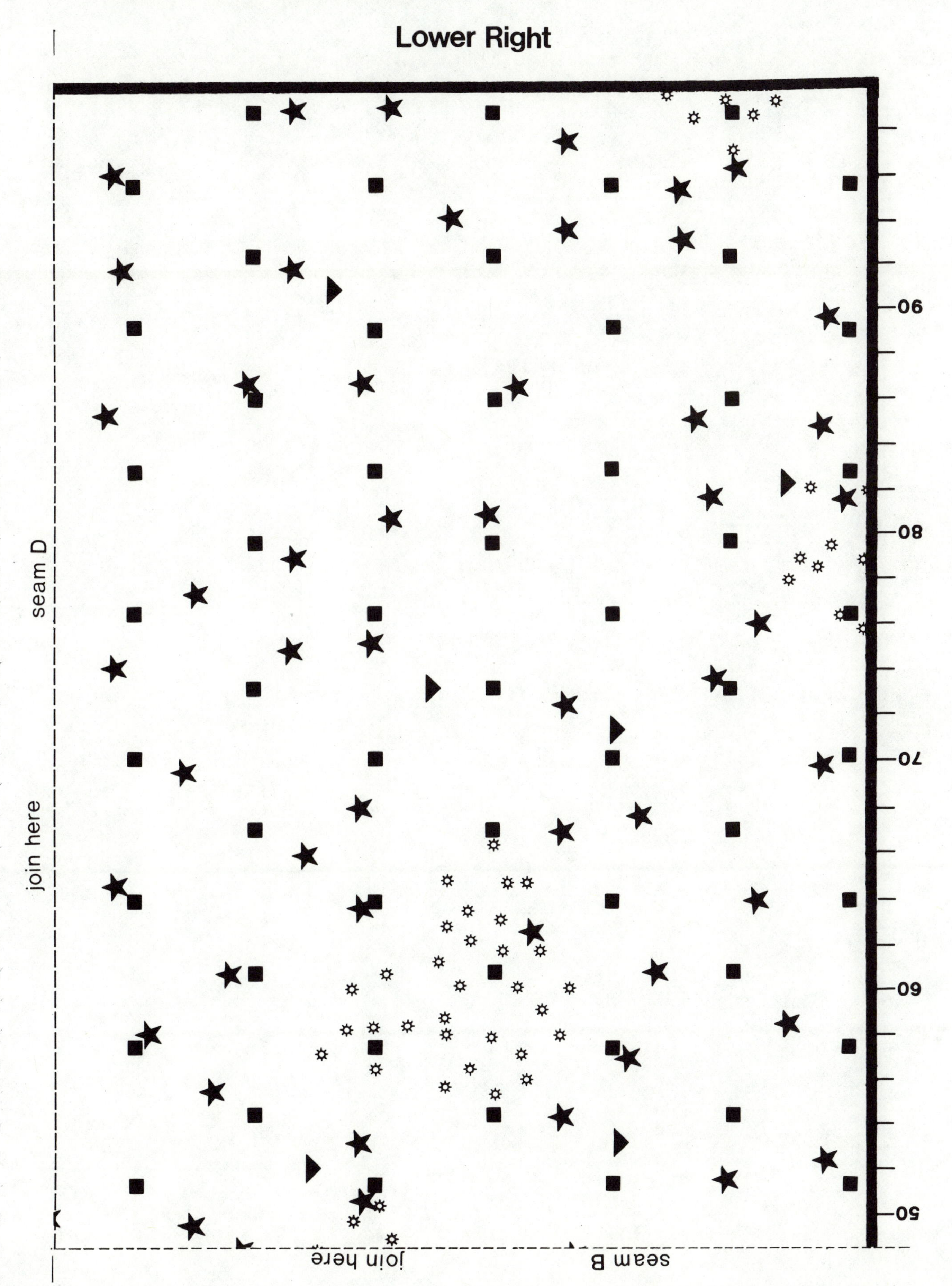
Lower Right
seam D
join here
join here
seam B
90
80
70
60
50

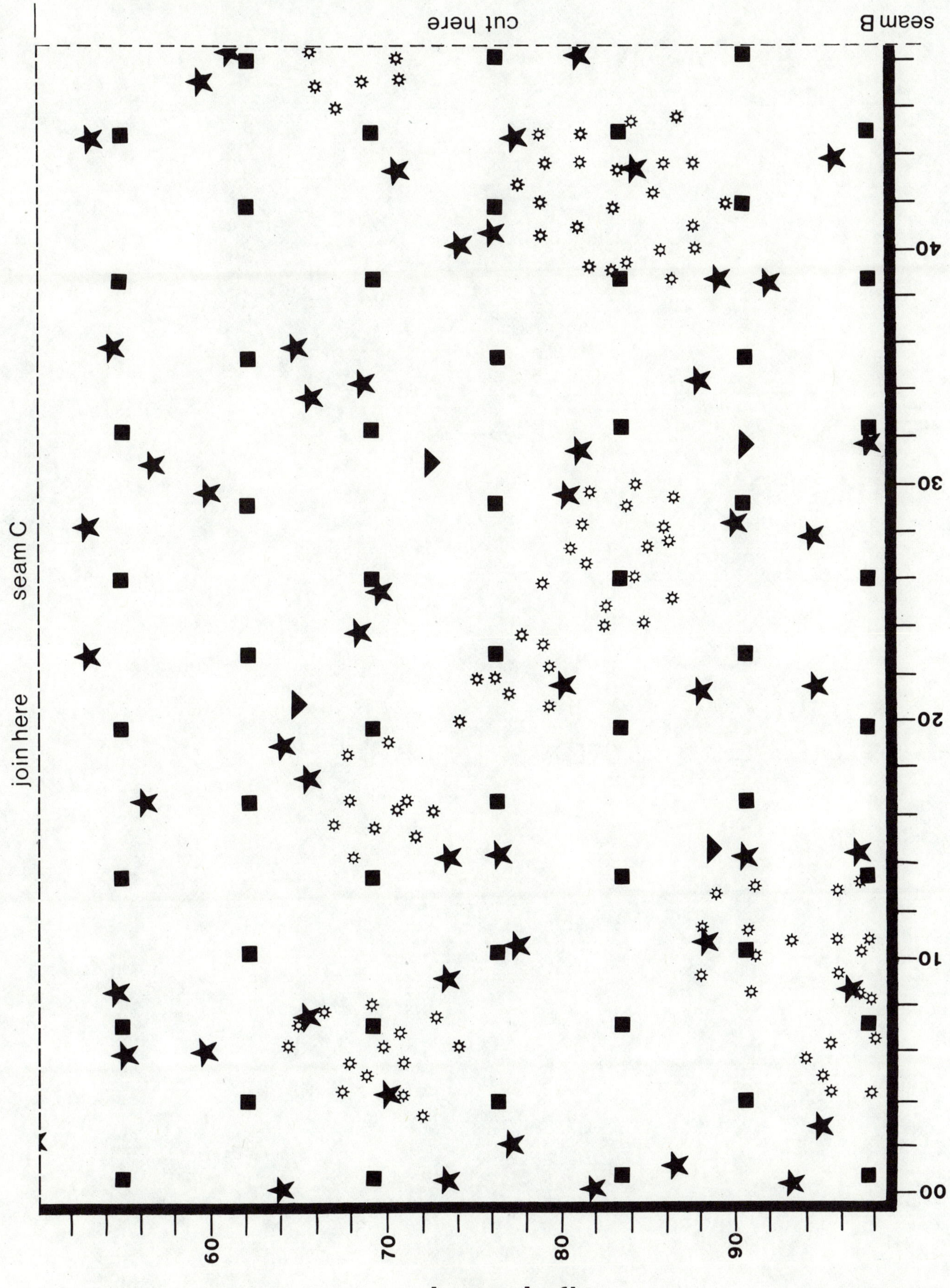

cut here
seam B
seam C
join here
00
10
20
30
40
60
70
80
90
Lower Left

Data Sheet for Quadrat Sampling

Quadrat	Species			
	■ Mint	▲ Sedge	* Sunflower	☆ Starflower
1				
2				
Estimated Density after Two Quadrats				
% Error				
3				
4				
5				
6				
7				
8				
9				
10				
11				
12				
Estimated Density after 12 Quadrats				
% Error				

TABLE 5–1. Sample Data: Plants Found in Square-Meter Quadrats on a Pioneer Area about Three Years Old, 9 October 1969, Kalamazoo County, Michigan

Species	Quadrat															
	1	2	3	4	5	6	7	8	9	10	11	12	13	14	15	16
Setaria lutescens	x	x	x	x	x	x	x	x	x	x	x	x			x	x
Digitaria sanguinalis	x				x	x	x	x	x	x	x	x	x	x	x	
Conyza canadensis	x					x			x				x	x	x	x
Poa compressa	x	x	x	x	x	x	x	x	x	x	x	x	x	x	x	
Cyperaceae, unidentified	x	x			x			x				x				
Cerastium sp.	x	x			x	x	x	x	x	x	x	x			x	
Melilotis alba	x	x	x	x	x	x	x	x	x	x	x	x	x	x	x	x
Medicago lupulina				x							x			x		x
Trifolium pratense		x	x	x	x		x						x	x		
T. repens	x		x		x	x	x	x			x					
Potentilla recta	x	x	x		x	x	x	x	x		x		x	x	x	x
Daucus carota	x	x	x	x	x	x	x	x	x	x	x	x	x	x	x	x
Plantago lanceolata	x	x	x	x	x	x	x	x	x				x		x	
P. major	x		x	x		x			x	x					x	x
Ambrosia artemisiifolia	x	x		x	x	x	x	x	x	x	x	x	x	x	x	x
Barbarea vulgaris	x															
Oxalis stricta	x		x	x	x	x		x								
Oenothera biennis					x											
Sporobolus cryptandrus		x														
Achillea millefolium		x	x	x				x								
Taraxacum officinale			x	x		x		x				x				x
Phleum pratense	x	x	x	x	x	x	x	x		x						
Erigeron strigosus			x	x	x	x	x				x					
Hypericum perforatum							x									
Sonchus asper															x	
Solidago sp.					x				x							
Aster pilosus	x	x	x	x	x	x		x	x	x			x	x	x	x
Verbascum thapsus						x		x								
Panicum capillare					x	x	x	x	x		x	x	x	x	x	
Aster sp.								x								
Prunella vulgaris					x	x	x								x	x
Bromus sp.					x											
Agropyron repens									x				x	x	x	x
Cichorium intybus									x	x						

accuracy of results of sunflower compared with starflower? Explain fully. (3) Why was there a difference in accuracy of sedge compared with starflower?

Frequency. Percentage frequency is often used as a measure of abundance, especially vegetation studies. Percentage frequency is calculated as follows:

$$\text{percentage frequency of species A} = \frac{\text{total no. of samples containing A}}{\text{total no. of samples}} \times 100.$$

Record below the percentage frequencies you determined for the four plant species in 12 quadrats.

Species	Percentage Frequency	Actual Density
Mint		
Sedge		
Sunflower		
Starflower		

Discussion. (1) Is percentage frequency correlated with density? (2) Is the correlation perfect? (3) Based on this exercise, on what besides density does frequency depend? (4) If you took another

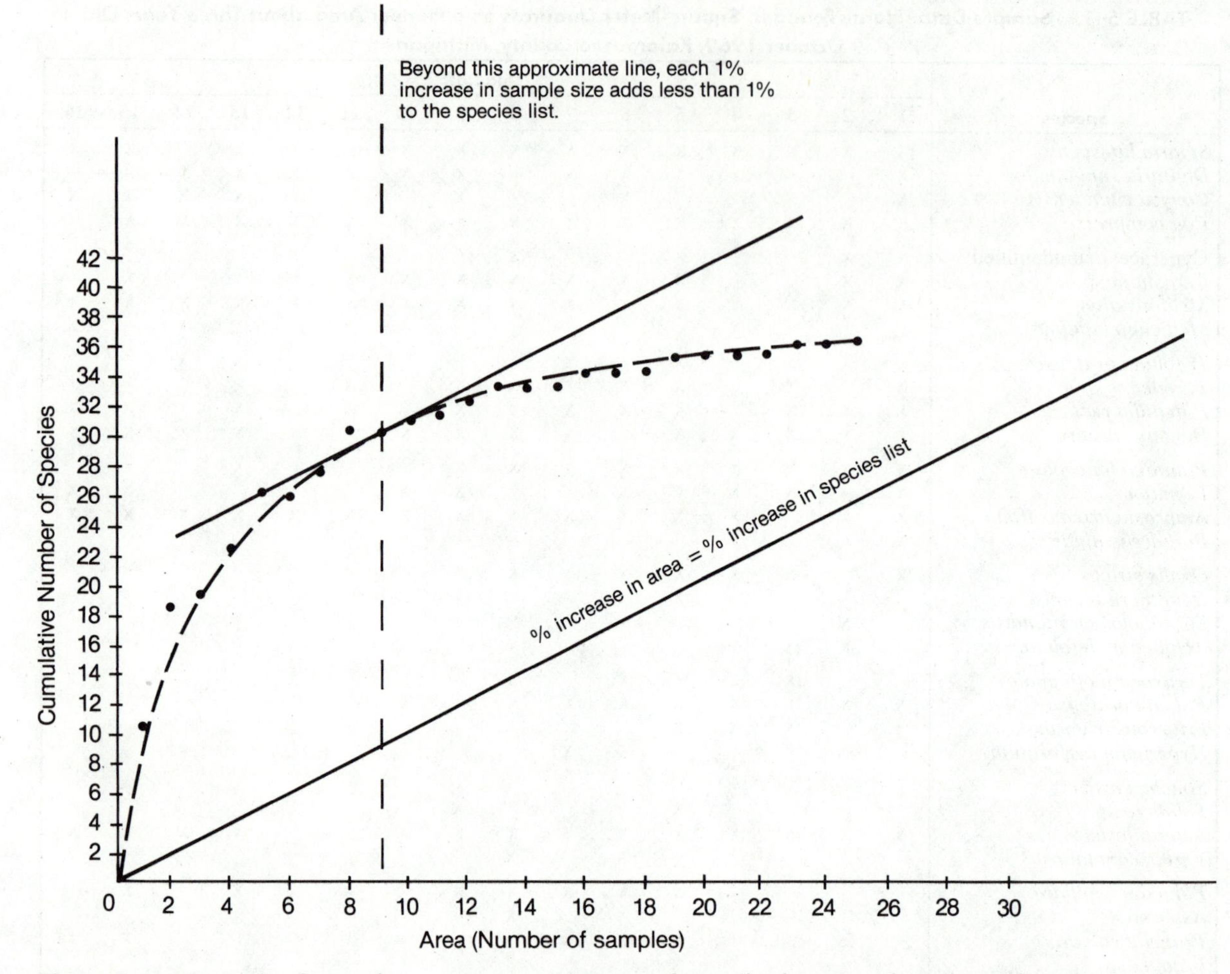

Figure 5–1. Example of a species: area curve. Adequacy of sampling in terms of species registration is reached at about 10 quadrats, where a 1% increase in sample size no longer adds as much as 1% to the species list.

set of samples, using quadrats of 2 m² rather than 1 m², how would percentage frequency change? Why?

ADEQUACY OF SAMPLING

It is not possible to specify ahead of time how large a sample is needed to be adequate. In general, the larger the sample size (that is, the more quadrats or other sample units) the more likely it is to be adequate; however, taking a sample far larger than necessary is a waste of time and money. In general, a simple, homogeneous area will require a smaller sample than a complex or heterogeneous one. Samples adequate for some purposes may not be adequate for others; a few quadrats may suffice to give us a reasonable approximation of the density of the very common species but may not even detect some of the rare ones. We will deal here with three ways of judging adequacy: (1) the species:area curve, (2) the performance curve, and (3) two-step sampling.

The Species:Area Curve. A plot of the accumulated number of species against the size of the sample (for example, the number of quadrats) is called a *species:area curve* (Fig. 5–1). It has some theoretical interest in ecology but we use it here as an indication of sampling adequacy. The specific aspect of adequacy to which it applies is whether our sample has been large enough to include most of the species of the community we are studying. The idea is that in our first quadrat we will have all new species. In the second quadrat we will probably repeat some species but will have several new ones. As we proceed we will reach a point where our samples have included all the common species and many of the rarer ones, so that each new quadrat adds little to the accumulated species total. Our plot of cumulative number of species begins to level off. There is no level at which species registration will be adequate for every purpose; a fairly widely used rule is that species sampling will be assumed to be adequate when a given percentage increase in sample size produces less than the same percentage increase in number of species. A mechanical way of determining this point is described below.

Procedure and Discussion. Using either data from Table 5–1, which gives the plant species found in square-meter quadrats on a three-year-old pioneer site, or similar data that your own class gathers, fill in the following table.

Quadrat Number	Number of Species	Number of New Species	Cumulative Number of New Species
1			0
2			
3			
4			
5			
6			
7			
8			
9			
10			
11			
12			
13			
14			
15			
16			
17			
18			

Now plot cumulative number of new species (the last column) on the vertical (y) axis against area (square meters in the case of the data provided) on the horizontal (x) axis.

You will see the sharp rise and leveling-off characteristic of species:area curves. Now draw a line that starts at zero on both axes and runs through the 10 for both species and quadrats (square meters). Along this line the addition of one new sample adds one new species. For any portion of the species:area curve that rises more steeply than this line, species are being added at a greater rate than sample area is being added; for any portion of the curve that has a slope flatter than the line, species are being added at a lower rate. Draw a line parallel to this line and just touching the species:area curve. At this point (where the line is tangent to the curve) one new species is being added with each new sample. We will have achieved adequate sampling, using this criterion, with the next quadrat.

1. At what quadrat was sampling adequacy achieved?
2. If you wanted to be sure to include a larger fraction of the total species pool you might wish to sample to the point where a 1% addition of area added less than a 0.5% increase in species. How would you draw the line to indicate this slope? Using this criterion, with what quadrat was adequacy of sampling achieved?

The Performance Curve. A performance curve (Fig. 5–2) plots the mean value of some trait against sample size. This could be density, biomass, weight, or any of a variety of other measurements. The first one or two samples are not likely to show exactly the population mean, so that the early part of the performance curve tends to be jagged; as more samples are added, it flattens out. When the fluctuation produced by a new sample becomes very small we assume that we are estimating a mean that is close to the true population mean.

Procedure and Discussion. Return to the data used in Density. Calculate mean density of the sunflower after each quadrat and plot those estimates (mean density on y-axis, number of quadrats on x-axis).

1. Does 12 quadrats seem to give an adequate sample for estimating density in this case?
2. If you were to plot a performance curve for the starflower, how would it differ from the sunflower curve? What about the curve for the sedge? Plot the curves and see if your predictions were right.

Two-step Sampling. Although interpretation of the performance curve could be made objective, a better way to obtain a more rigorous statistical approach is to employ two-step sampling. Here we take one sample (ten quadrats, for example) and, then, based on how variable that sample is, calculate how many more quadrats we will need to achieve some certain level of statistical reliability. Doing the calculations, as outlined below, will be of value only to students who have had enough exposure to statistics to know what means and standard deviations are.

The method depends on the statistical concept of the confidence interval. This is the interval within which we are confident, with some specified probability, that a population trait lies. Here we are interested in specifying the confidence interval for the mean density of a population. The formula for calculating the lower limit of the confidence interval is

$$\bar{x} - K\,(S/\sqrt{N})$$

and for calculating the upper limit is

$$\bar{x} + K\,(S/\sqrt{N})$$

where $\bar{x}$ is the sample mean, S the sample standard deviation, N the sample size (number of samples), and K the normal curve variate for a particular probability. For example, suppose we have a sample with the following traits: $N = 200$, $\bar{x} = 31.7$, and $S = 1.81$. We wish to be 90% certain that the true mean of the population lies within limits we select. The K value for the 90% level is 1.645. This is looked up in a table; however, the only probabilities that we are likely to want, and the corresponding K values, are

90%	1.645
95%	1.960
99%	2.576.

Using a K value of 1.645, we have confidence limits of

$31.7 - 1.645\,(1.81/\sqrt{200})$, or 31.49, and

$31.7 + 1.645\,(1.81/\sqrt{200})$, or 31.91.

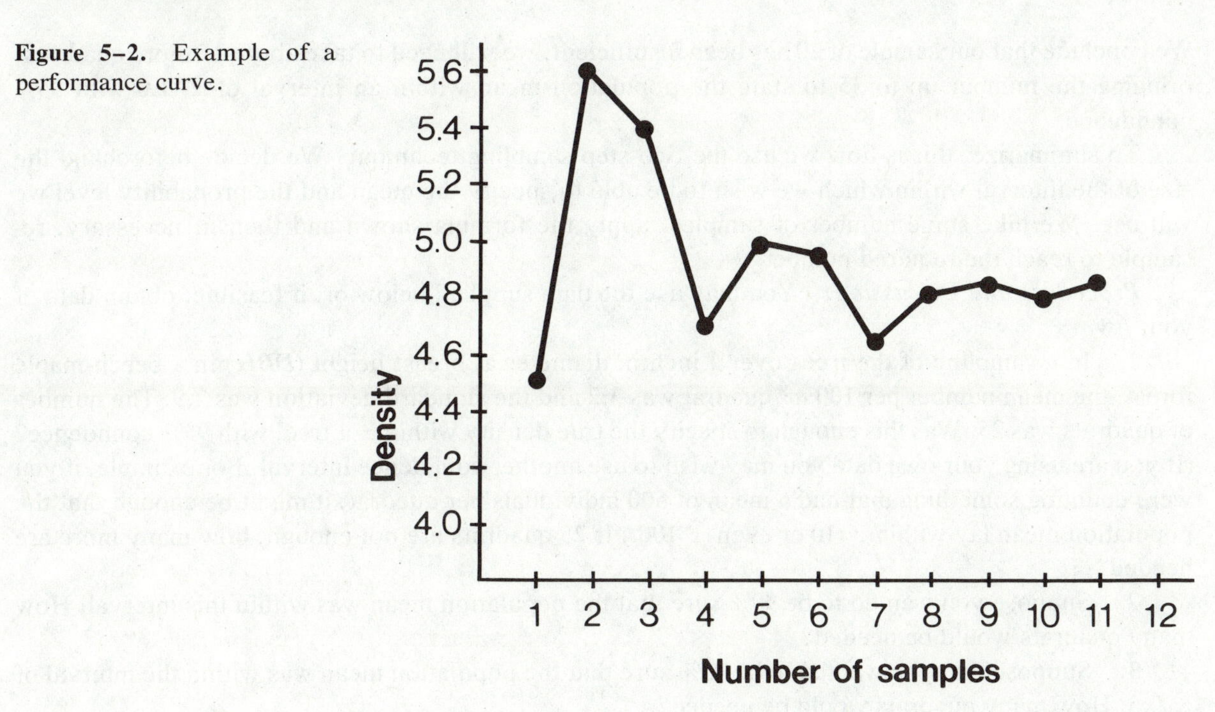

Figure 5–2. Example of a performance curve.

We say that we are 90% confident that the population mean lies within this range. In other words, when we make the statement that the population mean lies in this interval, we will expect to be right 90 times out of 100. (Some further explanation of confidence intervals is given in Appendix 4, on statistics.)

The important feature of the confidence interval for this method of determining adequacy of sampling is that any increase in sample size will decrease the size of the interval. (This should be clear from the fact that we divide by sample size in the calculations.) We can use this relationship to assess the number of samples we need to attain a given level of precision.

Suppose we have taken 20 quadrats and obtained a mean of 8.0 and a standard deviation of 3.00. We want to be sure with 95% confidence that we are within ± 1.0 of the population mean; that is, we want to be sure that the true mean is no more than 9.0 and no less than 7.0. Are the 20 samples we have taken enough to specify the mean within an interval of ± 1.0 and, if not, how many more do we need?

We want a sample size that satisfies the relationship

$$1.0 = 1.96\,(3.00/\sqrt{N}).$$

(1.96 is the K value.) Putting this in general terms we can represent it

$$L = K\,(S/\sqrt{N})$$

where L is the confidence interval and the other symbols have the same meaning as before. To solve for N, we may write this as

$$N = \frac{S^2 \times K^2}{L^2}.$$

For our example

$$N = \frac{3^2 \times 1.96^2}{1^2} = \frac{9 \times 3.84}{1} = 34.6.$$

We conclude that our sample of 20 has been insufficient; we will need to take about 15 more quadrats, bringing the number up to 35 to state the population mean within an interval of ± 1.0 with 95% confidence.

To summarize, this is how we use the two-step sampling technique: We decide beforehand the size of the interval within which we wish to be able to specify the mean and the probability level we will use. We take some number of samples, apply the formula shown and then, if necessary, resample to reach the required number.

Procedure and Discussion. You may use the data supplied below or, if feasible, obtain data of your own.

1. In a sampling of the trees over 1 inch in diameter at breast height (*DBH*) in a beech-maple forest, the mean number per 100 m^2 quadrat was 3.2 and the standard deviation was 2.9. The number of quadrats was 25. Was this enough to specify the true density within ± 1 tree, with 95% confidence? (If you are using your own data you may wish to use another confidence interval. For example, if you were counting something that had a mean of 500 individuals per quadrat, it might be enough that the population mean lay within ± 10 or even ± 100.) If 25 quadrats are not enough, how many more are needed?
2. Suppose you wanted to be 99% sure that the population mean was within this interval. How many quadrats would be needed?
3. Suppose that you wanted to be 95% sure that the population mean was within the interval of ± 0.5. How many quadrats would be needed?
4. Suppose you are dealing with a sample having a mean of about 50. Under some circumstances you might wish to specify a confidence interval of ± 5 whereas in other cases you might require a confidence interval of ± 1 or even smaller. What considerations dictate this choice?

ESTIMATING DENSITY BY THE MARK-AND-RECAPTURE METHOD (LINCOLN INDEX)

The mark-and-recapture method is based on an elegant idea, introduced by C. G. J. Petersen in the late nineteenth century and re-introduced for studying bird populations by F. C. Lincoln in 1930. The idea is that we somehow put a mark on some members of a population and then sample the population to find out what proportion of the sample bears the mark. We can then, by a simple proportionality, estimate total population size. The proportionality is

$$\frac{\text{number marked in the sample } (r, \text{for recapture})}{\text{total in the sample } (n)} = \frac{\text{marked in the population } (M)}{\text{total size of the population } (N)}.$$

Our estimate of N is usually designated $\hat{N}$ (called "N hat"). Rearranging the proportionality, we have the formula we will use for computation:

$$\hat{N} = (n/r)\,M.$$

The estimate is for the date of marking, not the date of recapture.

Although the idea behind the mark-and-recapture method is simple and logical, the results in practice are often not very accurate. This is for two basic reasons: (1) $\hat{N}$ is not a very precise estimator of N unless the sample size is large or a high proportion of the population is marked. As a rule of thumb, estimates are unlikely to be reliable if the product $r \times M$ is smaller than four times the population size (N). (2) The method makes several assumptions that may or may not be valid in a real population of fish, birds, or mammals. The necessary assumptions are: (a) Marks must not be lost; (b) there must be no recruitment into the population by reproduction, growth (for example, small fish getting large enough to be caught by a seine), or immigration; (c) marked and unmarked animals must behave alike; in particular, their mortality rates, activity, and response to traps must be the same; and (d) capture for marking and subsequent recapture must be random.

Many modifications of the mark-and-recapture method have been devised to correct some of the deficiencies of the method (examples are Bailey 1951 and Jolly 1965). If a mark-and-recapture method were to be used in serious research, one of these modifications should probably be used.

Procedure. In this exercise, we will use a plastic bucket containing several hundred poker chips to simulate a pond containing fish. The object is to estimate the number of white chips, which are meant to simulate a particular species of fish (possibly the round whitefish, *Prosopium cylindraceum quadrilaterale*). The other colors of chips are other species that we are not interested in. If the fish have not yet been marked, catch about 100 of the white chips and mark them with a press-on label or a piece of masking tape (simulating a jaw tag or a clipped fin). Make sure the marked individuals are thoroughly mixed with the rest. Each student should use a plastic cup with a capacity of 30–60 chips to simulate a seine and scoop up your recapture sample. Thoroughly mix the fish between samples.

Enter your figures in the spaces below. The instructor will supply the figure for the number of marked chips in the population. Compute $\hat{N}$. Also enter your classmates' estimates as they become available.

Number marked in sample (r) = ________________

Sample size (n) = ________________

Total marked in population (M) = ________________

Your estimate of total population size ($\hat{N}$) = ________________

Other estimates of total population size: ________________________________

Discussion

1. Get the correct value for N from your instructor. How far off was the worst estimate? Calculate the median of all the estimates. Was it a close approximation?
2. Why is the Lincoln index estimate for the time of marking rather than the time of recapture?
3. List several methods of capturing fish, birds, and mammals.
4. List several methods of marking fish, birds, and mammals.
5. Pick out a specific organism in a specific habitat (fish in a stream, mice in a woodlot, etc.) and try to assess which of the assumptions necessary for the validity of the mark-and-recapture method might be violated and, specifically, how.
6. What are trap-shy and trap-prone mice? How will they affect your estimates of mouse numbers?
7. Although it is necessary to assume that no reproduction occurs, the similar assumption of no mortality is not necessary. Explain the difference.
8. Indicate whether your estimate, $\hat{N}$, would be too high or too low, in each of the following circumstances:
   a. Some animals lose their marks.
   b. Immigration occurs.
   c. Marked animals have higher mortality rates.
   d. Marked animals seek out traps.
   e. Marked animals are sluggish.
9. In using the Lincoln index there are sometimes difficulties in defining exactly the population being studied (e.g., does it include all ages and sexes?) and also in defining the boundaries of the area to which the estimate pertains. Would these difficulties be serious in the following situations? Why or why not?
   a. Fish in a pond
   b. Rats on a ship
   c. Mice on the middle 10 acres of a 40 acre woods
   d. Geese wintering at a lake
10. It is desirable when using the Lincoln index to calculate a confidence interval, the interval within which you are confident (with a specified probability) that the true population total lies.* In

*Additional discussion of confidence intervals is given in Appendix 4.

general, the value we add to and subtract from (for the upper and lower confidence limits) any estimate is given by $K\sqrt{S^2}$ where S^2 is a variance and K is the normal curve variate. Unfortunately, the variance is not straightforward for the Lincoln index population estimate. The formula we will use, which is one of several that have been suggested, is

$$S^2 = \frac{M^2\, n\, (n - r)}{r^3}$$

This would be tedious if you had to do it by hand but is easy on a calculator.
Enter here your estimate of S^2 ____________

Take the square root and enter here your value of $\sqrt{S^2}$ ____________
The useful normal curve variates (K) are given on page 52. Here we will use the value for 95% probability, 1.96. Multiply 1.96 times S. ____________
This is the value you will add to and subtract from your estimate of N to give the upper and lower limits of the confidence interval. Give those limits here ____________

ESTIMATING DENSITY BY CATCH PER UNIT EFFORT (REMOVAL METHOD)

This method for determining total population size uses the law of diminishing returns. It was first used in ecology by P. H. Leslie and D. H. S. Davis (1939); DeLury's (1947) version has been widely used in fisheries research. Basically, the method consists of repeated samplings of an area during which members of the population are removed. As the population is depleted, the catch per unit effort drops. The diminishing returns can be projected to zero to estimate total population size. As an example of the situations to which the method has been applied, consider estimating the number of trout in a given stretch of stream. The number caught per fisherman-hour (the unit of effort) can be determined by creel censuses carried out on several days starting with the opening of trout season. As a rule of thumb, the method is not likely to be reliable unless more than a third of the population is removed.

Procedure. In this exercise, we will again attempt to estimate the number of white poker chips (corresponding to a particular species of fish) in the pond (the plastic bucket). Use as your sampling device a plastic cup that will hold 20–30 poker chips. Scoop up one sample, sort out the white chips and keep them; throw the others back. Record the number of white chips on the Work Sheet. Take nine more scoops (ten in all), following the same procedure.

When catch per unit effort (in this case, chips per scoop) reaches zero, we will have removed the whole population. All that we would have to do, if we reached that point, would be to add up our total catch and we would have the total population. Generally, though, we cannot continue until we have caught every last fish, grasshopper, or mouse. Instead, we will plot catch per unit effort against accumulated catch as far as we got in depleting the population and will then project the line. At the point where catch per unit effort would equal zero, the (projected) cumulative total catch should equal the total population.

Graph catch per unit effort (that is, white chips per scoop) on the y-axis against cumulative total catch on the x-axis. Calibrate the x-axis so that it goes up to slightly higher than the probable number of white chips in the bucket (for example, to 700 if you think there are about 650 chips in the bucket). Note that at the first sampling period you have not yet removed any chips, so cumulative total catch for the first sampling period is zero. By eye, determine the best-fitting straight line through the points; draw it in and figure what cumulative total catch would be where the line crosses the x-axis. This number is your estimate ($\hat{N}$) of total population size. Also obtain the estimates of your classmates.

A more objective way to estimate N is to do a least squares regression. In this process we calculate a straight line that best fits the points mathematically, in a particular sense described in Appendix 4. The equation for a straight line has the form

$$y = \alpha + \beta x.$$

Work Sheet for Catch per Unit Effort (Removal Method)

Removal Period	Catch per Unit Effort (y)	Cumulative Catch* (x)	Optional, for Least Squares Regression x^2	xy
1		0		
2				
3				
4				
5				
6				
7				
8				
9				
10				
Optional, for least squares regression				
Totals	$\Sigma y =$	$\Sigma x =$	$\Sigma x^2 =$	$\Sigma xy =$
Means	$\bar{y} =$	$\bar{x} =$		

*Cumulative catch for any given removal period is the total number caught up to, but not including, that removal period.

In this case, y is catch per unit effort, x is cumulative catch, α is the intercept of the line with the y-axis (catch per unit effort at the start), and β is the slope of the line (the rate at which catch per unit effort drops). Our estimate of the total population (that is, cumulative total catch when catch per unit effort is zero) is given by

$$\hat{N} = -(\alpha/\beta).$$

It remains only to calculate these two regression constants. This is done by completing the table and plugging the values into the following equations. Alternatively, if you have access to a pocket calculator that does automatic regressions simply by entering the x and y values, use it to obtain α and β if your instructor does not mind.

$$\beta = \frac{\Sigma xy - \dfrac{(\Sigma x)(\Sigma y)}{N}}{\Sigma x^2 - \dfrac{(\Sigma x)^2}{N}}$$ (N is the number of removal periods)

$$\alpha = \bar{y} - \beta\bar{x}$$ ($\bar{y}$ and $\bar{x}$ are means of y and x, respectively)

DISCUSSION

1. How close was your estimate to the actual number of white chips? How much variation within the class was there?
2. Considering both accuracy and time, is the mark-and-recapture method or the removal method better?
3. In what situations would the removal method be most useful?
4. What might be some effects of removal on the population remaining and the rest of the ecosystem?
5. Suppose that for the reasons given in question 4 or for other reasons, it was desirable not to remove organisms. Devise a way of using the catch-per-unit-effort method without removal.
6. The essential assumptions of the removal method are that the population be closed (no births, deaths, immigration, or emigration) and that catchability be uniform. That is, every fish, bird, or mouse has to be as easily caught, shot, or trapped as every other member of the population and, further, catchability must not change from one removal period to another. How likely are these conditions to be met?
7. Suppose that you are using the removal method to estimate the number of wild turkeys inhabiting an area of state forest. Your data will consist of the number of turkeys shot on the seven days of a special turkey season. How do you suppose "catchability" of the turkeys might change during the seven days. If this did happen, would it make your estimate of population size too high or too low?
8. Suppose that for one of your removal periods you used the cup that removes 50 rather than 25 chips. How should you enter this data in the table (and on the graph)?

Teaching Notes

Having students work in pairs, taking turns doing the work and recording data, works well. Three hours is enough time if students do not have to wait for a turn at the poker chip artificial population and if some of the calculations are done outside of class.

BIBLIOGRAPHY

Anderson, D. R. 1972. Bibliography on methods of analyzing bird banding data with special reference to the estimation of population size and survival. *U.S. Fish and Wildlife Service Special Scientific Report Wildlife* no. 156: 1–13.

Bailey, N. T. J. 1951. On estimating the size of mobile populations from recapture data. *Biometrika* 38: 293–306.

Begon, M. 1980. *Investigating Animal Abundance: Capture-recapture for Biologists*. University Park Press, Baltimore.

Cain, S. A., and G. M. de O. Castro. 1959. *Manual of Vegetation Analysis*. Harper, New York.

Caughley, G. 1977. *Analysis of Vertebrate Populations*. Wiley, London.

Cormack, R. M. 1968. The statistics of capture-recapture methods. *Oceanog. Mar. Biol. Ann. Rev.* 6: 455–506.

DeLury, D. B. 1947. On the estimation of biological populations. *Biometrics* 3: 145–167.

Giles, R. H., Jr., ed. 1971. *Wildlife Management Techniques*. 3rd ed. The Wildlife Soc., Washington, D.C.

Greig-Smith, P. 1964. *Statistical Plant Ecology*. Butterworth's, London.

Jolly, G. M. 1965. Explicit estimates from capture-recapture data with both death and immigration—stochastic model. *Biometrika* 52: 225–247.

Leslie, P. H., and D. H. S. Davis. 1939. An attempt to determine the absolute number of rats on a given area. *J. Anim. Ecol.* 8: 94–113.

Mueller-Dombois, D., and H. Ellenberg. 1974. *Aims and Methods of Vegetation Study*. Wiley, New York.

Pielou, E. C. 1974. *Population and Community Ecology: Principles and Methods*. Gordon and Breach, New York.

Pound, R., and F. E. Clements. 1898. A method of determining the abundance of secondary species. *Minn. Bot. Studies* 2: 19–24.

Tanner, J. T. 1978. *Guide to the Study of Animal Populations*. Univ. Tennessee Press, Knoxville.

Exercise 6

SPATIAL RELATIONS IN PLANTS

Key to Textbooks: Brewer 77–79; Barbour et al. 53–59, 79–82; Benton and Werner 66–69, 365; Krebs 295–296; McNaughton and Wolf 82–89; Odum 205–211; Pianka 145; Richardson 301–302; Ricklefs (short) 235–237; Smith 433–437, 700–703; Whittaker 67–73.

OBJECTIVES

We learn what random distributions look like and practice methods of determining whether spacing is random, even, or aggregated. Distribution relative to another species is also studied. We try to relate non-random spacing to the underlying ecological processes.

INTRODUCTION

The basic assumption of studies of spatial distribution is that individuals of a species will be spaced randomly unless something is biasing the distribution. Non-random distribution, referred to as *pattern,* is a clue to physical or biotic factors important in the ecology of the organism. Organisms may be clumped, or aggregated, because of patchiness in the physical environment (growing in the wet spots, for example) or for biotic reasons (because vegetative spread is more important than seeds, for example). Even spacing is usually the result of competitive interactions, either by one individual depleting a resource in its vicinity to a point where another individual cannot survive (*exploitation*) or by *interfering* with any nearby individuals. Among plants the distinction between exploitation and interference in intraspecific competition is often vague but in animals even distributions as the result of interference are frequent.

Organisms may show pattern on the ground and they may also show patterns in their distribution relative to other species. Two species may be distributed over the ground just as though the other species were not there; they may tend to grow near one another (*positive association*); or they may seem to avoid one another (*negative association*). Positive and negative association in the statistical sense do not necessarily stem from coactions of the two species. They may, for example, occur together because both prefer wetter parts of a habitat or they may show a negative association because one grows in the wetter parts and the other in the drier parts. Sometimes, of course, direct interactions are involved. The co-occurring species may be mutualists or parasite and host. Species that do not occur together may be competitors.

The distribution of organisms in space is examined in the first section (Spacing) and distribution in relation to other species (Interspecific Association) in the second.

Key Words: spacing (dispersion), pattern, random distribution, aggregation, contagious distribution, even spacing, competition, Clark-Evans (nearest neighbor) method, variance:mean ratio, Poisson distribution, interspecific association.

SPACING

The Clark-Evans Nearest Neighbor Method. Several methods are available for studying spacing (also called *dispersion*) of plants or animals. The first method we will use was developed by Clark and Evans (1954). It consists of obtaining a ratio, R, between the actual mean distance from each individual to its nearest neighbor ($\bar{r}_A$) and the expected mean distance between neighbors in a random population of the same density ($\bar{r}_E$). When R (i.e., $\bar{r}_A/\bar{r}_E$) ≈ 1, the distribution is random. When the members of the population are evenly spaced, each individual will be equidistant from six other individuals and R is 2.1491. Under conditions of complete aggregation, the observed mean distance between neighbors would be 0, and R likewise would be 0. The mean distance to nearest neighbor in a random population, $\bar{r}_E$, is equal to

$$\frac{1}{2 \times \sqrt{\text{density}}}\ .$$

(The derivation of these is in Clark and Evans 1954.)

1. To obtain a population on which to practice the method, use the Blank Plot and enter 40 dots with a pen or pencil. Try carefully (rather than by closing your eyes) to place them randomly. Remember that in a random distribution every point has an equal and independent probability of being occupied.

2. The specific steps in testing whether you achieved a random distribution are as follows: Measure the distance from each dot to the one nearest it and record on the Work Sheet (see Fig. 6–1). Here, use millimeters (for other populations, any convenient unit). Add these individual distances together and enter them in the blank Σr. Obtain the mean value by dividing Σr by N, the total number of individuals. To obtain density (as defined for these calculations), divide the total number of dots by the total area of the plot (here expressed as mm^2). Figure $\bar{r}_E$ by the formula above. Then calculate R.

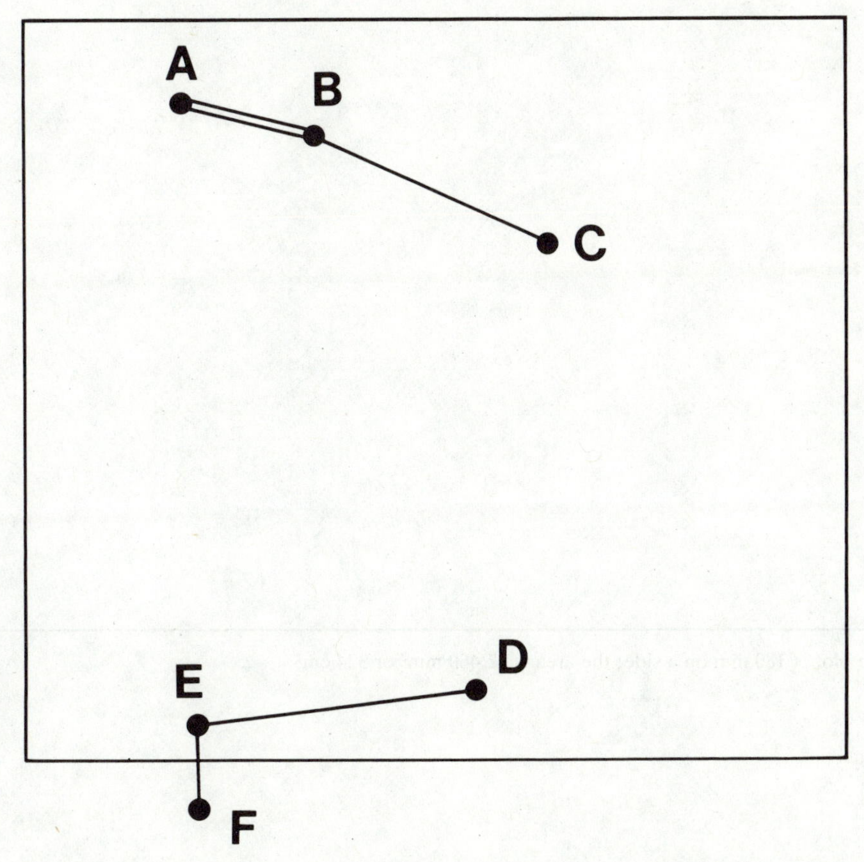

Figure 6–1. Nearest neighbor measurements. In this example, measurements of nearest neighbor would be made as follows: For A, measure to B; for B, measure to A; for C, measure to B; for D, measure to E; for E, measure to F.

Blank Plot for the Clark-Evans Nearest Neighbor Method

The plot is 180 mm on a side; the area is 32,400 mm² or 324 cm².

Work Sheet for the Clark-Evans Nearest Neighbor Method

Individual distances (r)

Number of distances measured	N 40
Sum of distances	Σr
Mean distance	$\Sigma r \div N = \bar{r}_A$
Density	$N \div$ area
$\bar{r}_E$	$\frac{1}{2 \times \sqrt{\text{density}}}$
Ratio	$\frac{\bar{r}_A}{\bar{r}_E} = R$
Spacing?	

Work Sheet for Variance:Mean Ratio Method

Number of individuals in each sample (x)

Total number of samples	N ______
Total number of individuals	Σx ______
Mean number of individuals per sample	$\Sigma x \div N = \bar{x}$ ______
Square Σx*	$(\Sigma x)^2$ ______
Square the individual x's and sum the squares	$\Sigma(x^2)$ ______
Calculate the variance, S^2, as $\dfrac{\Sigma(x^2) - \dfrac{(\Sigma x)^2}{N}}{N - 1}$	S^2 ______
Calculate the variance:mean ratio	$\dfrac{S^2}{\bar{x}}$ ______
Spacing?	______

*This step and the next two are used to calculate the variance. If calculators that compute the variance automatically are available, you may use them unless the instructor directs otherwise.

If this is about 1, the population is randomly distributed; if it is much smaller than 1, the population is aggregated; if it is much larger than 1, the population is evenly distributed. If it is larger than 2.1491, you have an error in your calculations. Be sure that nearest neighbor and plot measurements are in the same units.

Determining whether the calculated ratio is significantly different from 1 requires a statistical test which should be done only if the class has a satisfactory statistical background. Specifically, the significance of an apparent deviation from randomness can be determined by calculating K* as

$$K = \frac{\bar{r}_A - \bar{r}_E}{\sigma_{\bar{r}_E}}$$

where K is the standard variate of the normal curve and $\sigma_{\bar{r}_E}$ is the standard error of the mean distance to nearest neighbor in a randomly distributed population of the same density as that of the observed population. The value of $\sigma_{\bar{r}_E}$ is

$$\frac{0.26136}{\sqrt{N \times \text{density}}}.$$

Interpretation of the test is the same as with a t test. If the calculated value of K is greater than the table value of K for a given level of significance, then the hypothesis that $\bar{r}_A$ is the same as $\bar{r}_E$ is rejected. Table values of K are 1.96 for the 5% level of significance and 2.58 for the 1% level.

Discussion

1. Which distribution—random, even, or aggregated—predominated in the class? If it was something other than random, explain what defect in your fellow students' understanding of randomness may have produced the result.
2. What are some kinds of populations that could be studied using this method?
3. Suppose that for convenience in sampling, you restrict measurements to the nearest neighbor within your sample plots. That is, for some of the individuals in your quadrats their actual nearest neighbor lies just outside the quadrat but you measure instead to the nearest neighbor within the quadrat. In what direction will this bias your result? Why?

The Variance:Mean Ratio Method. Another approach to determining pattern is a quadrat method originally used by the Swedish botanist Svedberg (1922). The number of individuals per quadrat is counted and this information is compared with the results that would be expected if the population was randomly distributed. Specifically, the comparison is made with a Poisson (pronounced "pwa-sone") distribution. All that we need to know about the Poisson distribution is that it describes the probability distribution for random, fairly rare events; that is, it gives the frequencies with which we expect to find all the various numbers. If we know the average density per square meter of some plant or animal, the Poisson distribution will tell us how many square meters will have 0, 1, 2, etc. individuals if the distribution is random.

One method of studying spacing is to "fit" the observed data to a Poisson distribution (Blackman 1935). If the observed and expected numbers of quadrats with 0, 1, 2 . . . n individuals are similar (tested by χ^2) then the distribution is random. If not, the distribution is either aggregated (if the zero class is too big) or evenly distributed (if the zero class is smaller than expected on the assumption of randomness). A different method, which we prefer (because it avoids problems connected with certain requirements of the χ^2 test), makes use of the fact that in the Poisson distribution, the variance is equal to the mean. That is, variance ÷ mean = 1.

Count the number of individuals of anything in some 20 to 50 sampling units. Depending on opportunity you can count dandelion plants in square meter quadrats on a lawn, sowbugs under

*The symbol c, rather than K, is used in Clark and Evans (1954).

bricks, cowbird eggs in song sparrow nests, or whatever. In general, choose the size of your sampling unit so that mean density will come out to be less than 5 per sampling unit. Enter the data and calculate the mean, the variance, and the variance:mean ratio on the Work Sheet. These statistics are described in Appendix 4. If the ratio is about 1, the population is randomly spaced. If it is less than 1, the population tends toward even spacing and, if more than 1, toward aggregation. (Note that the interpretation is opposite to that of the nearest neighbor method.)

As with the nearest neighbor method, determining whether a given deviation from 1 is significant requires statistical testing. This can be done by calculating t as follows:

$$t = \frac{\text{calculated ratio} - 1}{\sqrt{\frac{2}{N-1}}}.$$

The resulting t value is compared with the values in a t table (p. 241) and the hypothesis that the two ratios are the same is rejected or accepted.* A general discussion of t tests is given in Appendix 4.

Discussion

1. Was the distribution you studied random, even, or aggregated?
2. If the distribution was not random, suggest what ecological factors might be involved in biasing the spacing.
3. If you had used a much larger (or smaller) quadrat, might the result have been different?
4. Clumped distributions are sometimes referred to as *contagious* distributions. Why is that?

INTERSPECIFIC ASSOCIATION

Several ways of measuring association among species have been devised. Unfortunately, most of them are influenced by the size of the sampling unit used and by various features of the distribution of the two species on the ground. We describe here one widely used method. It tests occurrence in sample units using χ^2 (chi square). The first application of χ^2 to study association among species was by Edmondson (1944) on the suggestion of G. E. Hutchinson.

In the method a series of random quadrats are taken and the presence of two (or more) species is noted. From such data in other exercises, we have calculated percent frequency. For example, species A may be present in 30 of 50 quadrats and so have a frequency of 60%, and species B may be present in 10 of the quadrats and so have a frequency of 20%. If the two species are distributed independently of one another, they should occur together in 20% of 60% of the 50 quadrats. That is, 30 (60%) of the 50 will have species A, and of those 30, 6 (20%) should have species B. The two will occur together then in 12% or 6 of the 50 quadrats, on the average. Suppose now that species A puts something in the soil that makes species B thrive. In this case species B might tend to grow near species A and the two might co-occur at a much higher rate. On the other hand, if the two were competitive then species B might grow only in the spaces where species A did not grow and they might occur together in few or no quadrats.

---

*Given certain peculiar circumstances, non-random distributions may yield variance:mean ratios near 1. If you obtain a ratio not significantly different from 1 but have some reason to suppose the distribution is not random, calculate the expected frequency of empty sampling units (zeros). This, the first term of the Poisson series, is given by

$$N\,(e^{-\bar{x}})$$

where N is the total number of sample units, $\bar{x}$ the mean number of individuals per sample, and e the base of natural logarithms. Compare this value with your observed number of zero samples. If the two differ substantially you may wish to consider further whether the population may be non-randomly distributed.

What we are interested in testing is whether the actual occurrences of the two species are statistically independent of one another or whether the distribution of one is somehow influenced by the distribution of the other. We could set this up for testing by chi square as follows:

	Number of Quadrats	
Species	**Observed**	**Expected**
A alone		24
B alone		4
A and B		6
Neither		16

All of the expected values are calculated as was shown for the joint occurrence of A and B. For example, the number of quadrats expected to have species B but not species A is derived as follows: The probability of a quadrat having B is 20%. The probability of a quadrat lacking A is 40% (since 60% of the quadrats have species A, 40% lack it). The joint probability of (species B present, species A absent), then, is 20% of 40%, or 8%, so out of 50 quadrats, 4 (8% of 50) should be in this category. We can fill in the actual observed values—suppose they are 28, 8, 2, and 12—and proceed to calculate χ^2 (as described in Appendix 4). There will be only a single degree of freedom because once we have specified the value for any one of the combinations, the values for all the rest are set.

Usally, however, χ^2 is calculated from data of this sort using a 2×2 contingency table, which is faster in that calculation of expected values is unnecessary. This is done as follows (the species with the higher frequency is used as species A).

		Species A	
		Present	**Absent**
Species B	**Present**	2 (a)	8 (b)
	Absent	28 (c)	12 (d)

The observed data given in the preceding paragraph are shown in the table. χ^2 is calculated from the contingency table as follows:

$$\chi^2 = \frac{(ad - bc)^2 N}{(a + b)(c + d)(a + c)(b + d)}$$

N is the total number of quadrats ($a + b + c + d$ should equal N). For the example, then χ^2 is

$$\frac{(24 - 244)^2\ 50}{(10)\ (40)\ (30)\ (20)} = \frac{2{,}000{,}000}{240{,}000} = 8.33$$

This value is compared with the table value of χ^2 (1 degree of freedom). The test is of the hypothesis that the distribution of one species is not dependent on the other. A level of significance (α) of 5, 1, or 0.5% would ordinarily be used; the corresponding table values of χ^2 with one degree of freedom are 3.84, 6.63, and 7.88. Since the calculated value exceeds the table value for any of these levels of significance, we would reject the hypothesis and would, in effect, conclude that the two species were associated. We could determine whether the association was positive or negative by going back and calculating the expected numbers in each category or, more simply, the association is positive if $ad > bc$ and negative if the reverse, as in this example.

If a significant value had not been found (for example if the 5% level of significance had been chosen beforehand and the calculated χ^2 value had been less than 3.84) then we could not reject the null hypothesis. We would conclude that the two species were distributed independently of one another.

The suggestions are sometimes made that the test should not be applied if the expected cell size (that is, the expected value for a, b, c, or d) is below 5, that if the sample size is below about 30, Yates's correction should be applied, or that if the sample size is below about 20, Fisher's exact method of calculating χ^2 should be used. It is not clear that any of these suggestions are correct (Pielou 1974, Cochran 1952, Snedecor and Cochran 1967) and, in any case, the usefulness of the test is ordinarily preliminary, for the detection of possible cases of association. These would then be studied by more direct methods to determine whether there was an association and what it depended on ecologically. Consequently, keeping the test simple and realizing that a significant χ^2 value is not very informative by itself seems to be the best approach.

If a significant χ^2 value is obtained and sample size is large, various coefficients have been devised that are supposed to measure the strength of association between species. For example, if in a forest there is a group of species that occurs in the low spots and another group in the high spots, members of the two groups should be strongly associated (positive) with one another, strongly associated (negative) with members of the other group, and perhaps weakly associated one way or the other with the species that occur on the middle ground.

1. Obtain a set of samples in which you can record the occurrence of two (or more) species. The simplest sort of data is probably the occurrence in a series of quadrats of various plants. Data from Exercise 19 or Exercise 16 could be used. The sample units need not be plots, however. If good data are available on the flora or fauna of the lakes and ponds of your region you could test for interspecific association among pairs of species of duckweeds or sunfish. The goldenrod plants in a large stand could be checked for two or more kinds of galls. Going further afield (or astray), blocks of strip development in a city could be checked to see if there were negative associations between, say, different brands of fast-food hamburger stands but no, or perhaps positive, associations between hamburger and fried chicken joints.

A sample size of 30 or more is desirable but 20 will do. The size of the sampling unit should be selected so that neither of the species to be considered has an extremely high or an extremely low frequency.

Be sure to record the data for each quadrat or sample unit separately.

2. Fill in the cells of the following contingency table. Make additional copies if you need to test more than one pair of species.

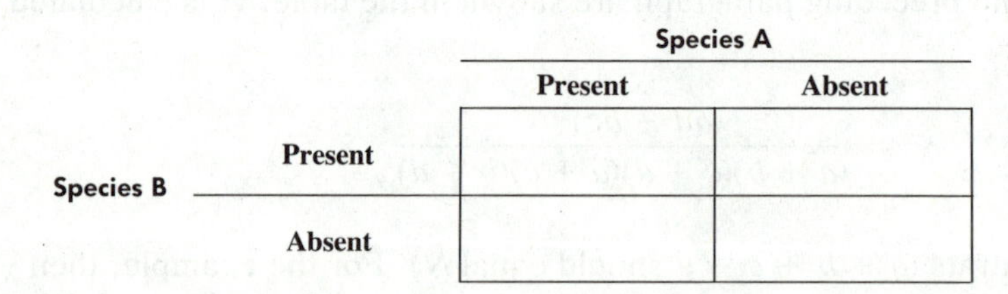

3. Calculate χ^2 using the formula given earlier.

$$\chi^2 = \underline{\hspace{6em}}.$$

Discussion

1. Do the two species show a significant statistical association at the 5% level? Was the association positive or negative? How can you tell? (If several tests were done, give answers for all pairs of species.)

2. Calculate for one or more of the species pairs the expected number of quadrats for each cell assuming the two species are distributed independently.

3. If you detected any significant associations, speculate on what the underlying ecological cause might be.

4. Suggest situations in which you are sure you would find a significant positive association between two species. Significant negative associations.

5. Under what circumstances might a predator and its prey show a positive association? A negative association?

6. Might two species be positively or negatively associated in some habitats or situations and independently distributed in others? How?

Teaching Notes

This exercise is readily done in three hours. Attempting to produce a random distribution and then testing it helps to understand the concept of randomness. One can use both nearest neighbor and variance:mean ratio methods on the same field population for an interesting comparison. If there is absolutely no way to use natural populations, the map of artificial plant populations (Exercise 5) can, of course, be used.

BIBLIOGRAPHY

Blackman, G. E. 1935. A study by statistical methods of the distribution of species in grassland associations. *Ann. Bot.* 49: 749–778.

Clark, P. J., and F. C. Evans. 1954. Distance to nearest neighbor as a measure of spatial relationships in populations. *Ecology* 35: 445–453.

———. 1979. Generalization of a nearest neighbor measure of dispersion for use in k dimensions. *Ecology* 60: 316–317.

Cochran, W. G. 1952. The χ^2 test of goodness of fit. *Ann. Math. Stat.* 23: 315–345.

Cole, L. C. 1949. The measurement of interspecific association. *Ecology* 30: 411–424.

Hurlbert, S. H. 1969. A coefficient of interspecific association. *Ecology* 50: 1–9.

Edmondson, W. T. 1944. Ecological studies of sessile Rotatoria. Part I. Factors affecting distribution. *Ecol. Monogr.* 14: 31–66.

Goodall, D. W., and N. E. West. 1979. A comparison of techniques for assessing dispersion patterns. *Vegetatio* 40: 15–27.

Greig-Smith, P. 1964. *Quantitative Plant Ecology*. 2nd ed. Butterworth's, London.

Kershaw, K. A. 1974. *Quantitative and Dynamic Plant Ecology*. 2nd ed. Elsevier, New York.

Pielou, E. C. 1974. Chapters 7 and 11 in *Population and Community Ecology*. Gordon and Breach, New York.

Dice, L. R. 1945. Measures of the amount of ecologic association between species. *Ecology* 26: 297–302.

Simberloff, D. 1979. Nearest neighbor assessments of spatial configurations of circles rather than points. *Ecology* 60: 679–685.

Snedecor, G. W., and W. G. Cochran. 1967. *Statistical Methods*. Iowa State College Press, Ames, Iowa.

Svedburg, T. 1922. *Ett bidrag till de statistiska metodernas användning inom växtbiologien. Svensk Bot. Tidskrift* 16: 1–8.

Exercise 7

SPATIAL RELATIONS IN HUMANS

Key to Textbooks: Brewer 69–72, 77–79; Colinvaux 457–470; Emlen 186–209, 280–281; Kendeigh 189–190, 198–200; McNaughton and Wolf 332–340, 618–621; Richardson 282–287, 310–326; Ricklefs 52–55, 216–229, 495–500; Ricklefs (short) 236; Smith 345–348, 428–430; Smith (short) 233–239.

OBJECTIVES

We will study the spacing achieved behaviorally in animals. We will use humans as our experimental species and speculate on the role of spatial organization and perception in human populations.

INTRODUCTION

Spacing in animals (or, more correctly, in motile organisms) is achieved behaviorally. At least among the more complicated animals, spacing may vary widely depending on stage of the life cycle, physiological state, and other factors. One species of bird may, for example, occur in evenly spaced territories in the breeding season, in loose flocks while feeding in the winter, and in very compact flocks when under attack in the air by a hawk. Humans, no less than other animals, show characteristic patterns of spacing. On a large scale, people may be clumped into cities. Here we consider human spatial arrangements on the scale of individuals, particularly *individual distance* or *personal space*. The study of human use of space has been called *proxemics* by Hall (1966). The two projects described below are based on work by him and by Sommer (1969).

PROCEDURE

Half the class should do one of the projects and half the other.

A. This is best done in pairs. Go to an area in a fair-sized library that has several reading tables, each with six or more chairs. Pick a time when there are few patrons so that there are plenty of chairs and, preferably, at least one totally empty table. One member of the pair should find a book or magazine and sit down directly alongside one of the patrons (or "victims," as Sommer called them). If the chairs are not closer than 0.5 m, the experimenter should hitch his or her chair closer to that of the victim until it is within 0.5 m.

The observer should have the victim and another person, preferably of the same sex, approximate age, etc., who is sitting alone, under observation at the same time. The second person is the

Key Words: spacing, behavior, individual distance, personal space, territoriality, dominance hierarchy, proxemics, crowding.

control. Observe the behavior of the victim and the control. Does the victim adopt any postures not seen in the control? Does the victim move his or her chair away from the experimenter? Does he or she get up and leave? Record behavior and times for ten minutes or until the victim leaves.

The observer-experimenter pair should obtain data on at least four subjects. Make sure that your activities are not obvious or conspicuous (for example, move to another part of the library or come back at another time). Preferably the two should change roles so that each gets a chance to function as both observer and experimenter.

Analyze the data by comparing the behavior of the victims with that of the controls.

B. This could be done in pairs also, facilitating estimating distance and providing someone to corroborate explanations, if necessary. However, it can be done alone if each observer practices estimating distances from himself or herself to other persons. Someone close enough to put your arm around is less than 0.3 m away. Arm's length is about 0.7 m. Choose subjects, or victims, from as many of the following categories as possible. Include at least six subjects in at least three categories.

1. An adult with a close genetic connection—parent, sister, or brother.
2. A child with a close genetic connection—offspring, brother, or sister.
3. Wife, husband, boyfriend, or girlfriend.
4. A longtime close friend of the same sex.
5. A co-worker or fellow student of the same sex.
6. A co-worker or fellow student of the opposite sex.
7. A male clerk in a store.
8. A female clerk in a store.
9. A male stranger standing still (for example, waiting for a bus or elevator, in a movie ticket line, etc.).
10. A female stranger standing still, as in 9.

Approach the subject from the front (or three quarters) so that your face is within about 0.2 m of his or hers. Ask some relevant question, such as a request for directions or some other information. Observe (or, better, have a partner unobtrusively observe) any changes in expression or posture following your close approach. Maintain a conversation for at least 15 seconds and longer if possible. Take note and later record the distance at which the conversation is conducted, as resulting from movements by the subject (you should avoid moving).

Analyze the observations, comparing results in the different situations. Were facial or postural responses evident and did they differ in the different categories? What was the individual distance that developed? Was it different in the different categories?

DISCUSSION

1. The distinction is sometimes made between "contact" and "non-contact" species. Contact species, such as the walrus and hippopotamus, tend to huddle together, with much body contact. Are humans contact or non-contact species?

2. Do humans appear to have an individual distance that they tend to maintain around them? Is it the same in all situations? If not, list some factors that tend to lower individual distance and some that tend to raise it.

3. What are some ways people behave to lessen the impact of being very close to others, such as in crowded elevators?

4. Is minimum distance, or personal space, in the same situation (for example, strangers talking) similar for all persons? Does it, for example, seem to be different between persons of different nationalities?

5. If such differences seem to exist, how would you account for them? Are they the result of learning? If so, how are they learned? Could genetic factors be important?

6. Some authors have suggested a reciprocal relationship between dominance and territoriality, with a more definite hierarchy where space is not controlled. Did anything in your observations support (or refute) this?

7. Sommer listed four major zones of personal distance in man. They are, with approximate distances:

Intimate 0–0.45 m

Personal 0.45–1.2 m

Social-consultive 1.2–3 m

Public 3 m or more

Do these categories seem reasonable? Which of the situations in project B fit in the various categories?

8. What connection, if any, is there between personal space and territoriality? Give some human examples of each.

9. High density and crowding (or overcrowding) are sometimes used almost synonymously. Are they necessarily synonymous? On what factors besides density does crowding (or overcrowding) depend?

10. Is there any evidence that high human population densities are tolerated less well in some places or at some times than others? If so, what are some possible ways of making high densities more tolerable?

Teaching Notes

A few minutes can be used to assign this exercise to be done out of class; later, an hour or so can be used to discuss the results. It may be necessary to clear this exercise with a Human Subjects Research Committee if your school has one. A movie, *Invisible Walls* (black and white, 12 minutes, Univ. California, Berkeley, Extension Media Center, 1968), goes well with part B (it is more enjoyable after the observations have been made). Its suggestion that somehow circumventing personal space requirements to allow for crowding is desirable for the future is, at least, debatable.

BIBLIOGRAPHY

Calhoun, J. B. 1962. Population density and social pathology. *Sci. American* 206 (2): 139–147.

Hall, E. T. 1966. *The Hidden Dimension*. Doubleday, Garden City, New York (Anchor Books, 1969).

Morris, D. 1978. *Man-Watching*. Abrams, New York.

Newman, O. 1972. *Defensible Space*. Macmillan, New York.

Sommer, R. 1969. *Personal Space: The Behavioral Basis of Design*. Prentice-Hall, Englewood Cliffs, New Jersey.

Wilson, E. O. 1975. Chapter 12: Social spacing, including territory; Chapter 13: Dominance systems. In *Sociobiology: The New Synthesis*. Harvard Univ. Press, Cambridge.

Exercise 8
TERRITORIALITY

Key to Textbooks: Brewer 69–72, 79–80; Benton and Werner 347–348, 364; Colinvaux 443–455; Kendeigh 199–200, 214–215; Krebs 60, 324–326; McNaughton and Wolf 68, 321–322; Odum 195–197, 209–211; Pianka 144–148, 159–164; Richardson 310–335; Ricklefs 216–225, 237–240, 272–275, 491–492; Ricklefs (short) 310–335; Smith 399–432, 485–491; Smith (short) 233–238; Whittaker 22.

OBJECTIVES

To determine the approximate size of territories of a species of bird and to observe how the territory is defended; also to determine the mating system and to attempt to define some essential habitat elements of territories of the species.

INTRODUCTION

Many kinds of animals defend an area against other animals, usually just those of the same species. Such defended areas are called *territories*. The *home range* of an animal is the area over which it routinely travels. An animal may defend no part of its home range if it is not territorial; it may defend a part of it; or it may defend the whole of it, in which case its home range and territory coincide.

Territorial defense may involve actual combat to determine who occupies an area or where the boundary between adjacent territories lies; more often defense is by various kinds of ritualized behavior, called *displays*. Among birds, song often advertises that a territory is occupied by an individual that will fight for it, if need be.

A frequent situation among birds is for the males to occupy territories early in spring. Somewhat later, after the territories are relatively well established, the females arrive, are courted by males, and, if a pair bond develops, settle on a male's territory and eventually lay eggs and rear young. If the usual situation in a population is for only one female to be associated with a male on his territory, the population is said to be *monogamous*. If a fair number of males have more than one female, the population is said to be *polygynous*. In the very rare case where one female has several male consorts, the term *polyandry* is applied.

For this exercise we recommend using the red-winged blackbird *(Agelaius phoeniceus)*. The red-wing is a widespread species of passerine bird, formerly occurring mainly in cattail marshes but now occupying hayfields and many other kinds of upland herbaceous communities. Males are black with red patches bordered with yellow on what appear to be their shoulders (but are actually their wrists).

Key Words: territory, home range, display, mating system (monogamy, polygyny, polyandry), group selection, individual selection, sexual selection, habitat requirements.

Females are very different. They are smaller, brownish, striped birds; they lack the red and yellow epaulets and look a little like a large sparrow.

This exercise is designed as a simple observational study with theory left to the discussion; however, depending on the inclination of individual classes, various sorts of experimental approaches could be tried. For example, vegetation or food supply could be sampled quantitatively in different territories. Models (lacking epaulets, with blue epaulets, miniature, round rather than bird-shaped, etc.) could be made and put out in the field to determine what features of the male red-wing other males respond to.

MATERIALS

dittoed maps of the study area with scale, landmarks, and north indicated; 1–3 per student

binoculars for each student are useful, as is a compass to allow more accurate plotting of locations

watches

a poncho or piece of plastic to sit on

PROCEDURE IN THE FIELD

This exercise should be done in spring or early summer from early to mid-morning. It is best done in an upland herbaceous habitat such as a hayfield; a marsh is satisfactory provided the students can move about in it or, failing that, have a good view over it from adjacent high ground.

The instructor should supply each student with a map having a scale not much larger than 100 m to 5 cm (about 150 feet to the inch). If such an area has been carefully mapped using transit and steel tape, that is ideal; however, much less precise mapping will suffice. For example, tracing landmarks from a large-scale aerial photograph (see Exercise 18) will do; so will pacing using a hand-held compass. In the latter case, mark coordinate intersections at 50 m (or 200 feet) intervals with plastic flagging. The map should show the grid of coordinates or the landmarks or both.

If possible, the size of the area used should be chosen so as to have slightly more red-wing territories than students; if a smaller area must be used, having students in pairs or threes on a single territory also works well.

1. The instructor should orient the students so that they can plot observations accurately. The instructor should then space the students in pairs 125–200 m (400–600 feet) apart over the area.

2. Each pair of students should make some preliminary observations to determine approximate locations of male red-wings around them. After a few minutes each should choose a different male red-wing, move apart 30 or 40 m (so as to provide a less concentrated disturbance), and center his or her observations for the rest of the period on that bird.

3. One major aim will be to map approximate territorial boundaries. Accordingly, attempt to plot each location of the male you have chosen. Every time the male moves to a substantially new position (more than 5 m), indicate the position with a dot and record the time next to it. If it stays in one spot for a long period, simply record the time next to the same dot every five minutes. Write small.

4. As time permits, record approximate locations of the males with bordering territories. If some of these have not been assigned to other students, your observations, though not intensive, will allow a better picture of the occupancy of the area. They will also help you keep better track of the boundaries of your own male's territory.

5. At some specified interval (one to five minutes depending on the volume of data wanted), record what the male is doing at that instant. Use the following categories: territorial defense (singing, displaying, chasing, etc.), resting and maintenance (sitting quietly, preening, feeding, etc.), out of sight, and other (specify).

6. Observe and record descriptions of vocalizations, postures, and movements that males use in territorial defense.

7. Once you have an idea of the approximate boundaries of the territory, keep track of how many females are within it. Try to rough out the areas they occupy within the male's territory. Note what role, if any, the females have in territorial defense. Do they defend sub-territories against one another within the male's territory? Do they defend their territories against bordering females in other males' territories? Do they help defend the male's territories against other males?

8. Note whether the male leaves his territory. If he does, record the time he leaves and returns. Note how he behaves when he is off his territory, if feasible. Do the females leave the territory?

9. At the end of the observations and before going back to the laboratory, the whole class should visit a few territories to compare territories of males having the fewest and males having the most females. You will have an opportunity to compare the two kinds of territories as to size later. Do you see any obvious, consistent differences in other features? Is the vegetation shorter or sparser in territories with more or fewer females? Are there more or taller singing perches in one kind of territory than the other? Are territories with more females in the middle of the field or at the edges? Do you see other differences?

PROCEDURE IN THE LABORATORY

1. Outline the territory of the male you studied as carefully and accurately as possible. Using the random dot planimeter in Exercise 18 calculate the area of the territory. Do this by determining how large an area on the ground corresponds to one square centimeter of your map. Then set up the proportionality

$$\frac{\text{10 dots}}{\text{on-the-ground equivalent of 1 cm}^2} = \frac{\text{numbers of dots in territory}}{\text{size of territory}}.$$

2. Combine all the territories on a single map. In each write the size of the territory and the number of females it contained.

3. To assess temporal variation in territorial defense divide the time you were in the field approximately into equal thirds. Calculate the percentage of time the male red-wings spent in territory defense and in maintenance during each of the thirds.

4. Also calculate the time spent off territory and see if it was greater in the first, second, or third third of the observation period.

RESULTS AND DISCUSSION

1. Give the mean size of territory for male red-wings on the study area. Also give the size of the smallest and largest territories.

2. Describe the behavior you observed that seemed to be part of territorial defense. Could some of the behavior patterns also serve other functions?

3. Was there a trend in the percent of time spent in territory defense from early to later in the morning?

4. What other sorts of behavior did the male engage in while on territory?

5. Did the male leave his territory? Was there a trend in the percent of time he spent off the territory? What was the male doing while he was off the territory? What were his postures and vocalizations?

6. Did the female show territorial defense? Elaborate.

7. Many ecologists believe that territoriality is an important mechanism regulating local population size. A few ecologists believe that it was this advantage of territoriality to the population

(group) that led to the evolution of territoriality. Outline, in more detail, a *group selection* hypothesis for the evolution of territoriality.

Other ecologists believe that although territoriality may regulate local population size, the evolution of territoriality can be explained in terms of *individual selection*. Suggest several advantages to the individual animals displaying territoriality.

8. Suggest some additional roles, not directly related to population size, that territoriality plays at the population level.

9. What happens to males that do not manage to set up territories in good habitat? How could you determine if there were non-territorial "floater" males around?

10. Prepare a list of habitat features that seem essential for red-wing territories.

11. Presumably the genetic fitness of males will usually be increased by polygyny or, at least, by mating with as many females as possible. Polygyny will not evolve, however, unless it is also advantageous for the female. Specifically, it must sometimes be advantageous to a female to be the second or third mate in a harem rather than becoming the sole mate of an unmated male. Suggest habitat and population features that might predispose a species to polygyny.

12. Are the females making their choices on settling down based on individual traits of the male (plumage, voice, vigor, etc.) or habitat traits of the territory? Suggest benefits of each. What habitat features might make an especially good territory for a female red-wing (favoring her survival and nesting success)? How can the female red-wing measure such habitat features?

Teaching Notes

About one and one-half hours of field observations will suffice, although two to two and a half will show changes in activity better. An additional half hour is needed to get started and another half hour to exchange data and make measurements after the field work. Red-wings can be used for this exercise during the spring term over much of the U.S.; however, other bird species can be used with some slight modifications. If the observations have to be made on campus, song sparrows *(Melospiza melodia)* and chipping sparrows *(Spizella passerina)* are widespread, strongly territorial—but monogamous—possibilities.

BIBLIOGRAPHY

Albers, P. H. 1978. Habitat selection by breeding red-winged blackbirds. *Wilson Bull.* 90: 619–634.

Davies, N. B. 1978. Ecological questions about territorial behavior. Pp. 317–350 in J. R. Krebs and N. B. Davies (eds.). *Behavioral Ecology, an Evolutionary Approach*. Sinauer, Sunderland, Massachusetts.

Kendeigh, S. C. 1948. Bird populations and biotic communities in northern lower Michigan. *Ecology* 29: 101–114.

Nero, R. W. 1956. A behavior study of the red-winged blackbird. *Wilson Bull.* 68: 5–37, 129–150.

Orians, G. H. 1961. The ecology of blackbird *(Agelaius)* social systems. *Ecol. Monogr.* 31: 285–312.

———. 1969. On the evolution of mating systems of birds and mammals. *Am. Nat.* 103: 589–603.

Partridge, L. 1978. Habitat selection. Pp. 351–376 in J. R. Krebs and N. B. Davies (eds.). *Behavioral Ecology, an Evolutionary Approach*. Sinauer, Sunderland, Massachusetts.

Searcy, W. A. 1979. Female choice of mates: a general model for birds and its application to red-winged blackbirds. *Am. Nat.* 114: 77–100.

Verner, J. 1977. On the adaptive significance of territoriality. *Am. Nat.* 111: 769–775. Also see replies by G. Tullock, 1979, *Am. Nat.* 113: 772–775 and others.

Wilson, E. O. 1975. Chapter 12: Social spacing, including territory; Chapter 15: Sex and society. In *Sociobiology: The New Synthesis*. Harvard Univ. Press, Cambridge.

Wynne-Edwards, V. C. 1962. *Animal Dispersion in Relation to Social Behavior*. Oliver and Boyd, Edinburgh.

———. 1978. Intrinsic population control: an introduction. Pp. 1–22 in F. J. Ebling and D. M. Stoddart (eds.). *Population Control by Social Behavior*. Praeger, New York.

Exercise 9

PREDATION

Key to Textbooks: Brewer 90–97, 152–154, 162–164; Colinvaux 397–442; Emlen 157–185; Kendeigh 98, 156–168; Krebs 241–263; McNaughton and Wolf 268–310; Odum 220–226; Pianka 147–150, 200–221, 260–262; Richardson 359–364; Ricklefs 529–582; Ricklefs (short) 299–332; Smith 129, 520–538; Smith (short) 247–260; Whittaker 28–36.

OBJECTIVES

We determine the diet of a predatory bird and try to view this both in the context of optimal foraging by the owl and in broader terms of predator-prey relationships and energy flow. Nocturnality is discussed.

INTRODUCTION

Predation is one of the most important interactions between organisms. From the standpoint of the individual predator, predation is largely a matter of optimizing its foraging—looking in the right places and taking the right food items to maximize its energy (or perhaps nutrient) intake. From the standpoint of the individual prey, it is a matter of predator defense and avoidance and, when these fail, it is a mortality source important in determining such demographic traits as age structure, longevity, and sex ratio. At the population level, predation sometimes limits or regulates both predator and prey numbers, and, of course, food is very often the object of competition, both intra- and inter-specifically. The enhanced fitness of the more rapacious predator or the swifter or wilier prey leads to evolutionary changes that, in some cases, fit predator and prey into a tight, coacting system. At the level of the ecosystem, predation is the basis of energy transfer.

We examine here some aspects of predation by owls, using an indirect method, the examination of pellets they regurgitate. Owls swallow their prey whole or in large chunks, but they do not digest the bones, fur, or feathers. The digestive tract of the owl is constructed such that this undigested material is formed into a mass and regurgitated. Most owls are nocturnal, so it is difficult to observe their hunting directly. By day the owl roosts in some protected spot and produces one or two pellets with remains from the prey of the preceding night. Any individual owl usually has one or a few roosts that it habitually uses, so it is often possible to locate these and pick up many pellets over the course of a few months.

A great deal can be learned about the food of the owl by identifying and counting bones in the pellets. They give a virtually complete record of the owl's vertebrate food. If an owl eats a mouse and a sparrow, the bones of the mouse and sparrow will show up in the next day's pellet. If we know the number of pellets the owl produces in a day, we can calculate the actual quantity of food, and thereby

Key Words: predation, optimal foraging, nocturnality, carnivore, food chains, energy flow.

energy, it took in. Invertebrates in the diet are difficult to deal with, so for the few species of owls in which invertebrates are important as prey, the estimates of food and energy intake will be too low.

MATERIALS

owl pellets, 1–3 per student

finger bowls or similar containers, 1 per student

forceps, 1 per student

keys, such as Glass (1973), to mammal skulls, several (ideally 1 per student)

skeleton of mammal (e.g., cat) and bird (e.g., pigeon)

reference collections of mammal skeletons and of bird skeletons, if possible

magnifiers or low-power dissecting microscopes are helpful

PROCEDURE

The exercise requires a supply of owl pellets, enough that each student can analyze one or more. The pellets should be from a single species of owl. The best way to determine what owl is producing the pellets is to see it on the roost from which the pellets are being collected; however, owl ranges are virtually exclusive, so simply seeing or hearing an owl in the area will usually be diagnostic. If the instructor does not know of an owl roost, someone in the community may be able to tell of a source of owl pellets. They are an interesting aspect of the natural history of predation, and many ornithologists, wildlife biologists, and amateur nature enthusiasts take note when they find an accumulation of pellets.

Ideally, the pellets should be gathered at intervals of a few days and stored dry until use. An accumulation can perhaps be divided into a few age categories (new, older, oldest) on the basis of how weathered the pellets are.

Some comparative element, between species or between different times of the year or different habitats for the same species, may be added but is unnecessary.

1. Put a pellet into a fingerbowl (culture dish) or similar container. Tear it apart with fingers and forceps, separating the fur from the bones. Treat the bones reasonably gently; bird bones can be crushed and teeth can be lost from mammal skulls. Clean enough of the adhering fur from the bones for identification; the bones do not need to be absolutely clean.

2. Identify the species of prey in the pellet and how many individuals are represented.

Mammals. Bring together all the parts of mammal skulls from the pellet and try to match upper and lower jaws. Determine how many individuals are represented. For small prey, each skull probably represents an individual consumed. Larger prey, such as cottontail rabbits *(Sylvilagus floridanus)*, however, may provide several meals, so that rabbit bones in one pellet may represent only a fraction of a rabbit. For large prey, try to get an approximate idea of the percent of the skeleton in the pellet and assume that a similar fraction of the meat was eaten.

Mammal skulls can usually be identified, at least to genus, using Glass (1973) or a state or regional mammal book such as those listed in the bibliography. A reference collection (preferably disarticulated bones rather than mounted skeletons) that includes each of the mammals that occurs locally up to about woodchuck size is useful for checking the results of keying. The collection is also useful in cases where some of the bones do not seem to belong with any of the skulls. Identification of the odd elements may be possible by direct comparison with skeletons of appropriately sized mammals.

Birds. Identification of the skeletal parts of birds is considerably harder than for mammals. A little familiarity with mammalian bones should allow you to sort out bird bones as something different. They tend to be hard and shiny with the long parts hollow. No keys to bird skulls exist, and it will probably be too time consuming to identify the bird bones in the pellets to genus or species even if a good reference collection is available. It is probably best just to determine the number of individual birds in each pellet by pairing up some particular bone, such as humeri or femora, and then to place the individual birds into size categories by comparison with a small reference collection. The following species would make a good collection. All are species that can be legally obtained and kept without federal and state permits (the game species[*] must be taken in season by someone with a hunting license).

house sparrow, *Passer domesticus*—28 g

starling, *Sturnus vulgaris*—80 g

*bobwhite, *Colinus virginianus*—190 g

rock dove (pigeon), *Columba livia*—330 g

*Am. crow, *Corvus brachyrhynchos*—500 g

*mallard, *Anas platyrhynchos*—1200 g; or

*ring-necked pheasant, *Phasianus colchicus*—1200 g

*Canada goose, *Anser canadensis*—3800 g

It is desirable to have an additional size class of smaller-than-house-sparrow (= chickadee) with a weight of about 12 g; however, no non-protected species this small is readily available.

3. Enter the number in each mammal species and in each bird size category (use the eight above plus intermediates if necessary) in the Pellet Analysis Sheet. If comparisons are to be made, you will need to fill out separate sheets for each different situation (species, season, etc.).

4. If possible, fill in an average weight per individual of each prey item. For mammals, use local data if available and, if not, figures from regional mammal books (see Bibliography). For birds, use the weights listed above but feel free to interpolate between categories.

5. Multiply weight per individual by number of individuals to obtain the total weight of each species (or bird size category) consumed. Then obtain a daily weight consumption for each prey (last column of Analysis Sheet). In making this calculation we are assuming that large owls (such as barred and great horned) produce one pellet per day and that smaller ones (such as long-eared or burrowing) produce about one and one-half pellets per day.

RESULTS AND DISCUSSION

1. Which species made up the greatest part of the owl's diet numerically? Which species made up the greatest part by weight? Here and in the following questions make appropriate comparisons between species, season, etc.

2. Are all of the animals that the owl caught nocturnal? If not, account for their presence in the owl's diet.

3. Based on the habitats of the prey, where was the owl searching for food?

4. Complete the table (page 81) for the mammalian prey. You might add any mammal species that are in the habitats the owl is hunting but that did not show up in the pellets. Use information from the literature, local data, and your experience, as appropriate. Take season, habitat, and the species of owl into account. Easily found animals are those that will venture out into the open rather than almost always being underground or otherwise inaccessible. Easily captured animals are those that, when the owl has a chance at them, are not too large nor able to put up a strong defense; very slow animals would also be easily captured. A mole would not be easily found by a barred owl, but when occasionally the mole did appear above ground, it would be easily captured. It might, however, be a difficult prey for a small owl to take. Raccoons would be easily found by owls, but because of their

Owl Pellet Analysis Sheet

Prey Item	A Number	Percent by Number	B Live Weight (g) per Individual	C Total Live Weight (g) ($A \times B$)	Percent by Weight	Daily Weight Consumed (g)*
Mammals						
Birds (Size Categories)						
Totals		**100%**			**100%**	

*Daily weight consumed is equal to $C \div$ total number of pellets for large owls. For small owls it is $C \div$ (total number of pellets $\times$ 0.7).

size, teeth, and claws would not be easily captured even by large owls. Density should be individuals per hectare or some similar unit. If local studies are available, use them; otherwise use your best estimates.

Species	Average Weight	Density	Ease of Finding	Ease of Capture

The weight column is a rough index of the benefit (in energetic terms) to the owl of taking one individual of each prey species. The other three columns together are a rough measure of the energetic expense of taking the item. A scarce, inaccessible, elusive, and/or vicious prey will be the most expensive. We might hypothesize that an animal seeks to optimize its foraging (not necessarily consciously, of course) by running a rough cost/benefit analysis and then foraging in such a way as to concentrate on the food items with the best rate of return. On this basis, predict which should be the most frequent items in the owl's diet. Were they, in fact, the most frequent? What additional factors, not considered above, might be entering into the owl's foraging or choice of prey?

5. Suggest some advantages that owls gain by being nocturnal (and, thus, factors probably influential in their original and continuing adaptation to hunting in the dark).

6. What are some features of owls' structure, physiology, and behavior that allow them to be successful nocturnal predators?

7. What was the average daily live weight of food eaten by the owl? What was this as a percentage of the owl's weight? Weights for seven widespread species of owls are given below. For other species see Snyder and Wiley (1976), from which these weights were derived.

Species	Males	Females
barn owl, *Tyto alba*	440 g	490 g
screech owl, *Otus asio*	160 g	180 g
great horned owl, *Bubo virginianus*	1140 g	1510 g
snowy owl, *Nyctea scandiaca*	1640 g	1960 g
barred owl, *Strix varia*	630 g	800 g
long-eared owl, *Asio otus*	245 g	280 g
short-eared owl, *Asio flammeus*	315 g	380 g

8. Obtain approximate daily gross energy intake (kilocalories) for the owl as follows: Multiply daily live weight (bottom right of the Owl Pellet Analysis Sheet) by 0.3 to obtain dry weight; multiply dry weight by 5 to obtain kilocalories.

9. Construct a food chain in which the owl is a part. Is the owl a top carnivore (does anything prey on it)? Is it a secondary consumer or a tertiary consumer or both?

10. To what degree do hawks and owls compete? Include both exploitation and interference in your considerations.

Teaching Notes

The laboratory work can be completed in three hours; some computations and questions will have to be done beyond that. Store pellets in egg cartons to lessen breakage. Skeletons may be prepared using dermestid beetles (see Anderson 1948: 133–145) or enzymatic laundry "pre-soaks" (Ossian 1970). The key to skulls is available from Bryan P. Glass, Department of Zoology, Oklahoma State Univ., Stillwater, Oklahoma 74074. Whether or not reference collections are available, having one mammal and one bird skeleton (e.g., cat and pigeon) on display is useful in deciding whether a given bone is mammalian or avian.

BIBLIOGRAPHY

Anderson, R.M. 1948. Chapter VI, Collecting skeletons. Pp. 133–145 in *Methods of Collecting and Preserving Vertebrate Animals*. 2nd ed. Canada Dept. Mines and Resources, Mines and Geology Branch, Natl. Museum of Canada Bull. no. 69.

Baumgartner, A.M., and F.M. Baumgartner. 1944. Hawks and owls in Oklahoma 1939–1942: food habits and population changes. *Wilson Bull.* 56: 209–215.

Bent, A.C., ed. 1938. Life histories of North American birds of prey, part 2. *U.S. Natl. Mus. Bull.* 167: 1–482 (reprinted 1961, Dover, New York).

Burton, J.A., ed. 1973. *Owls of the World: Their Evolution, Structure, and Ecology.* Dutton, New York.

Craighead, J.J., and F.C. Craighead, Jr. 1956. *Hawks, Owls, and Wildlife.* Stackpole, Harrisburg, Pennsylvania (reprinted 1969, Dover, New York).

Errington, P.A. 1932. Technique of raptor food habits study. *Condor* 34: 75–86.

Graham, R.R. 1934. The silent flight of owls. *J. Roy. Aeronautical Soc.* 38: 837–843.

Krebs, J.R. 1978. Optimal foraging: decision rules for predators. Pp. 23–63 in J.R. Krebs and N.B. Davies (eds.) *Behavioral Ecology: An Evolutionary Approach.* Sinauer, Sunderland, Massachusetts.

Marti, C.D. 1973. Food consumption and pellet formation rates in four owl species. *Wilson Bull.* 85: 178–181.

Ossian, C.R. 1970. Preparation of disarticulated skeletons using enzyme-based laundry pre-soakers. *Copeia* 1970: 199–200.

Sparks, J., and T. Soper. 1970. *Owls: Their Natural and Unnatural History.* Taplinger, New York.

Snyder, N.F.R., and J.W. Wiley. 1976. Sexual size dimorphism in hawks and owls of North America. *Amer. Ornith. Union Monogr.* no. 20: 1–96.

Wallace, G.J. 1948. The barn owl in Michigan: its distribution, natural history and food habits. *Mich. State Agric. Expt. Sta. Tech. Bull.* 208: 1–61.

Sources for Mammal Identification and Weights*

Banfield, A. W. F. 1974. *The Mammals of Canada*. Univ. Toronto Press, Toronto.

Burt, W. H. 1957. *Mammals of the Great Lakes Region*. Univ. Michigan Press, Ann Arbor.

______, and R. P. Grossenheider. 1976. *A Field Guide to the Mammals; Field Marks of All North American Species Found North of Mexico*. 3rd ed. Houghton Mifflin, Boston.

Godin, A. J. 1977. *Wild Mammals of New England*. The Johns Hopkins Univ. Press, Baltimore.

Glass, B. P. 1973. *A Key to the Skulls of North American Mammals*. 2nd ed. Published by the author, Stillwater, Oklahoma.

Golley, F. B. 1962. *Mammals of Georgia: A Study of Their Distribution and Functional Role in the Ecosystem*. Univ. Georgia Press, Athens.

Ingles, L. G. 1954. *Mammals of California and its Coastal Waters*. 2nd ed. Stanford Univ. Press, Stanford.

Lechleitner, R. R. 1969. *Wild Mammals of Colorado: Their Appearance, Habits, Distribution, and Abundance*. Pruett Publ., Boulder, Colorado.

Lowery, G. H., Jr. 1974. *The Mammals of Louisiana and its Adjacent Waters*. Louisiana State Univ. Press, Baton Rouge.

*The reference list in Burt and Grossenheider includes additional state and regional mammal guides.

Exercise 10

LIFE TABLES

Key to Textbooks: Brewer 51–54; Barbour et al. 59–70; Benton and Werner 352–356; Colinvaux 369–379; Emlen 242–244; Kendeigh 204–208; Krebs 150–158; McNaughton and Wolf 194–200; Odum 170–179; Pianka 97–104; Richardson 302–309; Ricklefs 392–395, 449–452; Ricklefs (short) 238–242; Smith 438–459; Smith (short) 204–222.

OBJECTIVES

We will use mortality and survivorship data to construct life tables and compare vital statistics in human populations in historic and modern times.

INTRODUCTION

A summary of mortality, survivorship, and expectation of further life, by age, is called a *life table*. The most straightforward type of life table starts with a *cohort* of young organisms and follows their fortunes through their lives, until the last one dies. Partly because cohort data are difficult to obtain, most life tables are calculated from other sorts of information. If we can obtain the current mortality rates by age of a population we can, after the appropriate assumptions and calculations, construct a life table (called a *time-specific* table versus the *age-specific* cohort table). In the rare case in which neither population size nor mortality rates change with time, all we need for a time-specific table are the sizes of the different age classes. A frequent approach, and the one used here, is to use age at death to estimate mortality rates and calculate the other vital statistics from that. Tables produced in this way are age-specific, even though the cohort is composite, made up of individuals that started life in different years.

MATERIALS

arithmetic graph paper, 4 sheets per student

semi-log graph paper, 4 sheets per student

calculators

death notice sections of newspapers

Key Words: life table, age-specific, time-specific, cohort, mortality, survivorship, life expectation, longevity, dependency ratio, age structure, sex ratio.

PROCEDURE

We will obtain two sets of data giving numbers of deaths by age. The first set, to represent vital statistics of your region in historic times, will be obtained from one or more early graveyards. The second set, to represent current mortality statistics, will be obtained from the obituary section of your local paper.

1. For the historic data, divide the cemetery into sections. In each section, one student or pair should, from information on the gravestones, figure the age at death of every person who died before 1900. Use the Life Table Data Sheet: Pre-1900 (Cemetery); record males and females separately. Each student should record as many gravestones as necessary to give a total class sample of at least 300 of each sex; 500 is better. (Use common sense in collecting the graveyard data. If you are unable to tell from the name or other evidence whether an individual was male or female, omit him or her. If only initials are used rather than given names, the individual was probably male. "Infants" or children without names can be put in the 0–0.99 category. Occasionally infants or stillborn children are recorded simply as "baby" or some other designation from which sex cannot be determined. Tally these for the class and put half in one sex and half in the other.)
2. Pool the class data and record on the Data Sheet.
3. For the current mortality data, use the death notices in your local paper. It is best to spread the sample over a year to avoid seasonal trends. This could be done in class, or students can be assigned specific months to cover using microfilm editions in the library. Obtain a sample of the same size as was obtained from the graveyard.
4. Enter the pooled class data on the Life Table Data Sheet: Current (Newspaper).
5. Analyze all four sets of data, filling in the appropriate Life Tables for Humans (it may be desirable to divide the calculations among individuals or groups).

In the life table, each line represents mortality and survival for an age interval. The columns, specifically, are

x, the beginning age of the interval;

l_x, the number alive (survivors) at age x;

d_x, the number dying within the interval;

q_x, mortality rate (that is, $d_x \div l_x$); and

e_x, the average life span still left to individuals alive at x.

In the analysis we will, in effect, treat the sample as a cohort. Enter the total number of individuals dealt with in the sample in the zero age class of the l_x column. This, then, is the number of individuals that started out at age zero (although they did so at different times). In the d_x column, enter the figures obtained for number dying by age. The figures are mortality in each age interval.*

Calculate l_x for age 1 by subtracting the number dying in the zero age class from the number alive at the beginning, that is,

$$l_1 = l_0 - d_0 .$$

Repeat the operation (subtract d_1 from l_1 to give l_6, etc.) for the rest of the l_x column.

Complete the q_x column by dividing the number dying in the interval by the number alive at the beginning of the interval; that is,

$$q_x = d_x \div l_x .$$

---

*As applied here to an increasing population, this method will underestimate mortality rates for older age classes. The error is not serious for the purposes of this exercise; correcting the d_x column is discussed by Caughley (1977).

Life Table Data Sheet: Pre-1900 (Cemetery)

Age at Death (yr.)	Male		Female	
	Your Data	Class Total	Your Data	Class Total
0–0.99				
1–5.99				
6–10.99				
11–15.99				
16–20.99				
21–25.99				
26–30.99				
31–35.99				
36–40.99				
41–45.99				
46–50.99				
51–55.99				
56–60.99				
61–65.99				
66–70.99				
71–75.99				
76–80.99				
81–85.99				
86–90.99				
91–95.99				
96–100.99				
101 and over				
Total				

Life Table for Humans: Pre-1900 (Cemetery)*

x(yr.)	Male					Female				
	l_x	d_x	q_x	e_x	L_x No. %	l_x	d_x	q_x	e_x	L_x No. %
0										
1										
6										
11										
16										
21										
26										
31										
36										
41										
46										
51										
56										
61										
66										
71										
76										
81										
86										
91										
96										
101†										

*The L_x column is optional; see question 8.

†For the midpoint of this age class, 101 up, either use 103.5 or devise some defensible way of calculating a midpoint for it.

Life Table for Humans: Current (Newspaper)*

	Male					Female				
x(yr.)	l_x	d_x	q_x	e_x	L_x No. %	l_x	d_x	q_x	e_x	L_x No. %
0										
1										
6										
11										
16										
21										
26										
31										
36										
41										
46										
51										
56										
61										
66										
71										
76										
81										
86										
91										
96										
101†										

*The L_x column is optional; see question 8.

†For the midpoint of this age class, 101 up, either use 103.5 or devise some defensible way of calculating a midpoint for it.

Life Table Data Sheet: Current (Newspaper)

Age at Death (yr.)	Male: Your Data	Male: Class Total	Female: Your Data	Female: Class Total
0–0.99				
1–5.99				
6–10.99				
11–15.99				
16–20.99				
21–25.99				
26–30.99				
31–35.99				
36–40.99				
41–45.99				
46–50.99				
51–55.99				
56–60.99				
61–65.99				
66–70.99				
71–75.99				
76–80.99				
81–85.99				
86–90.99				
91–95.99				
96–100.99				
101 and over				
Total				

(Note that the mortality rate for the 0–0.99 age class is *annual* mortality, mortality rate per year. For the other classes, mortality is per 5 years. To make the figures directly comparable, you must convert the later mortality rates to annual rates by dividing each by 5.)

The e_x column gives expectation of further life. This is somewhat tedious to calculate and the figures are not widely used in ecology. The value is, however, used in describing human populations, so understanding how it is derived and thereby what it means is worthwhile.

For age zero, expectation of further life is the same as average life span (or mean longevity), and if we were dealing with exact ages, it would be easy to calculate. We would simply add up all the ages of death and divide by the number of individuals so as to obtain an average. In life tables, however, we are always dealing with grouped data and, also, we want expectation of further life at every age, not just at birth, so the mechanics are a little more complicated. The reasoning is this: Suppose we have 100 animals at age zero; 90 die during the course of the first year and 10 die during the course of the second. What was the expectation of further life at age zero? Ninety of the animals lived about 0.5 years more (starting at zero); some of these, of course, lived only a month or two and others lived nearly the full 12 months, but we will use the midpoint of the age class as the average that these 90 animals lived. For the portion that died between 1 and 2 years of age, we will use 1.5 as the average years lived. We have, consequently,

$$(90 \text{ animals} \times 0.5 \text{ year}) + (10 \text{ animals} \times 1.5 \text{ years}) = 60 \text{ animal-years.}$$

To calculate e_0, we divide by the number of animals

$$\frac{60 \text{ animal-years}}{100 \text{ animals}} = 0.6 \text{ year.}$$

An average animal has a life expectation of 0.6 year at age zero. To obtain life expectation at age 1 in the same example, we reason as follows: the 90 animals that have already died are of no further concern in the calculations. At age 1, 10 members of the cohort were left. All of these died sometime during the next one year, consequently,

$$e_1 = \frac{10 \text{ animals} \times 0.5 \text{ year}}{10 \text{ animals}} = 0.5 \text{ year.}$$

That is, the animal that has survived to age 1 can expect to live 0.5 year longer, on the average.*

RESULTS AND DISCUSSION

1. Plot survivorship curves (l_x against x) for the four sets of data both arithmetically and on semi-log paper (l_x on the log scale). How does the form of the semi-log plots differ from the arithmetic plots? Are there straight-line portions of the semi-log plots? What does this indicate about survivorship rate during that period of life?

2. Can you divide the tables up into a juvenile segment (high but declining mortality), an adult segment (low mortality), and a senile segment (rising mortality)? What ages are included in the three segments? How do the historical and current data differ?

3. How do the sexes differ in survivorship and longevity? Is the pattern different now compared with pre-1900? At what ages, specifically, have the greatest changes in mortality rate occurred? Account for the differences.

---

*An alternative method of calculating e_x by constructing L_x and T_x columns, generally given in textbooks, is less satisfactory because it requires equal age intervals.

4. What are the biggest differences overall in mortality rates for the current data compared with the pre-1900 data? What did the fact that mean life span was lower then mean to an infant? To a person of 25? Account for the differences.

5. In Rome during the 1st–4th centuries, the expectation of further life was about 22 years at birth and about 18 years at age 50. Account for this.

6. Why is the life table you have constructed age-specific rather than time-specific?

7. Under what circumstances would a table such as you have prepared from the newspaper data be the same as a cohort table that began with the year of birth of the oldest person included in your sample? Have these circumstances prevailed?

8. Construct a new column, L_x, giving age distribution in the population:

$$L_0 = \frac{l_0 + l_1}{2},\ L_1 = \frac{l_1 + l_6}{2},\ L_6 = \frac{l_6 + l_{11}}{2},\ L_{11} = \frac{l_{11} + l_{16}}{2},\ \text{and so on.}$$

This gives the number alive at the middle of each age interval: the whole L_x column approximates the age distribution that would develop in a population having the schedule of mortality found. Comparisons are easier if numbers are expressed in percentages. Do this by adding up the L column and dividing each L_x by this total and multiplying the result by 100%.

9. How does the current age distribution compare with the pre-1900 one? What are the dependency ratios (assume that persons under 16 or over 65 are "dependent" on the production of persons 16–65)?

10. Your calculations of age distributions (in question 8) do not picture the age structure as it is but as it would be if the population were stationary in size. In fact, in both pre-1900 and current data, there were higher percentages of younger ages because the population was expanding. What differences do you see in American society if an age structure such as you calculated, which is to be expected under the current mortality schedule, comes to prevail? Try to make some general projections and also discuss specifically politics, education, marriage and family, health care, and popular culture. What will the sex ratio be like? Do you foresee any effects from this?

Teaching Notes

If the cemetery is nearby and newspapers are available in the lab, this exercise can be done in three hours with completion of calculations done outside of class. The instructor may wish to have students work through the calculations for one set of data and then use the BASIC program "LIFE" given in Appendix 5 for the other three sets.

BIBLIOGRAPHY

Caughley, G. 1977. *Analysis of Vertebrate Populations*. Wiley, New York.

Deevey, E. S., Jr. 1947. Life tables for natural populations of animals. *Quart. Rev. Biol.* 22: 283–314.

Harper, J. L. 1977. *Population Biology of Plants*. Academic Press, London.

Hutchinson, G. E. 1978. *An Introduction to Population Ecology*, Chapter 2. Yale Univ. Press, New Haven.

Sharitz, R. M., and J. F. McCormick. 1973. Population dynamics of two competing annual plant species. *Ecology* 54: 723–740.

Southwood, T. R. E. 1978. *Ecological Methods with Particular Reference to the Study of Insect Populations*. Chapters 10 and 11. Chapman and Hall, London.

Exercise 11

POPULATION GROWTH

Key to Textbooks: Brewer 55–65; Benton and Werner 340–343; Colinvaux 309–328; Emlen 233–238, 328; Kendeigh 211–213; Krebs 180–193; McNaughton and Wolf 179–204; Odum 179–188; Pianka 113–118; Richardson 300–301, 336–344; Ricklefs 385–390; Ricklefs (short) 244–251; Smith 460–475; Smith (short) 222–228, 244; Whittaker 14–19.

OBJECTIVE

To demonstrate logistic population growth in the laboratory.

INTRODUCTION

When an organism is introduced into a favorable habitat, its population increases. If the new habitat were unlimited, the population could grow *exponentially:* 2, 4, 8, 16, 32 (Fig. 11–1). If, however, the number of new organisms that habitat can accommodate is limited, the population growth may be *logistic:* the rate of increase slows, then levels off, as the population approaches the carrying capacity (Fig. 11–2). *Carrying capacity (K)* is defined as the maximum number of a species that can be supported indefinitely by the habitat. Notice that the first part of the logistic curve looks like the exponential growth curve. This is because the number of organisms is so far below the carrying capacity that the habitat is essentially unlimited then.

Both the exponential and the logistic growth curves are conceptual models. Populations do not necessarily grow like these models, but they are often fair approximations of reality. The mathematics will be explained more later; for now, realize that a formula for exponential growth is

$$\frac{dN}{dt} = r_m N$$

which can be read as "the population growth rate equals the increase per individual (r_m) times the number of individuals in the population (N)." A formula which describes logistic population growth is

$$\frac{dN}{dt} = r_m N \left(\frac{K-N}{K}\right).$$

This is the same as the formula for exponential growth with the addition of the $(K-N)/K$ term, which is a measure of environmental resistance.

In this exercise you will be working with a duckweed, *Lemna minor*. Duckweeds are small, free-floating, freshwater plants in the monocot family Lemnaceae. They can be found in great numbers,

Key Words: population, exponential population growth, logistic population growth, carrying capacity (K), biotic potential (also called intrinsic rate of natural increase, r_m), r selection, K selection.

Figure 11–1. Exponential population growth.

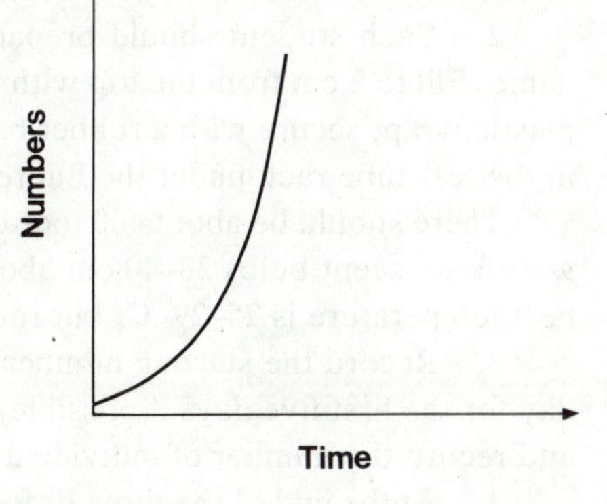

often covering the surface of ponds or arms of lakes protected from waves or currents. Although some species, including *Lemna minor,* have roots, they are not rooted to the substrate. Reproduction is nearly always vegetative, by budding, but they can flower.

MATERIALS

Lemna minor (put it all in a net and rinse with tap water)

filtered water collected from where *Lemna* grows, about 100 ml per student

test tubes, 20 cm × 2.5 cm, 1 per student

grease pencils or labels

plastic wrap

rubber bands, 1 per student

rack for test tubes

fluorescent light

class Data Sheet, taped to table near the light and test tubes

calculators with e^x and *ln* ($\log_e$) keys are very helpful for the calculations

graph paper, arithmetic, 1 sheet per student

PROCEDURE

1. Familiarize yourself with the growth habit of *Lemna minor.* Each frond is counted as an individual; there are usually three or four individuals in a cluster. Not every frond has a root. Count

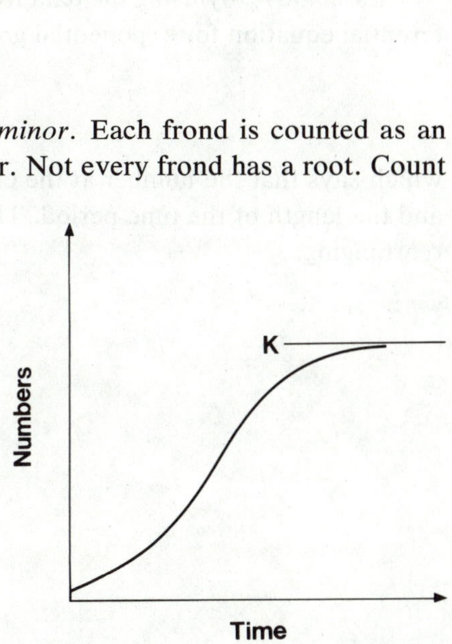

Figure 11–2. Logistic population growth.

buds when they are big enough to see; if you are not sure, wait until the next count. Do not include dead (not green) fronds in your count.

2. Each student should prepare one test tube. Label the tube with the date, class, and your name. Fill to 2 cm from the top with the filtered water. Add a cluster of three duckweeds. Cover with plastic wrap, secure with a rubber band, and poke five small airholes through the plastic wrap. Place in the test tube rack under the fluorescent light.

There should be about 400 foot-candles of light (about 4,300 lumens m^{-2}). This is given by two 40 watt fluorescent bulbs 38–40 cm above the top of the test tubes. They may be on continuously. The best temperature is 25–29° C, but room temperature is satisfactory.

3. Record the starting number (3) on the class Data Sheet. At least three times a week (every day for the first five days if possible) at about the same time each day for three or four weeks, count and record the number of individual live fronds in your test tube.

4. At the end of the three or four weeks (as instructed), pool the class data. Use the daily mean in all graphs and calculations.

RESULTS AND DISCUSSION

1. Graph the number of duckweeds versus time (on the horizontal axis). Does the curve level off? Does it look like the exponential or the logistic model (or neither)?

2. When is the growth rate the highest? To demonstrate the changing growth rate, fill in the ΔN and $\Delta N/t$ columns on your Data Sheet and graph $\Delta N/t$ against time. For example, if there were 5 fronds one day and 7.2 fronds at the next count two days later, $\Delta N/t$ would be $(7.2-5)/2 = 1.1$. During which interval was growth rate highest? From day ______________ to day ______________. This marks the point of inflection of graph 1.

3. The population growth rate per capita (that is, per individual or per head) in uncrowded conditions (if the age distribution is stable) is the *biotic potential,* or r_m, of that population in that habitat. We will use the fastest per capita growth rate the population of duckweeds showed in the laboratory as an approximation of r_m for this species of duckweed.

First, fill in the per capita growth rate $(\Delta N/t)/N$ column on the Data Sheet. This is the growth rate which you figured above divided by the number on the earlier day. To use the same example, with 5 fronds then 7.2 fronds two days later, the per capita growth rate is $1.1/5 = 0.22$. During which interval is the per capita daily growth rate the highest? From day ______________ to day ______________.

Estimate r_m by using the data from these days. The formula used is an integrated form of the differential equation for exponential growth:

$$N_t = N_o e^{r_m t}$$

which says that the number at the end of a time period depends on the number at the beginning, r_m, and the length of the time period. This form will be used in the next question, but to find r_m it needs rearranging:

$$N_t = N_o e^{r_m t}$$

$$\frac{N_t}{N_o} = e^{r_m t}$$

$$ln\left(\frac{Nt}{N_o}\right) = r_m t$$

$$\frac{ln\left(\frac{N_t}{N_o}\right)}{t} = r_m$$

N_o is the number at the earlier day, N_t is the number at the later day, and t is the number of days between counts.

Solve for an estimate of r_m. Result: r_m = _______________.

4. If the duckweeds had grown at an exponential rate for three weeks, how many would there be? N_{21} = _______________. At day 5? N_5 = _______________. How does this compare with the data? Do you think the growth follows the exponential model?

5. Did the growth curve level off? What seems to be the carrying capacity (K) of the test tube? List some possible limiting factors. What experiments could be done to identify the factor that is actually limiting?

6. Try to fit the data to the logistic model. One form for the logistic equation is

$$N_t = \frac{K}{1 + e^{a - r_m t}}.$$

To "fit" the data, you will solve for several N_t (numbers at various times, t) and compare them to the numbers you actually had. (Later we will use another method of figuring r_m to grind out better N_t values.) Perhaps you will find the data fit a logistic curve, and perhaps not.

To use the logistic equation, you must know K, the carrying capacity, which was estimated in question 5. K = _______________.

For this time through, use the r_m from question 3: r_m = _______________.

To account for the starting number so that the curve is at the right place on the axis, calculate a by rearranging the formula to

$$ln\left(\frac{K-N}{N}\right) = a - r_m t.$$

Let $t = 0$, which makes N the number at day 0, which was 3. Solve for a.* a = _______________.

Now, solve for N_t (use $N_t = K/[1 + e^{a - r_m t}]$) when t is about 2, 4, 7, 11, and 16 (adjust as necessary to match days you actually counted). Fill in the Summary Table column and add the calculated points to the graph of the data points. Are the calculated values close to the data? Is K reached at about the same time? Is the r_m value you used too high or too low?

7. In number 3, r_m was estimated from only two data points and by assuming exponential growth during that time. Now, r_m will be estimated using all the data and assuming logistic growth. To do that, calculate $ln([K-N]/N)$ for each count, record in the Summary Table, and plot those numbers (which range from negative to positive) against time. Draw a best-fit straight line by eye for the data points. The slope of that line is r_m.*

$$r_m = \frac{\left[ln\left(\frac{K-N}{N}\right) \text{ value of the line at day 0}\right] - \left[ln\left(\frac{K-N}{N}\right) \text{ value of the line at some later day, for instance, day 10}\right]}{\text{time, say, 10 days}}$$

r_m = _______________

8. As you did in number 6, grind out some N_t values, but use the r_m from number 7. Record in the Summary Table and graph alongside the actual data. How do they compare? Is K reached at about the same time?

---

*Note that $ln([K - N]/N) = a - r_m t$ is the equation for a straight line. Accordingly, you may, if you have the background or the calculator, do a least squares regression to estimate both a (the y intercept) and r_m (the slope of the line). The y values you will enter will be the natural logs of $(K - N)/N$ and the corresponding x values will be t. On some calculators, negative numbers can't be used in regression. Overcome this problem by adding a number to all the $ln([K - N]/N)$ values to make them positive, then subtract it from the resulting intercept value to get a.

Data Sheet for *Lemna* Population Growth

Day	Number in Your Test Tube	Mean Number, All Test Tubes, N	Change in Numbers, All Test Tubes, ΔN	Change in Numbers per Day, $\frac{\Delta N}{t}$	Per Capita Growth Rate, $\frac{(\Delta N/t)}{N}$
0	3	3	—	—	—
1					
2					
3					
4					
5					
6					
7					
8					
9					
10					
11					
12					
13					
14					
15					
16					
17					
18					
19					
20					
21					
22					
23					
24					
25					
26					
27					
28					

Summary Table for *Lemna* Population Growth

Day	N	Logistic Fit from Question 6; r_m=_____	$ln\ \frac{K-N}{N}$ for Question 7	Logistic Fit from Question 8; r_m=_____	$\frac{K-N}{K}$ for Question 9	$\frac{K-N}{N}$ for Question 9
0	3					

9. Figure and graph $(K - N)/K$ and $(K - N)/N$. State in words their meanings. Biologically, what causes them to vary during the course of population growth?

10. How many days would it take four *Lemna minor* to cover a 100 m^2 pond if growth were exponential? Use r_m from number 7. (Hint: Estimate the number of *Lemna minor* covering 1 cm^2.) On what day would the pond be half covered?

11. How many weeks would it take four *Lemna minor* to cover a 100 m^2 pond if growth were logistic? (You will need to estimate K for the pond.) How does this compare with the time calculated assuming exponential growth?

12. In nature, when might populations be growing exponentially?

13. After a catastrophe, which population would rebound faster—one with a high r_m, or one with a low r_m? What would be some other advantages of having a high r_m? Why don't all organisms have a high r_m?

14. There has been some effort lately to use the nutrients in sewage as fertilizer for plants to be used as animal feed. Could duckweed be grown in sewage treatment lagoons? Do you think it would be good at taking up nutrients? Would the duckweed be readily digestible?

15. Is the number of fronds the best thing to count in this experiment? What else could be counted or measured which might give a better picture of population changes?

16. Some population growth curves look like Figure 11–3. They have an overshoot period. Did the duckweeds show an overshoot? What natural factors could cause such an overshoot? Don't limit your thoughts to duckweeds.

17. What life history factors determine r_m in *Lemna minor?* What determines r_m in, say, trees? In animals?

18. Look up and list r_m values for several other organisms. What animals have an r_m close to that for *Lemna minor?* (r_m per week is 7 times r_m per day, and so forth.)

Teaching Notes

The *Lemna* cultures take only a few minutes to set up and a minute or so for each count. Run the cultures until there has been no net increase for about a week. Many variations, such as replenishing the water regularly, are possible. To mimic the natural situation more closely, a time switch can be used to give a 12-hour photoperiod rather than continuous light. In this case, raise the light intensity to about 8,000 lux. We recommend collecting local *Lemna minor* and pond water. You can, however, purchase the duckweed from Carolina Biological Supply Company, Burlington, North Carolina 27215, and grow it in a nutrient solution such as Hoagland's solution:*

$Ca(NO_3)_2 \cdot 4H_2O$	1.180 g/l H_2O
$MgSO_4 \cdot 7H_2O$	0.493 g/l
KNO_3	0.506 g/l
KH_2PO_4	0.136 g/l
$Fe_2(C_4H_4O_6)$	0.005 g/l
H_3BO_3	2.86 mg/l
$MnCl_2 \cdot 4H_2O$	1.81 mg/1
$ZnSO_4 \cdot 7H_2O$	0.22 mg/l
MoO_3	0.07 mg/l
$CuSO_4 \cdot 5H_2O$	0.08 mg/l

The computer program GROW given in Appendix 5 calculates values of N_t for various intervals according to exponential and logistic models (see questions 6 and 8).

*From Gorham, P.R. 1945. Growth factor studies with *Spirodela polyrrhiza* (L.) Schleid. *Am. J. Bot.* 32: 496–505.

Figure 11–3. Population growth curve with an overshoot.

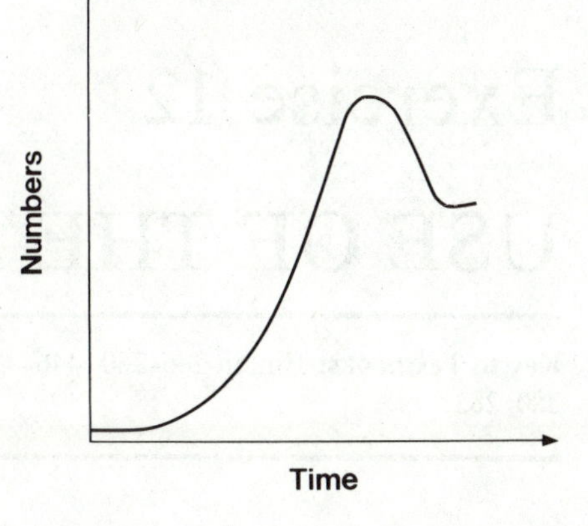

BIBLIOGRAPHY

Brewer, R., and Swander, L. 1977. Life history factors affecting the intrinsic rate of natural increase of birds of the deciduous forest biome. *Wilson Bull.* 89: 211–232.

Clatworthy, J. N., and J. L. Harper. 1962. The comparative biology of closely related species living in the same area. V. Inter- and intraspecific interference within cultures of *Lemna* spp. and *Salvinia natans*. *J. Exp. Bot.* 13: 307–324.

Cole, L. C. 1954. The population consequences of life history phenomena. *Quart. Rev. Biol.* 29: 103–137.

Gadgil, M., and O. T. Solbrig. 1972. The concept of *r*- and *K*-selection: evidence from wild flowers and some theoretical considerations. *Am. Nat.* 106: 14–31.

Gause, G. F. 1934. *The Struggle for Existence*. Williams & Wilkins, Baltimore.

Harper, J. L. 1977. *Population Biology of Plants*. Academic Press, London.

Hillman, W. S. 1961. The Lemnaceae, or duckweeds: a review of the descriptive and experimental literature. *Bot. Rev.* 27: 221–287.

———, and Culley, D. D., Jr. 1978. The uses of duckweed. *Amer. Scientist* 66: 442–452.

Keddy, P. A. 1976. Lakes as islands: the distributional ecology of two aquatic plants, *Lemna minor* L. and *Lemna trisulca* L. *Ecology* 57: 353–359.

Pearl, R. 1930. *Introduction to Medical Biometry and Statistics*. Saunders, Philadelphia.

Wilson, E. O., and W. H. Bossert. 1971. *A Primer of Population Biology*. Sinauer Associates, Stamford, Connecticut.

Exercise 12

USE OF THE LESLIE MATRIX

Key to Textbooks: Emlen 246–260, 446–458; Krebs 616–619; Odum 280–283.

OBJECTIVES

The objective is mainly methodological, to give an introduction to the use of matrices in studying population growth. In addition, following the reasoning on which the matrix is based helps to understand the role of age structure in population growth and gives insight into some otherwise puzzling questions about population growth.

INTRODUCTION

It is sometimes desirable to make calculations of population growth without the simplifying assumptions of the exponential or logistic equations or to study the growth of various age classes separately. Use of matrix algebra (introduced to ecology in this context by P. H. Leslie 1945, 1948) allows this.

The basic terms are as follows:

$n_{x,t}$ the number of females alive in the age group x to $x+1$ at time t (here as in many computations involving populations we consider only females and assume that the males follow along)

P_x the probability that a female in the x to $x+1$ age group at time t will be alive in the age group $x+1$ to $x+2$ at time $t+1$

F_x the number of female offspring born in the interval t to $t+1$ per female aged x to $x+1$ at time t that will be alive in the age group 0 to 1 at time $t+1$

The sum of n's for all the age groups (all the x's) at a particular time t gives the total population (of females) for that time. Total population size we will represent as usual by N; thus

$$N_0 = \Sigma n_{x,0}.$$

Note that P_x is a survivorship and F_x is primarily a maternity (or fertility) term.

We can think of two successive periods in the growth (or decline or staying steady in size) of a population in the way shown in Figure 12–1. This shows the underlying basis of the matrix that Leslie developed. He realized that age distribution at any one time forms a column matrix (or column vector) that, when multiplied by another matrix containing survival and fertility rates, gives a second column vector consisting of the age distribution in the next time interval.

Key Words: population growth, fertility, survival, age distribution.

To use the method we need first to learn a little matrix algebra. The way that a square matrix (which will contain the survival and fertility rates) is multiplied by a column vector is shown by the example below.

$$\begin{bmatrix} 2 & 4 & 6 \\ 0 & 1 & 3 \\ 5 & 2 & 1 \end{bmatrix} \times \begin{bmatrix} 1 \\ 2 \\ 3 \end{bmatrix} = \begin{bmatrix} 2\times 1 + 4\times 2 + 6\times 3 \\ 0\times 1 + 1\times 2 + 3\times 3 \\ 5\times 1 + 2\times 2 + 1\times 3 \end{bmatrix} = \begin{bmatrix} 28 \\ 11 \\ 12 \end{bmatrix}.$$

By convention a capital boldface letter is used to denote a matrix having several rows and columns, so that we may use **M** for the square matrix shown above. A lower case boldface letter is used for a column vector, so we can refer to the column vectors above as $\mathbf{n}_0$ (for the first) and $\mathbf{n}_1$ (for the second). What we have done in the example above then is

$$\mathbf{M} \times \mathbf{n}_0 = \mathbf{n}_1.$$

That is, we multiplied the square matrix **M** by the column vector $\mathbf{n}_0$ to obtain the column vector $\mathbf{n}_1$. We could similarly obtain $\mathbf{n}_2$ in this way:

$$\mathbf{M} \times \mathbf{n}_1 = \mathbf{n}_2.$$

(We may also obtain $\mathbf{n}_2$ as $\mathbf{M}^2\mathbf{n}_0$; that is, we can multiply the square matrix by itself and then multiply the resulting square matrix by the column vector $\mathbf{n}_0$. This is not difficult but it does require knowing how to multiply a square matrix by itself, information contained in Emlen 1973, Poole 1974, or any algebra text.)

To connect this with populations (Fig. 12–1), we can obtain numbers by age at time 1 as the product of numbers at time 0 and a matrix containing age-specific survival and fertility rates. Likewise, we can obtain numbers by age at time 2 by multiplying numbers by age at time 1 by that same matrix, and so on. We can continue this as long as there are no changes in age-specific survival or fertility. If such changes occurred, we would have to produce new square matrices containing the new values.

The matrices that Leslie used for computation are as follows:

$$\begin{bmatrix} F_0 & F_1 & F_2 & F_3 & \dots & F_{n-1} & F_n \\ P_0 & 0 & 0 & 0 & \dots & 0 & 0 \\ 0 & P_1 & 0 & 0 & \dots & 0 & 0 \\ 0 & 0 & P_2 & 0 & \dots & 0 & 0 \\ 0 & 0 & 0 & P_3 & \dots & 0 & 0 \\ \cdot & \cdot & \cdot & \cdot & & \cdot & \cdot \\ \cdot & \cdot & \cdot & \cdot & & \cdot & \cdot \\ \cdot & \cdot & \cdot & \cdot & & \cdot & \cdot \\ 0 & 0 & 0 & 0 & \dots & P_{n-1} & 0 \end{bmatrix} = \mathbf{M}.$$

The symbols are as defined earlier.

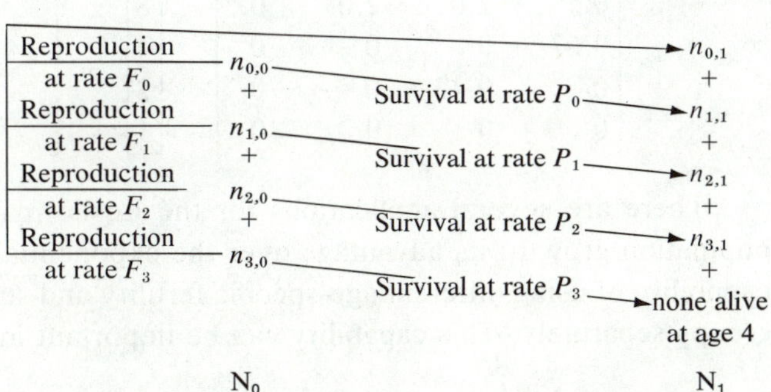

Figure 12–1. Biological basis of the Leslie matrix.

The column vector is $\begin{bmatrix} n_{0,0} \\ n_{1,0} \\ n_{2,0} \\ n_{3,0} \\ \cdot \\ \cdot \\ \cdot \\ n_{n,0} \end{bmatrix} = \mathbf{n}_0.$ Multiplying these together gives the column vector. $\begin{bmatrix} n_{0,1} \\ n_{1,1} \\ n_{2,1} \\ n_{3,1} \\ \cdot \\ \cdot \\ \cdot \\ n_{n,1} \end{bmatrix} = \mathbf{n}_1.$

Given ordinary life table data with l_x and m_x, how do we produce the matrices?

x	l_x	m_x	L_x	P_x
0	1.0	0.5		
2	0.8	1.0		
4	0.4	2.0		
6	0.1	0.5		
8	0	—		

F_x is usually taken as equivalent to the m_x column. P_x is usually found by constructing an L_x column (see question 8, Exercise 10) where (as an approximation) $L_x = (l_x + l_{x+1})/2$. P_x is then calculated as $P_x = L_{x+1}/L_x$. Fill in the L_x and P_x columns.

For the life table above then we will have this square matrix:

$$\begin{bmatrix} 0.5 & 1.0 & 2.0 & 0.5 \\ 0.67 & 0 & 0 & 0 \\ 0 & 0.42 & 0 & 0 \\ 0 & 0 & 0.2 & 0 \end{bmatrix}.$$

We can deal with any age distribution we choose. For the purposes of illustration we will use 8 in age class 0, 4 in 1, 2 in 2, and 1 in 4. The column vector then is

$$\begin{bmatrix} 8 \\ 4 \\ 2 \\ 1 \end{bmatrix}.$$

Total population size is 15. Now multiply and thereby obtain population size and age distribution at time 1.

$$\begin{bmatrix} 0.5 & 1.0 & 2.0 & 0.5 \\ 0.67 & 0 & 0 & 0 \\ 0 & 0.42 & 0 & 0 \\ 0 & 0 & 0.2 & 0 \end{bmatrix} \begin{bmatrix} 8 \\ 4 \\ 2 \\ 1 \end{bmatrix} = \begin{bmatrix} \quad \\ \quad \\ \quad \\ \quad \end{bmatrix} = \begin{bmatrix} \quad \\ \quad \\ \quad \\ \quad \end{bmatrix}$$

There are several applications for the Leslie matrix. It is, of course, one way of calculating population growth; its advantage over the exponential and logistic equations we used earlier is the capability of using different age-specific fertility and survivorship rates and of following different age groups separately. This capability can be important in planning. For human societies, for example, we can determine when or if the supply of school teachers or of geriatric social workers will need to be increased at some time in the future, given the current age structure of the population. We can determine how much longer and to what level human populations in the U.S. will continue to grow, despite current fertility rates being below replacement levels.

The Leslie matrix can be used in modeling problems having to do with predation or hunting such as the following: Suppose we wish to know the consequences on some game animal of allowing hunting (which will decrease survivorship) on a certain age group or at a certain period of the year. By setting up different matrices incorporating these different effects we can estimate the consequences on population size, including the consequences by age group. The Leslie matrix can also be used for such purposes as determining the stable age distribution and the intrinsic rate of natural increase of a population, although it offers few advantages over other methods.

Similar matrices may be useful in other branches of ecology. The use of transition matrices in studying succession is described in Exercise 20.

Computations involving small matrices used in the examples and problems can and should be done by hand or with a pocket calculator for the sake of understanding the processes involved. As the size of the matrix increases, however, the number of operations required increases at a high rate, and it is impractical to attempt manipulations of large matrices except by means of computers.

PROBLEMS

1. Calculate population size by age and total population size at time 2 and time 3 for the sample population given earlier.
2. Calculate L_x and P_x for the life table data that follow.

x	l_x	m_x	L_x	P_x
0	1.0000	0.8000		
1	0.8700	1.2000		
2	0.5300	1.8000		
3	0.1850	2.0000		
4	0.0080	1.7000		
5	0.0000	——		

3. Set up the Leslie matrix that would be derived from such a set of data.
4. Explain in biological terms exactly what is happening when you multiply each n_x by each F_x. Explain in biological terms exactly what is happening when you multiply each n_x by each P_x.
5. The following table gives l_x and m_x data for a particular species, possibly the hitherto undescribed Unfortunate Reindeer (*Rangifer inauspicius*), in a particular environment.

x	l_x	m_x	L_x	P_x
0	1.0000	0		
2	0.0900	2		
4	0.0100	6		
6	0.0004	6		
8	0	—		

Suppose that a population is introduced into such an environment and that the population consists of age 0 = 0, 2 = 1, 4 = 3, 6 = 3.

Using the Leslie matrix, calculate population size and age distribution for times 1 through 5. Enter total population size here:

Time 0 7
1 ______
2 ______
3 ______
4 ______
5 ______

What seems to be the eventual fate of the population?

See if you can find actual cases in which introductions of exotic species met with initial success but later failed or dropped to low population levels.

Teaching Notes

Since a main point is to learn how to do the computations, it is probably worth working through much of the exercise in class and checking results as you go. This can easily be done in three hours.

BIBLIOGRAPHY

Jeffers, J. N. R. 1978. Chapter 4 in *An Introduction to Systems Analysis: with Ecological Applications*. University Park Press, Baltimore.

Klein, D. R. 1968. The introduction, increase, and crash of reindeer on St. Matthew Island. *J. Wildl. Manage*. 32: 350–367.

Leslie, P. H. 1945. On the use of matrices in certain population mathematics. *Biometrika* 33: 183–212.

———. 1948. Some further notes on the use of matrices in population mathematics. *Biometrika* 35: 213–245.

Michod, R. E., and W. W. Anderson. 1980. On calculating demographic parameters from age frequency data. *Ecology* 61: 263–269.

Pielou, E. C. 1974. Pp. 18–25 and 367–369 in *Population and Community Ecology: Principles and Methods*. Gordon and Breach, New York.

Poole, R. W. 1974. Pp. 21–41 in *An Introduction to Quantitative Ecology*. McGraw-Hill, New York.

Scheffer, V. B. 1951. The rise and fall of a reindeer herd. *Sci. Monthly* 73: 356–362.

Searle, S. R. 1966. *Matrix Algebra for the Biological Sciences including Applications in Statistics*. Wiley, New York.

Williamson, M. H. 1967. Introducing students to the concepts of population dynamics. Pp. 169–176 in J. M. Lambert (ed.). *The Teaching of Ecology*. Brit. Ecol. Soc. Symp. 7.

Exercise 13
POPULATION PROBLEMS

Key to Textbooks: Brewer 51–65, 72–77, 110–115; Barbour et al. 89–90; Benton and Werner 339–343, 348–350; Colinvaux 309–339, 369–379; Emlen 223–246, 309–313; Kendeigh 201–217; Krebs 150–222; McNaughton and Wolf 179–256; Odum 168–188; Pianka 97–118, 175–182; Richardson 336–343; Ricklefs 385–406, 461–473, 502–528; Ricklefs (short) 240–251; Smith 439–466, 505–509, 711–715; Smith (short) 204–230; Whittaker 6–28.

OBJECTIVES

Although some intuitive feeling of what happens in logistic population growth or under what conditions extinction occurs according to the Lotka-Volterra models may be good enough for some purposes, actually working with these and other equations, putting numbers to them, and seeing what happens is necessary for a thorough grounding in ecology. We here provide some problems that require using some of the mathematical procedures and simpler models involved in population ecology.

PROCEDURE

The problems should be worked in connection with lectures and readings on the various topics.

Some Comments on r and r_m. Growth rate per capita in equations of population growth is usually represented by the symbol r. There are two ways of looking at r. We can consider it a constant as in the logistic equation

$$dN/dt = r_m N\ ([K - N]/K).$$

So conceived it is a population parameter that expresses the intrinsic capacity of the population for increasing. It is the biotic potential; however, the potential is realized only during times when the population is very low relative to K, the carrying capacity. At other times the actual per capita growth rate is below the value of r_m.

A simpler equation for population growth is

$$dN/dt = rN.$$

Key Words: populations, exponential population growth, per capita growth rate, r and r_m, the intrinsic rate of natural increase, crude birth and death rates, doubling time, immigration and emigration, mortality and survival rates, life tables, survivorship curve, logistic population growth, carrying capacity, net reproductive rate (R_0), generation length, Euler-Lotka equation, natality, delayed reproduction, Lotka-Volterra competition model, competition coefficients (α and β).

In this equation if r is a constant we have, of course, an expression for exponential growth. If, however, r is a variable, curves of other shapes may be obtained and, in particular, if r varies in the course of population growth from a high value to zero, we can obtain a sigmoid curve like the logistic.

In this book, r as a constant, the intrinsic rate of natural increase, is always shown as r_m. Note that the value of r_m is a function both of the life history traits of the species and of the effects of the environment; thus, r_m for a given species will differ from habitat to habitat. As any per capita growth rate, r is shown as just r, without a subscript. The high value that r has when N is very low will be equal to r_m if there is a stable age distribution. (In Brewer's *Principles of Ecology*, r always refers to the intrinsic rate of natural increase, r_m, but for simplicity it was used without the subscript.)

PROBLEMS

Exponential Population Growth, $N_t = N_o e^{rt}$

(Here r represents any per capita rate of increase and not necessarily the intrinsic rate of natural increase, r_m.)

1. Assume that two populations both begin with two members and that each grows exponentially for 60 days. Assume r per day to be 0.1 for one population and 0.2 for the other. What will be the population sizes for the times listed below?

	Population Size	
	$r = 0.1$	$r = 0.2$
Initial	2	2
1 day		
5 days		
20 days		
59 days		
60 days		

2. The crude birth rate in the U.S. is about 14.8 and crude death rate about 8.4 (crude birth and death rates are births and deaths per thousand in the population). (a) What is the annual percentage increase? (b) What is r? Note that r is an instantaneous per capita rate of increase. Approximating r by dividing the percentage growth rate by 100 is generally good enough when the growth rate is low; however, it is best to calculate r as

$$r = ln \text{ finite rate.}$$

The finite rate of population change is given as

$$\text{finite rate of population change} = \frac{\text{population size at end of time period}}{\text{population size at beginning of time period}}$$

(c) If this rate of increase were to continue, how long would it take for the U.S. population to double (ignoring immigration and emigration)?

3. The yearly percentage increase for Mexico is about 2.8%. (a) Calculate r for the Mexican population. (b) How many years will it take to double?

4. The U.S. population in April 1980 was 226,500,000. Considering only births and deaths, you calculated rates of increase in problem 2. However, populations may change size by other factors.

Current legal immigration is about 650,000 per year. No one has good figures for illegal immigration but 1,000,000 per year is probably conservative. (Emigration is negligible.) Calculate a new per capita rate of increase and a new doubling time taking immigration into account.

5. The crude birth rate in Nigeria is 49 births per thousand; the percentage growth rate is 3.3%. Give the crude death rate.

6. In West Germany, crude birth rate is about 10 per thousand and crude death rate about 12 per thousand. Assuming that the rates are maintained, what event may be expected about 347 years hence?

7. In 1962 15 white-tailed deer were released on South Fox Island in Lake Michigan, where the species had not previously occurred. In 1969 it was estimated that the population was about 400. (a) If we assume that population growth was exponential for this whole period, estimate r per year. (b) Also give r per day.

Life Tables

8. The data given in the following life table are for white-tailed deer in a section of Michigan. The data (except for l_0) are derived from the ages of deer shot or accidentally killed. (a) Why are the data entered in the l_x column rather than the d_x column? Is this an age-specific or a time-specific life table? (b) Complete the life table (assume that the midpoint of the last age class is 14.5). (c) Plot survivorship against age on ordinary graph paper and semi-log paper.

x (years)	l_x	d_x	q_x	e_x
0	479			
1	324			
2	241			
3	178			
4	114			
5	81			
6	56			
7	42			
8	32			
9	19			
10 or older	15			

Logistic Population Growth, $N_t = \dfrac{K}{1 + e^{a-r_m t}}$

9. Following are data for a culture of *Paramecium aurelia* studied by Gause. Numbers are per 0.5 ml. (a) Assume K is 570 (or some other value if you prefer). Complete the table and plot N_t, $\Delta N/\Delta t$, and $(K-N)/K$ all against t on one graph. (b) On another graph plot $ln[(K–N)/N]$ against t (or else plot $(K–N)/N$ on semi-log paper). The straight line you obtain is for the expression $ln[(K–N_t)/N_t] = a–r_m t$. Thus, a in the integrated form of the expression for the logistic curve is the value of $ln[(K–N_t)/Nt]$ when t is 0. What is the meaning of $(K–N_t)/N_t$? (c) Estimate r_m per day from the expression above. Depending on your statistical background and the sophistication of the calculator you have available (1) you may simply use the a value as entered in the table (i.e., $ln[(K–N)/N]$ for day 0) and calculate r by plugging in the appropriate figures for day 6 in the expression $ln[(K–N)/N] = a–r_m t$ or (2) you may run a least-squares regression on the data (with $ln[(K–N)/N]$ as x and time as y) to obtain a and r. For directions see Appendix 4 or the manual for your calculator. (d) Using the values calculated in (c), draw the expected logistic curve and compare it with the actual data.

Time (days)	N_t	$\frac{\Delta N}{\Delta t}$	$K-N$	$(K-N)/N$	$ln[(K-N)/N]$	$(K-N)/K$
0	2					
2	17					
4	39					
6	185					
8	267					
10	510					
12	650					
14	575					

Calculation of the Intrinsic Rate of Natural Increase, r_m

Populations under uncrowded conditions (and having a stable age distribution) will grow at a per capita rate called the *intrinsic rate of natural increase* (r_m). It is possible to estimate this rate in nature (as we did in question 7 under Exponential Population Growth). It is also possible to obtain estimates by keeping individual females under uncrowded laboratory conditions and recording their survivorship and fertility. One obtains, then, an age-specific schedule of survivorship and fertility. From these data, it is possible to estimate r_m using one of two methods. The first involves calculation of the net reproductive rate and generation length. The second involves trial-and-error solution of the Euler (pronounced "Oiler")-Lotka equation.

Calculating r_m from Net Reproductive Rate and Generation Length. This method starts with the equation

$$N_T = N_0 e^{r_m T}.$$

This is the same as our previous equation for exponential growth except that t is replaced by T, which is mean generation length, the average time between the birth of parents and the birth of their offspring. Dividing through by N_0 and representing N_T/N_0 as R_0 we have

$$R_0 = e^{r_m T}.$$

R_0 is given the name *net reproductive rate*. It is the ratio of individuals in one generation to the individuals in the preceding. (Actually, it is the ratio of daughters to mothers. In all of our subsequent calculations we will deal only with the female segment of the population and assume that the males follow along.)

To solve for r_m in the preceding expression, we take the natural log of both sides and divide through by T, obtaining

$$r_m = (ln\ R_0)/T.$$

This expression is useful because it illustrates the dependence of r_m on reproductive rate and generation length. Because of the difficulty of defining generation length in organisms with overlapping generations, an estimate obtained from this relationship is not precise.

Life table and maternity data are used to calculate r_m as follows. Below are standard age and survivorship data. Note that midpoints of age classes are used. The m_x column is maternity (specifically, live births).

x (years)	l_x	m_x	$l_x m_x$	$l_x m_x x$
0.5	0.800	0.750	0.600	0.300
1.5	0.500	2.000	1.000	1.500
2.5	0.100	3.000	0.300	0.750
3.5	0.005	2.000	0.010	0.035

To put things on a per head basis, we assume a cohort of 1. (For example, if we observed 20 females from birth, the table says that by 0.5 years of age, there were 16 left and that in the span from 0 days through 365 days, the total production of female offspring was 12.)

R_0 is the ratio of births of daughters to mothers. The number of daughters produced by one of these females, on the average, is obtained by multiplying survivorship by maternity for each age and adding them up; that is, $\Sigma l_x m_x$. The number of mothers is 1, since we are using a cohort of 1. R_0, then, is given by $R_0 = \Sigma l_x m_x$. For generation length, we wish to estimate an average time between the birth of the mothers and the birth of the (female) children. It is a populational value. It is calculated as

$$T = \frac{\Sigma l_x m_x x}{\Sigma l_x m_x}$$

Accordingly, $R_0 = 1.910$, $T = 2.585/1.910 = 1.353$, and $r_m = 0.647/1.353 = 0.478$ per year.

Calculating r_m from the Euler-Lotka Equation. The other method of calculating r_m from survivorship and maternity data is tedious, involving the trial-and-error solution of the following equation, called the Euler-Lotka equation. (Although it looks different, it too can be derived from the equation $N_t = N_0 e^{r_m t}$.)

$$1 = \Sigma l_x m_x e^{-r_m t}$$

This equation says that you have obtained the correct value of r_m when the summation is equal to 1. If the sum is below 1 you have overestimated r_m; if it is over 1 you have underestimated r_m.

We can demonstrate this method using the same data as before, but we now need another column giving $l_x m_x e^{-r_m t}$. It is simplest to set this up as below (using a trial r_m of 0.478).

x (years)	l_x	m_x	$l_x m_x$	$r_m t$	$e^{-r_m t}$	$l_x m_x e^{-r_m t}$
0.5	0.800	0.750	0.600	0.239	0.787	0.472
1.5	0.500	2.000	1.000	0.717	0.488	0.488
2.5	0.100	3.000	0.300	1.195	0.303	0.091
3.5	0.005	2.000	0.010	1.673	0.188	0.001
						$\Sigma l_x m_x e^{-r_m t} = 1.052$

The simplest way to approach the calculations is to use a trial value that we know will be close to the real value. We may accordingly wish to try the approximate method

$$r_m = \frac{ln\ R_0}{T}$$

first. If so, we will use a trial value of $r_m = 0.478$. This produces a summation of $l_x m_x e^{-r_m t}$ of 1.052. It then slightly underestimates r_m. If we have computer time available, we can simply try more values until we hit one that sums to 1.000. A simpler method of getting a precise value is to do another summation with a slight overestimate of r_m and then use DeWitt's method (DeWitt, R. M. 1964. The intrinsic rate of natural increase in a pond snail [*Physa gyrina* Say]. *Amer. Nat.* 88: 353–359) to solve for r_m graphically. This is illustrated below, with a new trial value of $r_m = 0.600$.

x (years)	l_x	m_x	$l_x m_x$	$r_m t$	$e^{-r_m t}$	$l_x m_x e^{-r_m t}$
0.5	0.800	0.750	0.600	0.300	0.741	0.445
1.5	0.500	2.000	1.000	0.900	0.407	0.407
2.5	0.100	3.000	0.300	1.500	0.223	0.067
3.5	0.005	2.000	0.010	2.100	0.122	0.001
						$\Sigma l_x m_x e^{-r_m t} = 0.920$

This sums to only 0.920. The correct value of r_m, then, lies somewhere between 0.478 and 0.600. We can graph the relationship as in Figure 13–1 and read off the value for r_m when the summation equals

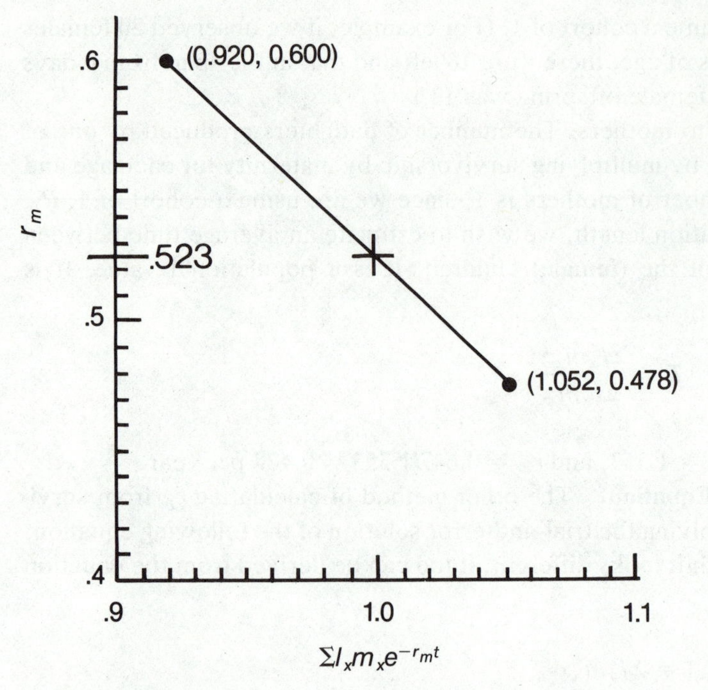

Figure 13–1. Graphical method of determining r_m given two trial summations; $r_m = 0.523$ sums to 1.001.

1.000. (Alternatively, with the sophisticated pocket calculators now available, we can readily do a linear regression of r_m (y) on the summation (x) and find r_m by obtaining $\hat{y}$ when $x = 1.000$.)

10. Calculate r_m using R_0 and T from the following data, which might be from a fictitious small mammal, the LaMont sea vole, *Microtus iterativus*.

x (years)	l_x	m_x	$l_x m_x$	$l_x m_x x$
0.5	0.80	1.0		
1.0	0.60	2.0		
1.5	0.25	3.0		
2.0	0.10	1.5		
2.5	0.02	0.0		

R_0 = ____________ T = ____________ r_m = ____________

11. Now use the value of r_m you just calculated as the first trial value in calculating r_m with the Euler-Lotka equation. Next, try a second value, and then obtain a more precise estimate graphically (or by interpolation or extrapolation using linear regression if you prefer).

		1st Trial Value r_m =			2nd Trial Value r_m =		
x (years)	$l_x m_x$	$r_m t$	$e^{-r_m t}$	$l_x m_x e^{-r_m t}$	$r_m t$	$e^{-r_m t}$	$l_x m_x e^{-r_m t}$
0.5							
1.0							
1.5							
2.0							
2.5							

$\Sigma l_x m_x e^{-r_m t}$ = ____________ $\Sigma l_x m_x e^{-r_m t}$ = ____________

Estimate of r_m = ____________

12. List several changes in life history traits of the LaMont sea vole that would have the effect of raising the value of r_m.

13. Suppose that two different populations of the LaMont sea vole encountered relatively permanent climatic shifts such that the number of offspring produced per female was reduced in one population and the age of reproduction was delayed in another population. To see which change would have the most drastic effect, calculate r_m using T and R_0 for the two life tables below. On the left reproductive age is delayed six months; on the right the number of offspring is halved but reproductive scheduling is not changed. In both cases, survivorship is unaltered from problem 10.

		Delayed Breeding			Reduced Litters		
x (years)	l_x	m_x	$l_x m_x$	$l_x m_x x$	m_x	$l_x m_x$	$l_x m_x x$
0.5	0.80	0.0			0.5		
1.0	0.60	1.0			1.0		
1.5	0.25	2.0			1.5		
2.0	0.10	3.0			0.75		
2.5	0.02	1.5			0.0		

$R_0 =$ ________ $R_0 =$ ________

$T =$ ________ $T =$ ________

$r_m =$ ________ $r_m =$ ________

14. Under a certain set of conditions a population of sparrows has a net reproductive rate (R_0) of 3.2. A population composed of ten females and ten males is introduced onto an island that has similar conditions and where the species did not hitherto occur. How many sparrows could you expect one generation later?

15. Suppose that you kept track of survivorship and maternity in a stable natural population of mice or birds and calculated R_0 and r. Approximately what values would they have?

Competitive Exclusion and the Coexistence of Species

The following problems make use of the Lotka-Volterra competition equations

$$\frac{dN_1}{dt} = r_{m1}N_1\frac{(K_1-N_1-\alpha N_2)}{K_1} \quad \text{and} \quad \frac{dN_2}{dt} = r_{m2}N_2\frac{(K_2-N_2-\beta N_1)}{K_2}$$

where N_1 refers to individuals of species 1, and N_2 to individuals of species 2, and α and β are competition factors involving respectively the effect of species 2 on species 1 and the effect of species 1 on species 2. Specifically, α says how many individuals (or what fraction of an individual) of species 2 is equal to one individual of species 1, insofar as population growth goes, and β does the same thing for the effect of species 1 on species 2.

Additional details on the Lotka-Volterra equations, which will be necessary for solving the following problems, are given on pages 110–115 in Brewer 1979, equivalent pages in some of the other texts, or pages 156–164 in *A Primer of Population Biology* (Sinauer Associates, Stamford, Connecticut) by E. O. Wilson and W. H. Bossert.

16. Determine, by drawing vectors, the outcomes of the situations in the graphs on the following page. (The exact lengths of the arrows are not important; however, if you wish a guide to relative lengths for the two species, make each proportional to the distance left to the line where its growth rate is zero.)

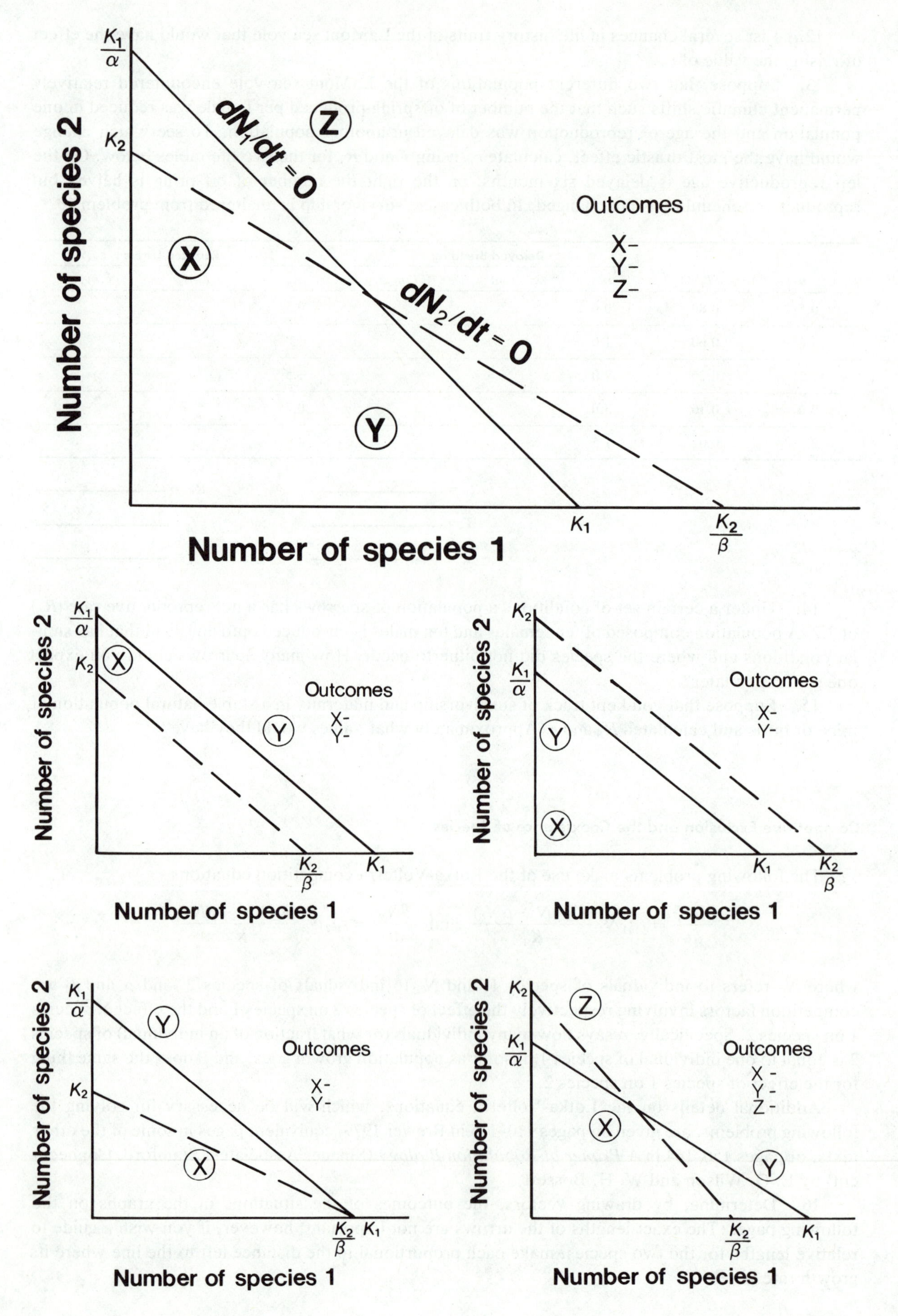
$\frac{K_1}{\alpha}$
K_2
$dN_1/dt = 0$
$dN_2/dt = 0$
Z
X
Y
Outcomes
X–
Y–
Z–
Number of species 2
Number of species 1
K_1
$\frac{K_2}{\beta}$
Outcomes
X-
Y-
Outcomes
X-
Y-
Outcomes
X-
Y-
Outcomes
X
Y
Z

17. What do you predict as the outcome of the following sets of conditions?
   (a) $K_1 = 175, K_2 = 120, \alpha = 1.7, \beta = 1.04$.
   (b) $K_1 = 44, K_2 = 120, \alpha = 4.50, \beta = 0.20$.
   (c) $K_1 = 100, K_2 = 125, \alpha = 0.62, \beta = 0.88$.

18. Graph the following situation: $K_1 = 110, K_2 = 125, \alpha = 0.62, \beta = 0.88$. What would be the predicted outcome of the competition between these two species? Suppose now that conditions changed, becoming more favorable for both species. An example might be two species of algae living in a lake to which fertilizer is added. The carrying capacity for both is increased, to $K_1 = 180$ and $K_2 = 150$. The competition coefficients are unchanged. What is now the predicted outcome?

19. Discuss the findings in question 18 in connection with the effects of organic pollution on species diversity.

Exercise 14

BIOMASS IN A GRASSLAND

Key to Textbooks: Brewer 122–135; Barbour et al. 518–532; Benton and Werner 101–120; Colinvaux 129–143, 164–165, 177–180; Emlen 341–351; Kendeigh 120–129, 164–168; Krebs 488–494, 549–557; McNaughton and Wolf 97–103, 546–549; Odum 63–85, 369–373; Pianka 271–283; Richardson 259–264; Ricklefs 643–680; Ricklefs (short) 128–155; Smith 26–30, 117–146, 282–299; Smith (short) 47–71, 380–403; Whittaker 213–231.

OBJECTIVES

The first objective of this exercise is to obtain estimates of biomass of some parts of an ecosystem. We use an ecosystem dominated by herbaceous vegetation rather than one dominated by woody plants because sampling is simplified. Another objective is to learn some standard terrestrial sampling techniques. The biomass relationships will be looked at from the standpoint of energy flow.

INTRODUCTION

Ecosystems have a trophic organization based on food chains. One organism feeds on another and is in turn eaten by a third. If we look at all of an ecosystem's food chains together, we can recognize *trophic levels* based on how many links each organism is from the entry of energy into the ecosystem as sunlight. One step away from sunlight are the green plants, forming the *producer* trophic level. The herbivore links in the food chains collectively form the *primary consumer* level and the carnivores that eat herbivores form the *secondary consumer* level. Carnivores that eat secondary consumers are *tertiary consumers*. It is usually true that when we measure either weight (biomass) or numbers or energy content (calories) in an ecosystem, we find a pyramid in which producers > primary consumers > secondary consumers > tertiary consumers.

Although the trophic level concept is often a useful way of thinking about ecosystems, there are both theoretical and practical problems in its use. In this exercise, consequently, we measure biomass in several compartments based primarily on convenience in sampling.

Key Words: ecosystem, trophic levels (producers, primary consumers, secondary consumers, tertiary consumers), herbivores and carnivores, biomass, biomass pyramid, energy flow, sweep-net sampling, trap-line (NACSM) sampling, Berlese-Tullgren and Baermann funnels, grazing and detritus-based food chains, standing crop, litter.

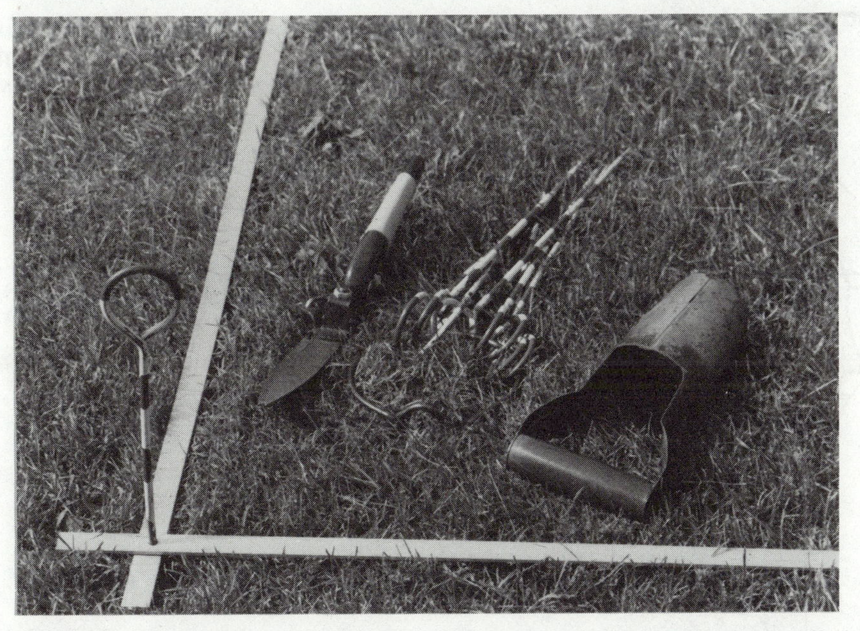

Figure 14–1. Quadrat frames are made of ¾ inch × ⅛ inch aluminum stock (sometimes called strapping). The ends are drilled with ¼ inch (or smaller) holes and the corners are secured with quadrat pins (surveyor's arrows) as shown or with short bolts and wing nuts. The inside measurement of the frame should be 1 m × 1 m; this may be produced by drilling the end holes on centers 101.9 cm (or 40⅛ inch) apart. Some of the other equipment used in the biomass exercise is also shown (photograph by Philip M. Brewer).

MATERIALS

1 m × 1 m quadrat frame (Fig. 14–1) or meter sticks, string, and stakes
forceps
grass clippers or scissors
3 sturdy, sharpened bulb planters (sharpen with a file)
knife or spade
4 paper sacks
3 sweep nets with muslin bags about 30 cm in diameter
anesthetic (ether or choloroform or a mixture)
several vials with lids
formalin
plastic flagging tape
metric tape at least 10 m long
120 break-back mousetraps
wire tent stakes and twine (optional)
peanut butter
rolled oats
freezer
meter stick or ¼ × ¼ m quadrat frame
trowels
12 plastic bags large enough to hold core from bulb planter

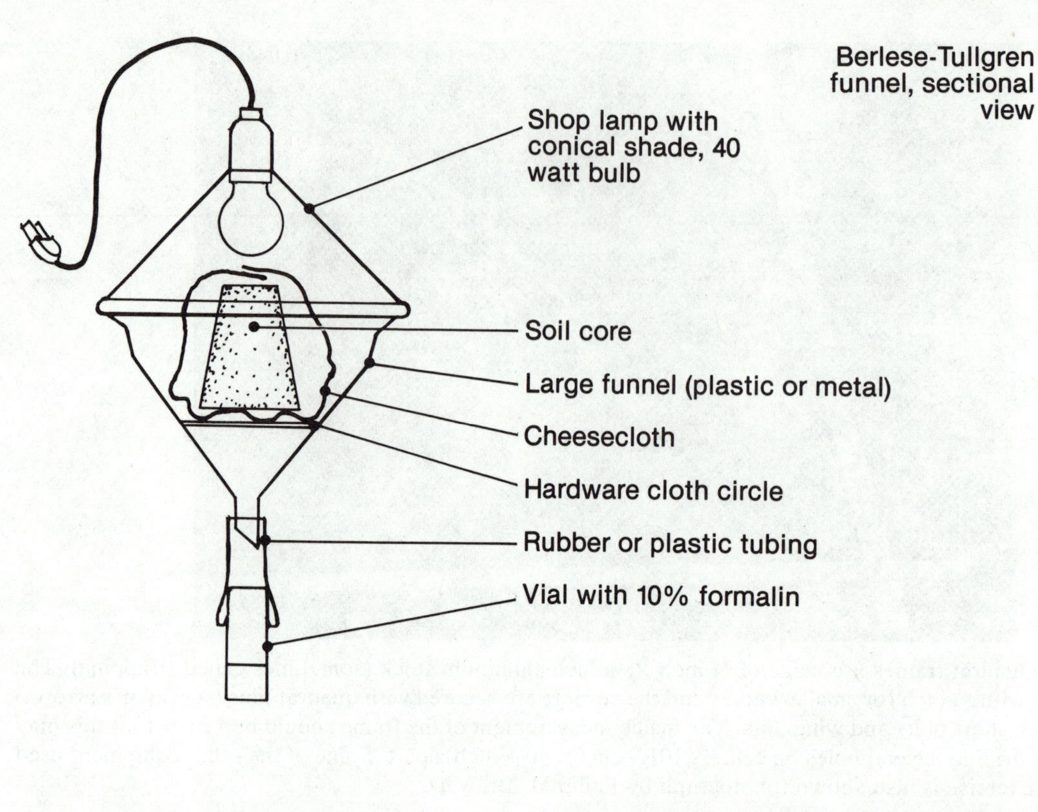

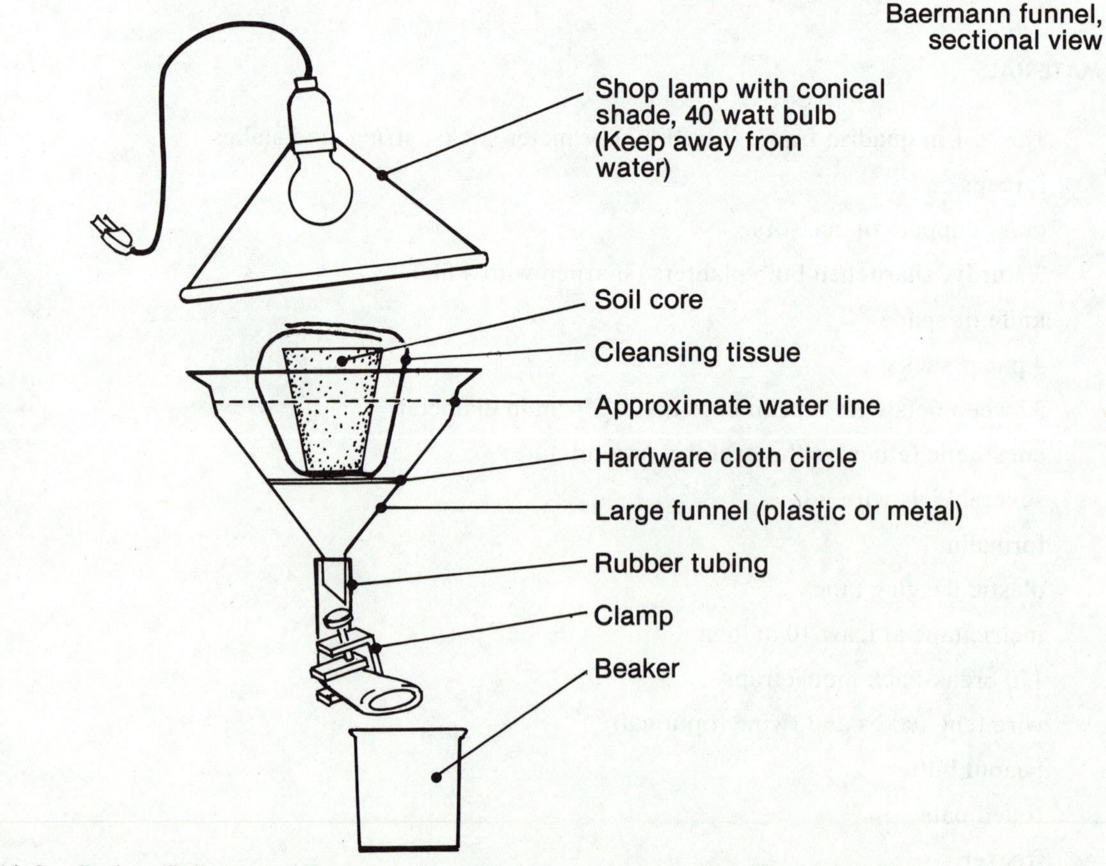

Figure 14–2. Berlese-Tullgren and Baermann funnels. Although they can be more elaborate, one easy way to construct them is to use a cheap shop lamp with a conical metal shade for the top heat/light source and a large plastic funnel (available in the automotive section of most discount stores) for the bottom funnel. If you do not want to build permanent supports, the funnels can be held on ringstands with rings and clamps. Use good Hoffman screw compressor clamps to close off the Baermann funnel tubing so that it won't dribble.

12 twist ties
4 Berlese-Tullgren funnels (Fig. 14–2)
2 Baermann funnels (Fig. 14–2)
binoculars (optional)
refrigerator
drying oven
sorting tray (white enamel)
sieve
bucket
dissecting scope
7 small petri dishes (about 5 cm diameter)
mammal identification book
books for identifying soil animals
balance
desiccator (optional)
aluminum foil

PROCEDURE IN THE FIELD

This exercise is best done about the end of the growing season; however, any season except winter is satisfactory.

Plant Material. Set up one or more square-meter quadrats. At the borders of the quadrat, part the standing vegetation so that all of the stems rooted in the square meter are inside the boundary markers (string or quadrat frame). Clip all of the standing leaves and stems as close to the ground as possible and put the material into a paper sack labeled "Standing Vegetation."

Now pick up all of the plant litter on the soil surface and put it in a sack marked "Litter." Cut around the borders of the quadrat with a knife or spade and take only the litter lying on the surface of the square meter. Try to avoid including soil with the litter.

Now, using a sturdy, sharpened, bulb planter take six cores evenly spaced over the surface of the square meter. Place the cores, soil and all, into a paper sack marked "Roots."

Foliage Invertebrates. Use a heavy sweep net with a muslin bag about 30 cm in diameter. Sweep the foliage, following these directions: Take a series of 48 strong sweeps as you walk briskly along. Make the sweeps 1 m long and turn the net sharply at the end of each sweep (to keep the insects trapped in the bag from flying out). At the end of the series, flip the bottom of the bag over the rim, so that the insects are trapped inside. Pour anesthetic on the bottom of the bag and then turn the net over so that the fumes rise through the bag. Bear in mind that the anesthetic is highly flammable and in enclosed places can form explosive mixtures with air. Keep sparks and flames away and avoid inhaling it.

We will assume that 48 sweeps (Shelford 1951) taken as described will sample as many insects as are contained in one square meter of foliage (from ground to top). Take several such sets of sweeps. When the insects are anesthetized, remove them from the bag with forceps, sorting through plant material as necessary. You may wish to empty the bag into a large white enamel tray for sorting. Transfer the insects into labeled vials filled with 4% formalin.

Small Mammals. Lay out a pair of North American Census of Small Mammals (NACSM) traplines, 190 m long, parallel to one another and at least 100 m apart. For each line (Fig. 14–3), put plastic flagging on the grass at the beginning and end and every 10 m between. At the 20 stations thus marked

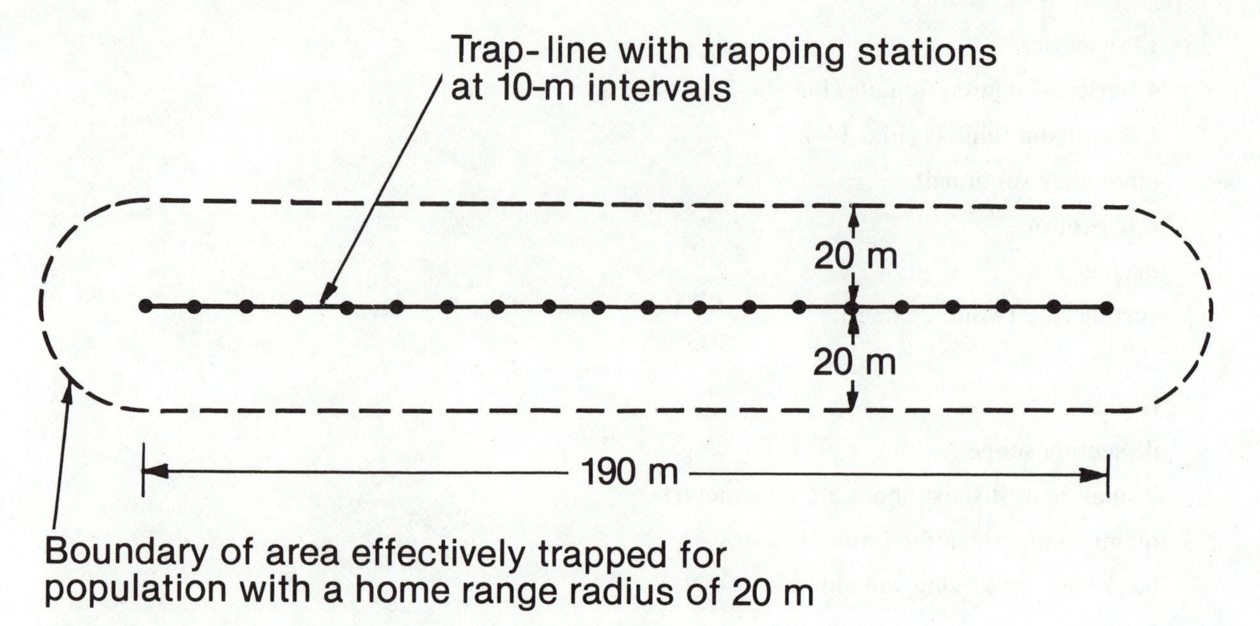

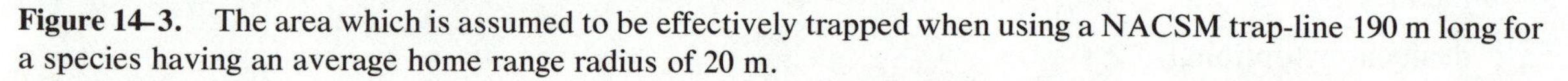

Figure 14–3. The area which is assumed to be effectively trapped when using a NACSM trap-line 190 m long for a species having an average home range radius of 20 m.

on each line, place three break-back mouse traps baited with peanut butter and rolled oats. Preferably, each set of three traps should be attached to a wire tent stake or similar anchoring device with half a meter of heavy twine threaded through a hole drilled in each trap. This is to prevent the traps from being dragged away. Move the litter aside to put the traps on the ground. If there are runways of small mammals in the vicinity, place the traps in runways, insofar as the 0.5 m tethers will allow.

Make arrangements for someone to run the trap-lines, collecting captured animals and rebaiting traps, for three successive mornings. It is considered that most residents will be removed in three nights and little influx into the depopulated area will have occurred. The animals should be frozen for later identification and weighing.

Soil Invertebrates. Set up one or more quadrats 25 cm on a side (0.0625 m²). (If time permits or if the soil fauna is very sparse, quadrats of 50 cm on a side [0.25 m²] can be used.) Work through the soil, breaking it up with trowels. Save the soil to replace in the holes. You may wish to put portions of the soil on white enamel trays for sorting. Excavate to the depth that you did with the bulb planter (probably 10–12 cm). Remove the larger, macroscopic invertebrates (for example, earthworms, snails, and beetles); the smaller forms are obtained from the funnel samples described below. Place these in labeled vials filled with 10% formalin.

From nearby take six bulb planter cores. Place them in plastic bags and seal. Try to keep the cores intact. These are for the Berlese-Tullgren and Baermann funnels.

Birds. If feasible, take a breeding-bird census using the spot-map method given in Kendeigh (1944), Pough (1947), or Hall (1964). If long-term population figures are available from the tract you are using that is even better.

Other Organisms. Make observations on other organisms that make use of the study area. Depending on the nature of the area, equipment available, and the interests of the class, it may be feasible and desirable to obtain estimates for larger mammals, reptiles and amphibians, garden spiders, slugs, or various other populations. Try to get an idea of what segments of the community are not represented in your quantitative sampling.

PROCEDURE IN THE LABORATORY

With the exception of the funnel samples and the bird and mammal data, the basic procedure is to prepare the various materials so that they can be dried in an oven at 80–105° C and then weighed.

Standing Vegetation. Open the tops of the sacks and put them in the oven for drying. If oven space is limited, air dry all the sacks, weigh them, and then oven-dry one (or a portion of one) to obtain a conversion factor for calculating the oven-dry weights of the rest. Oven-drying should be continued until a constant weight is obtained. This will take several days if a large amount of green vegetation is dried and one or two days for material that was first air-dried.

Litter. If little soil was included with the litter it can be handled exactly as described for the standing vegetation above. If appreciable amounts of dirt were included, do one or both of the following: (1) Agitate the material in the paper sack to cause the dirt to separate from the litter, collecting at the bottom of the sack. Transfer the litter to another sack. If it is now clean, put it in the oven for drying. (2) If the litter still includes a large amount of dirt, place the litter sample in a pan or dish of water and swirl it around. The dirt will sink but the litter will tend to float (do not pour the dirt down the drain). It can be skimmed out and put in another sack for drying. Drying time will be several days. Turn or stir material treated in this way occasionally while it is drying.

Roots. Place the soil cores from the ''Root'' sack into a sieve and wash away the dirt over a bucket. This should be done outside if possible. Put all of the plant material retained in the sieve into a paper sack for drying. Also stir the water in the bucket to bring any plant fragments to the surface. Add them to the sack before placing it into the oven for drying. Stir this material occasionally during the several days it will take to dry.

Earthworms, etc. Place the earthworms, snails, and other macroscopic soil invertebrates from the 0.0625 (or 0.25) m^2 samples into a preweighed aluminum foil pan and put the pan in the oven.

Berlese-Tullgren Funnels. From the set of six soil invertebrate core samples, use four for the Berlese-Tullgren funnels. Place a small amount of 10% formalin in a collecting vial and attach it to the bottom of the funnel. Hold the core over the funnel to remove it from the plastic bag. Place it upside down on a single layer of cheesecloth over the hardware-cloth platform in the funnel. Empty the plastic bag on top of the core before folding the cheesecloth over the top. Now replace the top of the funnel with the light turned on. Most kinds of soil animals will move away from the heat, light, or dryness and fall through the funnel into the collecting vial. As little as two days will suffice to obtain most of the potential organisms; however, the material can be left in place for a week if that is more convenient.

Baermann Funnels. Set up the funnels and make sure the tubing is clamped securely shut. Fill the funnels with lukewarm water (about 38° C). Wrap the other two soil cores individually in a facial tissue and place each gently on the support in a funnel. The water should come up about halfway on the core. Place a light over the top of the funnel. (The light is sometimes not used; if it is, take precautions to avoid shocks.) Nematodes will move downward and away from the heat and dryness and accumulate in the water above the clamp. A small amount of water should be run from the tube into a beaker after two hours. The class should have an opportunity to see the living nematodes at this time, if possible. Then add concentrated formalin to the beaker to produce a 4–5% solution. After another 24 hours, run more water from the funnel into the beaker and again add concentrated formalin.

Treatment of Berlese-Tullgren Funnel Material. Pour the contents of each collecting vial into two or three (or more depending on the number of organisms) small petri dishes (about 5 cm diameter). Rinse the vial with a small amount of distilled water and wash any remaining organisms into another petri dish. Examine the contents of the petri dishes under a dissecting microscope (about 10× for finding organisms and high powers for initial identifications). Count the numbers of each kind of organism and obtain totals, combining all the samples. Enter the totals for each kind of organism in the Berlese-Tullgren Worksheet. Convert these figures to oven-dry weights by multiplying them by the average dry weight per individual (these values, given in the table, are approximations mostly based on Table IV, page 155, of Kevan 1968 and converted to dry weight by assuming that dry weight = 0.25 × wet weight). The table includes the most common soil animals; for others, estimate dry weight per individual based on similarity in size to those for which weights are given. Enter the total dry weight in column *W* in Summary Table.

Treatment of the Baermann Funnel Material. Decant as much fluid from the beaker as feasible without losing worms. Wash the remaining fluid containing the nematodes into two or three (or more

Berlese-Tullgren Worksheet

Soil Animal	(A) Total Number, All Samples	(B) Dry Weight per Individual in g	(W) Dry Weight in Total Sample in g (A × B)
Nematoda		1.2×10^{-7}*	
Enchytraeidae		3×10^{-5}	
Tardigrada		5×10^{-7}	
Acarina		2.5×10^{-6}	
Small spiders		4×10^{-3}	
Pseudoscorpionida		2.5×10^{-4}	
Chilopoda and Diplopoda		7×10^{-3}	
Collembola and Thysanura		2.5×10^{-6}	
Diptera larvae		7×10^{-2}	
Medium ants		1.4×10^{-3}	
Total			

*Here are some examples to help you remember scientific notation. 3.2×10^6 means 3,200,000; the decimal is moved 6 places to the right. 3.2×10^{-6} means 0.0000032; the decimal is moved 6 places to the left. To divide, subtract the exponents:

$$\frac{6.3 \times 10^6}{3 \times 10^4} = \left(\frac{6.3}{3}\right) \times 10^{(6-4)} = 2.1 \times 10^2 = 210, \text{ and } \frac{7 \times 10^{-6}}{14 \times 10^3} = \left(\frac{7}{14}\right) \times 10^{(-6-3)} = 0.5 \times 10^{-9} = 0.0000000005.$$

To multiply, add the exponents: $(2 \times 10^9) \times (1.7 \times 10^2) = (2 \times 1.7) \times 10^{(9+2)} = 3.4 \times 10^{11} = 340{,}000{,}000{,}000$ and $(1.5 \times 10^{-3}) \times (2 \times 10^{-2}) = (1.5 \times 2) \times 10^{(-3+-2)} = 3 \times 10^{-5} = 0.00003$.
To raise a power to a power, multiply: $(10^3)^4 = 10^{12}$.

TABLE 14–1. Diets for Various Invertebrate Orders

Order	Diet Category: Herbivore	Diet Category: Carnivore
Odonata	—	Dragonflies and damselflies
Orthoptera	Walkingsticks, grasshoppers, katydids	Mantids
Thysanoptera	Thrips	—
Hemiptera	Many, including leafbugs and lacewings	Ambush bugs, assassin bugs
Homoptera	All	—
Neuroptera	—	Most
Coleoptera	Most	Ground, rove, and ladybird beetles
Lepidoptera	Most, including caterpillars	—
Diptera	Many	Mosquitos, blackflies, horseflies, deer flies, robber flies, long-legged flies
Hymenoptera	Many	Many wasps, which catch other insects as food for larvae
Stylommatophora	Slugs	—
Araneae	—	Spiders

depending on numbers of worms) small petri dishes (about 5 cm in diameter). Count the nematodes and add the individual counts together to obtain a total count for all the samples. Convert this figure to oven-dry weight by multiplying it by 1.2×10^{-7} (the dry weight in g of an average soil nematode, calculated as described above). Enter this figure in column W in the Summary Table.

Trophic Relations of the Soil Invertebrates. For simplicity, we will not attempt to separate the soil fauna into herbivores and carnivores. Most of the organisms feed on dead plant tissue (or on fungi that grow on it) or on living plant tissue, including roots. Others, such as maggots, are specialized for feeding on dead animals. There are a number of carnivores, including several kinds of beetles (both larvae and adults), a few collembolans, centipedes, spiders, and pseudoscorpions. Mites are very diverse in feeding habits, different species feeding on live plants, dead material, fungi, or other soil animals. Some nematodes attack the roots of plants; others eat bacteria or fungi and some are carnivorous. Earthworms eat mostly dead plant material.

Foliage Invertebrates. Sort the invertebrates from the foliage sweeps into herbivore (primary consumer) and carnivore (secondary or tertiary consumer) categories. Most of the invertebrates you are likely to find can be put in the right category using Table 14–1 (there are exceptions for several of the groups, and immature and adult stages may differ).

Combine the invertebrates from all of the sweeps and place them into two pre-weighed aluminum foil pans, one for herbivores and one for carnivores. After making sure all the animals are dead (pinch their heads with forceps if necessary) place the pans into the oven for drying.

Mammals. Identify any mammals caught. Weigh all individuals and obtain an average wet weight for each species by summing the weights of all individuals of each species and dividing by the number of individuals. To obtain an average oven-dry weight, multiply the average wet weight by 0.3. You will probably catch only rodents (mice, ground squirrels), which are largely herbivorous, and shrews, which are carnivorous. If you catch anything else, and for specifics of diets, consult a mammal book.

To express biomass on a per area basis, some further calculations are required. First, estimate separately for each mammal species the area sampled, A. The formula is $A = 2(2rL + r^2)$ where L is the length of each trap-line (190 m in this case) and r is the radius of an average home range (that is, r is the radius that would give a home range of the average size for the species, if the home range were a circle). We are, then, calculating the area of two plots, each shaped as in Figure 14–2.

If possible, you should obtain home range sizes from the same habitat, season, and geographical area as your study. If no local figures are available, use the following approximations of home range radii for U.S. grassland mammals.

Blarina brevicauda	35 m	*P. maniculatus*	27 m
Cryptotis parva	28 m	*P. nuttalli*	18 m
Microtus californicus	8 m	*P. polionotus*	21 m
M. ochrogaster	11 m	*Reithrodontomys megalotus*	26 m
M. pennsylvanicus	12 m	*R. humulis*	42 m
Onychomys leucogaster	70 m	*Sorex cinereus*	42 m
Otospermophilus beecheyi	25 m	*Spermophilus tridecemlineatus*	40 m
Peromyscus leucopus	24 m	*Zapus hudsonius*	34 m

For each species of mammal enter in the table the number caught for the entire trapping period *(N)*, the calculated trapping area *(A)*, and the average oven-dry weight ($\bar{X}$). Carry out the calculations shown in the table to obtain biomass expressed as oven-dry weight per square meter.

Mammal Species	(N) Number Caught	(A) Area Trapped	(D) Number per m² (N/A)	($\bar{X}$) Mean Dry Weight	(B) Dry Weight per m² ($\bar{X} \times D$)

Birds. If local population figures are available, convert them to (fractional) numbers of each species per square meter and carry out the calculations indicated in the table to obtain biomass. If no local wet weights are available, obtain them from the sources listed in the Bibliography.

Most of the smaller birds of grassland will eat both invertebrates and seeds during the breeding season, functioning thus as primary and secondary consumers (and when they eat spiders or carnivorous insects, they function as tertiary consumers). You may, if you wish, consult Martin, Zim, and Nelson (1951) or some other source and try to classify the individual species into the appropriate consumer category based on what the majority of their diet is.

If it is not feasible to obtain population figures for birds from your study area, it will suffice to assume a breeding season biomass as follows:

Eastern hayfield-type grasslands with red-winged blackbirds	0.175 g/m²
Tall and mid-grass prairies	0.0125 g/m²
Short-grass and other dry prairies	0.010 g/m²

These figures are for sparrows, meadowlarks, bobolinks, horned larks, etc., which eat both seeds and insects. For top carnivores, such as hawks, 0.0005 g/m² will usually be an overestimate.

Weighing. If feasible, cool the weighing pans in a desiccator before weighing. If this is not possible, transfer them from the oven directly to the balance and weigh immediately, since they will

Bird Species	(D) Number per m^2	($\overline{W}$) Mean Wet Weight	(J) Wet Weight per m^2 ($D \times \overline{W}$)	(B) Dry Weight per m^2 ($0.3 \times J$)

absorb moisture from the air. Do the same with the paper sacks. Since the sacks have not been preweighed, you will have to weigh them with their contents and empty in order to calculate the weight of the vegetation.

Calculation of Biomass for the Various Compartments. Fill in the Summary Table.

RESULTS AND DISCUSSION

1. What was the total weight of the vegetation per square meter? Of animal material? Weights can be converted to energy equivalents by multiplying by an appropriate conversion factor. Caloric content may vary by species, part of the body, season, and possibly other factors (Cummins and Wuycheck 1971) so that for precise work caloric determinations would have to be done for the specific material being worked on. Most plant material, nevertheless, has close to 4 kcal/g, and most animal material has about 5 kcal/g. Assuming those values, state plant and animal biomass in kilocalories per square meter.

2. Fill in the values on the energy flow diagram (Fig. 14–4). It will suffice to use weights rather than kilocalories. How does biomass in the grazing food chains compare with biomass in the detritus and underground food chains? How does energy flow probably compare?

3. Compare the biomass of nematodes or earthworms with that of birds or mammals. Some biologists have implied that this result means that birds or mammals are of less ecological importance than nematodes or earthworms. Evaluate this conclusion.

4. Based on your observations and reading, construct and show two grazing food chains and two detritus-based food chains. Be as specific as possible as to the identification of the organism. Is it likely that any of the chains link to form a food web? If so, where do the links occur?

5. Define standing crop. What is the relationship between the standing vegetation you sampled and the net primary prouction on the area?

6. Where does litter come from? What becomes of it? Suggest methods of determining how fast it disappears.

Summary Table for Biomass in a Grassland

Part of Ecosystem	(*G*) Oven-dry Weight of Container + Organisms	(*T*) Weight of Container	(*W*) Weight of Organisms ($G - T$)	(*A*) Total Area Included in Sample (in m^2)	(*B*) Biomass in g/m^2 (W/A)
Above-ground standing vegetation					
Roots					
Litter					
Total Vegetation					
Earthworms, snails, etc.					
Berlese-Tullgren samples					
Baermann samples					
Foliage primary consumers					
Herbivorous small mammals					
Birds					
Foliage carnivores					
Carnivorous small mammals					
Carnivorous birds					
Total Animals					

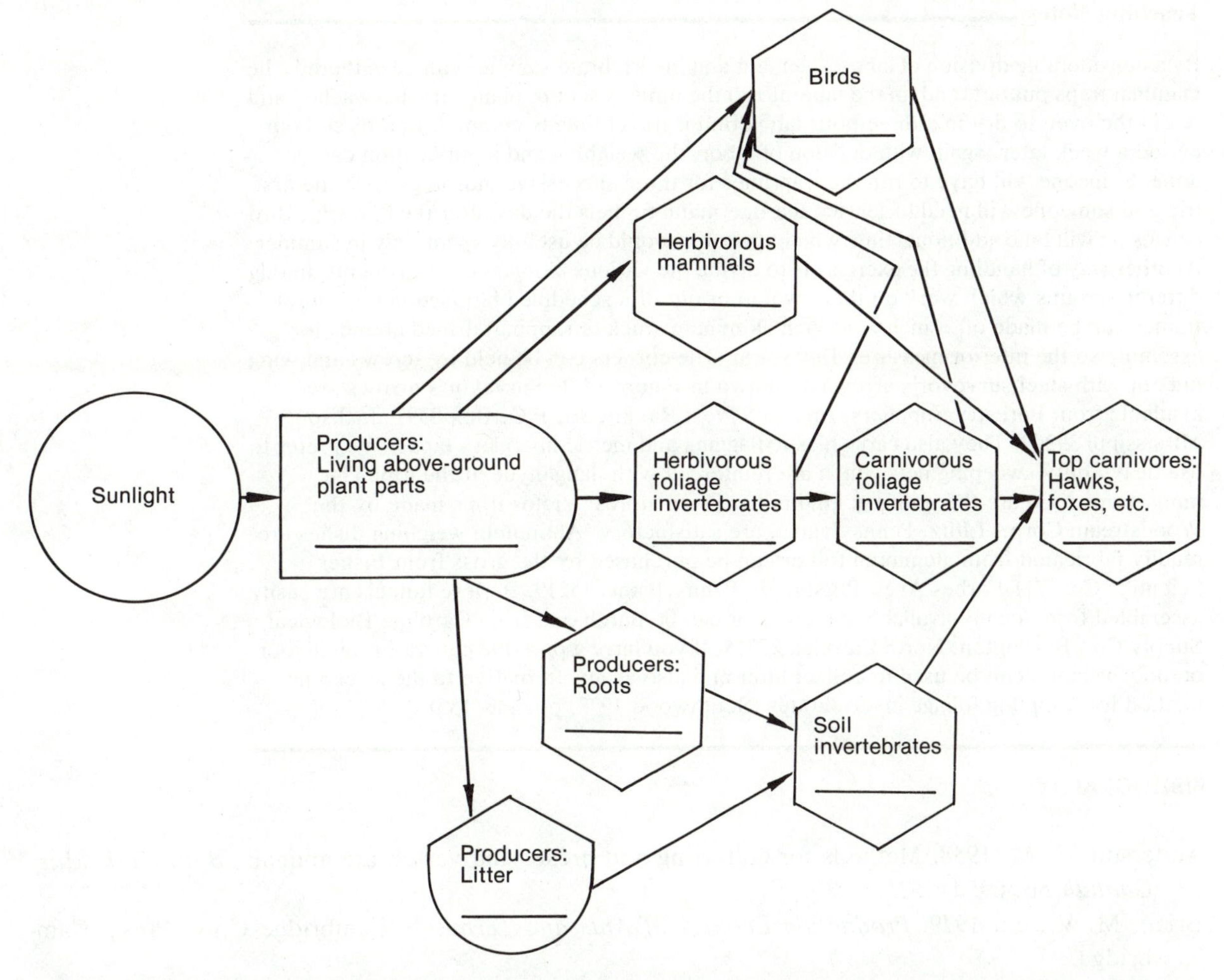

Figure 14–4. Energy flow diagram for a herbaceous ecosystem, using nine compartments. The symbols are those used in energy systems analysis (see diagrams and citations in Exercise 23).

7. What were some groups of organisms that were not sampled in this exercise? How could they have been sampled in such a way as to yield biomass figures?
8. What are some other sources of error in the exercise? How could they be corrected?
9. What are some difficulties in the trophic level concept?
10. In many eastern grasslands, the small mammal community of a grassland consists of three species, a vole *(Microtus),* a deer mouse *(Peromyscus),* and a shrew *(Sorex* or *Blarina).* They often have home ranges of about 0.1 ha, 0.3 ha, and 0.6 ha, respectively. Explain the differences in home range size on an energetic basis.
11. Evaluate the sweep-net method of determining foliage invertebrate densities. How could you calibrate it to see if 48 sweeps is equivalent to 1 m²? How might the insect population estimated in this way differ from those actually in a typical single square meter of grassland?
12. Outline the reasoning used in estimating the area sampled in the mammal trapping. Is the area sampled likely to be the same for all segments of the population of a species? Why or why not?
13. The average home range size of *Peromyscus eremicus* in one study was determined to be 1 acre. What home range radius does this give?
14. Using population and area figures for your state, calculate a rough estimate of human biomass per square meter. How does it compare with any single mammal species in the ecosystem you studied?

Teaching Notes

By a considerable division of labor, the plant and invertebrate samples can be gathered, the mammal traps put out, and (in the laboratory) the funnels set up, plant samples washed and put in the oven to dry in a three-hour lab period if travel time is minimal. In a three-hour period a week later, again with division of labor, the weighing and identification can be done. Someone will have to run the trap-lines for three successive mornings after the first trip and someone will need to service the Baermann funnels the day after the first trip. Bird censusing will take additional time which probably would be usefully spent only in summer. Another way of handling the exercise is to divide the various biomass compartments among different groups which work on their own in or out of a scheduled lab meeting. Quadrat frames can be made of ¾ inch × ⅛ inch aluminum stock (strapping) drilled at ends for assembly so the interior measures 1 m × 1 m. The corners can be held by screws and wing nuts or with steel surveyor's arrows as shown in Figure 14–1. Surveyor's arrows are available from Forestry Suppliers, Inc., 205 West Rankin St., P.O. Box 8397, Jackson, Mississippi 39204. They also carry plastic flagging and metric fiberglass tapes. We prefer to use heavy duty sweeping nets which are reinforced with naugahyde at the top. The mousetraps used are the ordinary kind in hardware stores; Victor traps made by the Woodstream Corp., Lititz, Pennsylvania, are satisfactory. Aluminum weighing dishes are readily fabricated from aluminum foil or can be purchased by the gross from Fisher Scientific Co., 711 Forbes Ave., Pittsburgh, Pennsylvania 15219. Berlese funnels are easily assembled from locally-available materials or can be purchased from Carolina Biological Supply Co., Burlington, North Carolina 27215. If you have a portable generator, an indoor-outdoor vacuum can be used to collect litter and also as an alternative to the sweep net method for sampling foliage invertebrates (Southwood 1978, pp. 148–155).

BIBLIOGRAPHY*

Anderson, R. M. 1958. Methods for collecting and preserving vertebrate animals. *Bull. Natl. Mus. Canada* no. 69: 1–162.

Brian, M. V., ed. 1979. *Production Ecology of Ants and Termites*. Cambridge Univ. Press, Cambridge.

Burt, W. H., and R. P. Grossenheider. 1976. *A Field Guide to the Mammals: Field Marks to All North American Species Found North of Mexico*. 3rd ed. Houghton-Mifflin, Boston.

Cockrum, E. L. 1962. *Introduction to Mammalogy*. Ronald Press, New York.

Cummins, K. W., and J. C. Wuycheck. 1971. Caloric equivalents for investigations in ecological energetics. *Mitt. int. Ver. Limnol.* no. 18: 1–158.

Fitch, H. S. 1958. Home ranges, territories, and seasonal movements of vertebrates of the Natural History Reservation. *Pub. Univ. Kansas Mus. Nat. Hist.* 11(3): 63–362.

French, N. R., R. K. Steinhurst, and D. M. Swift. 1979. Grassland biomass trophic pyramids. Pp. 59–87 in N. R. French (ed.). *Perspectives in Grassland Ecology*. Springer-Verlag, New York.

Giles, R. H. 1971. *Wildlife Investigational Techniques*. 9th ed. The Wildlife Soc., Washington, D.C.

Golley, F. B., J. B. Gentry, L. D. Caldwell, and L. B. Davenport, Jr. 1965. Number and variety of small mammals on the AEC Savannah River plant. *J. Mamm.* 46: 1–18.

Hall, G. A. 1964. Breeding bird censuses—why and how. *Aud. Field Notes* 11: 413–416.

Howell, J. C. 1954. Populations and home ranges of small mammals on an overgrown field. *J. Mamm.* 35: 177–186.

Jackson, R. M., and F. Raw. 1966. *Life in the Soil*. St. Martin's Press, New York.

Kendeigh, S. C. 1944. Measurement of bird populations. *Ecol. Monogr.* 14: 67–107.

Kevan, D. K. McE. 1968. *Soil Animals*. Witherby, London.

Martin, A. C., H. S. Zim, and A. L. Nelson. 1951. *American Wildlife and Plants: A Guide to Wildlife Food Habits*. McGraw-Hill, New York.

McFadyen, A. 1962. Soil arthropod sampling. *Advances in Ecological Research* (F. B. Cragg, ed.) 1: 1–34.

*Books for mammal identification are listed with Exercise 9.

Menhinick, E. F. 1963. Estimation of insect population density in herbaceous vegetation with emphasis on removal sweeping. *Ecology* 44: 617–621.

Phillipson, J., ed. 1971. *Methods of Study in Quantitative Soil Ecology: Population, Production and Energy Flow*. IBP Handbook no. 10. Blackwell Scientific, Oxford.

Pough, R. W. 1947. How to take a breeding bird census. *Audubon Mag*. 49: 290–297.

Shelford, V. E. 1951. Fluctuation of forest animal populations in east-central Illinois. *Ecol. Monogr*. 21: 148–181.

Smith, M. H., R. H. Gardner, J. B. Gentry, D. W. Kaufman, and M. H. O'Farrell. 1971. Density estimations of small mammal populations. Pp. 25–33 in F. B. Golley, K. Petrusewicz, and L. Ryszkowski (eds.). *Small Mammals: Their Productivity and Population Dynamics*. Cambridge Univ. Press, Cambridge.

Southwood, T. R. E. 1978. *Ecological Methods*. 2nd ed. Chapman and Hall, London.

Stonehouse, B., ed. 1978. *Animal Marking*. MacMillan, London.

Wiegert, R. G., and F. C. Evans. 1964. Primary production and the disappearance of dead vegetation on an old field in southeastern Michigan. *Ecology* 45: 49–63.

Identification of Terrestrial Invertebrates

Borror, D. J., D. M. DeLong, and C. A. Triplehorn. 1981. *An Introduction to the Study of Insects*. 5th ed. Saunders College, Philadelphia.

______, and R. E. White. 1970. *A Field Guide to the Insects of America North of Mexico*. Houghton Mifflin, Boston.

Chu, H. F. 1949. *How to Know the Immature Insects*. Wm. C. Brown, Dubuque, Iowa.

Eddy, S., and A. C. Hodson. 1961. *Taxonomic Keys to the Common Animals of the North Central States*. Burgess, Minneapolis.

Jaques, H. E. 1947. *How to Know the Insects*. Wm. C. Brown, Dubuque, Iowa.

Bird Weights

Amadon, D. 1940. Bird weights and egg weights. *Auk* 60: 221–234.

Clench, M. H., and R. C. Leberman. 1978. Weights of 151 species of Pennsylvania birds analyzed by month, age, and sex. *Bull. Carnegie Mus. Nat. Hist*. 5: 1–87.

Esten, S. R. 1931. Bird weights of 52 species of birds (taken from notes of Wm. Van Gordes). *Auk* 48: 572–574.

Graber, R., and J. Graber. 1962. Weight characteristics of birds killed in nocturnal migration. *Wilson Bull*. 74: 74–88.

Hartman, F. A. 1946. Adrenal and thyroid weights in birds. *Auk* 63: 42–63.

Johnston, D. W., and T. D. Haines. 1957. Analysis of mass bird mortality in October 1954. *Auk* 74: 447–458.

______, and R. A. Norris. 1958. Weights and weight variations in summer birds from Georgia and South Carolina. *Wilson Bull*. 70: 114–129.

Kendeigh, S. C. 1945. Resistance to hunger in birds. *J. Wildl. Manage*. 9: 217–277.

Murray, B. G., Jr., and J. F. Jehl, Jr. 1964. Weights of autumn migrants from coastal New Jersey. *Bird-Banding* 35: 253–263.

Poole, E. L. 1938. Weights and wing areas in North American birds. *Auk* 55: 511–517.

Skinner, R. W., and P. A. Stewart. 1967. Weights of birds from Alabama and North Carolina. *Wilson Bull*. 79: 37–42.

Stegeman, L. 1955. Weights of some small birds in central New York. *Bird-Banding* 26: 19–27.

Exercise 15

PRODUCTIVITY

Key to Textbooks: Brewer 133–135; Barbour et al. 237–262; Benton and Werner 118–120; Colinvaux 152–158; Emlen 341–351; Kendeigh 171–173; Krebs 517–540; McNaughton and Wolf 104–127; Odum 43–63; Pianka 47–53; Richardson 261; Ricklefs 626–642; Ricklefs (short) 110–120; Smith 117–126, 295; Smith (short) 47–71; Whittaker 192–230.

OBJECTIVES

We obtain an estimate of net primary production for a site dominated by herbaceous plants. In the process we examine various ways of measuring productivity and the reasoning on which they are based. This involves a consideration of the processes of production in individual, population, and community.

INTRODUCTION

Productivity is the rate (amount per time) at which new organic matter is added to an individual, a population, or an ecosystem. The term is sometimes applied both to plants (or other autotrophs and the new biomass is called *primary production*) and to animals and other heterotrophs (where the new biomass is called *secondary production*). We are concerned here with primary production.

The process underlying most of the production of new biomass in the world is photosynthesis (in a few habitats, chemosynthesis is an important consideration), generally summarized as

$$6\ CO_2 + 6\ H_2O \longrightarrow 6\ C_6H_{12}O_6 + 6\ O_2.$$

In the process, the energy of sunlight is stored as the energy of chemical bonds in the new organic molecules that form the added biomass.

Productivity as just defined is *gross productivity,* the total production of organic molecules in photosynthesis. The plant itself uses some organic matter in respiration. *Net productivity* is the rate of storage of new protoplasm after plant respiration is subtracted. Because of the great variability in water content in different tissue, production figures are generally given as oven-dry weight.

Key Words: productivity, primary production, secondary production, gross production, net production, respiration, biomass, harvest method, dimension analysis, light and dark bottle method, above-ground net production (*ANP*), below-ground net production (*BNP*), consumption, litter, humus.

Ways of Measuring Productivity

From the summary statement for photosynthesis it is clear that production can be measured in several ways. In this exercise we approach it directly by measuring added biomass. Because consistent relationships exist between the various terms of the equation it is also possible in principle to estimate production by measuring the depletion of any raw material or the increase in any product. Knowing, for example, how much carbon dioxide is used will allow us to calculate how much dry weight has been added.

The theoretical foundation for studying productivity is straightforward. Measuring productivity in practice is less so, and a very large number of methods and variants have been devised to deal with specific habitats or situations.

In this exercise we use variants of the *harvest technique* which is the most widely used method for studying herbaceous terrestrial ecosystems. A complicated relative of the harvest technique called *dimension analysis* is generally used for forests. Dimension analysis methods involve harvesting representative trees and shrubs, determining their annual production (how much biomass was added in circumference growth, in new twigs, in leaves, etc.), and then generalizing to the rest of the forest. This is done using regression equations relating productivity to one or more measures of tree size combined with a detailed stand table (Exercise 16) for the forest. Harvesting and dimension analysis both measure net productivity.

Another technique used in terrestrial situations is the measurement of CO_2 uptake (now generally by infrared gas analysis) in a transparent enclosure. Although the method is usually employed to study the photosynthetic rate of single leaves, some ecologists have enclosed large portions of an ecosystem in plastic bags. Potentially both gross and net productivity are measurable.

In aquatic ecosystems, O_2 changes are more often used. The familiar light and dark bottle method measures oxygen loss and gain in two bottles each containing a microcosm of the aquatic ecosystem to be studied. Knowing oxygen concentration at the start, the increase of oxygen in the light bottle gives a measure of gross primary production. The decrease in oxygen in the dark bottle gives a measure of respiration. Subtracting respiration from gross production gives net production.

The ^{14}C method injects radioactive carbon (in the form of bicarbonate) into an aquatic microcosm to obtain an estimate of the amount of carbon produced in photosynthesis. Since we can estimate biomass from carbon (1 g carbon $\approx$ 2 g dry weight), this provides us with a measure of production. This method is generally considered to estimate net productivity.

This list by no means exhausts the approaches that have been tried. In fact, some of the methods that have been devised are marvels of ingenuity. Carbon dioxide accumulation (as a measure of respiration) has been monitored in the air mass retained in a forest because of a meteorologic temperature inversion; and we have not even mentioned the dynamite method of obtaining tree roots for dimension analysis. Newbould (1967), Vollenweider (1969), and Part 2 of Lieth and Whittaker (1975) are useful reviews.

MATERIALS

quadrat sticks and pins, also called surveyor's arrows (see Fig. 14–1) (or meter sticks, quadrat pins, and string), 1 set per 2–4 students

grass clippers, 1 per 2 students

bulb planters (sharpen with a file), 1 per 4 students

drying oven

paper sacks, 2 or 3 per sample plot

bucket, 1 per 4 students

sieve, 1 per 4 students

balance accurate to 0.1 g

PROCEDURE

We here study production using a harvest method on an area dominated by herbaceous vegetation. Harvest methods are adaptations of crop harvests. If a farmer cuts and bales hay on an acre of alfalfa three times one growing season and gets 4 tons, that figure converted to oven-dry weight or kilocalories is a pretty good estimate of above-ground (or aerial) net primary production (*ANP*). The harvest method applied by ecologists does much the same thing.

A number of variations exist. In its simplest form, we try to pick the time when the standing crop of the area to be sampled is at its peak, clip the vegetation, weigh it, dry it, and call that an estimate of the year's *ANP*. It is, but it is generally an underestimate. The main reason for this is that most communities are composed of several species with different phenologies; one species may make its maximum production in May and then begin to die back, another may do the same in June, and so on. The overall peak may occur in August or September but many of the earlier species may be badly underestimated by then. To counteract this, another variant of the method samples repeatedly, say every month or two weeks throughout the growing season, and sorts and weighs each species separately. *ANP* is calculated by summing the peak weights for each species whenever they are reached. This gives a better estimate, much better if the community is phenologically complex but scarcely better at all if it is dominated by one species of annual plant.

A third variant measures the increase in biomass over fairly short intervals, two weeks or a month, but does not consider the species separately. All increases over the growing season are summed to measure the year's *ANP*. This method takes less time than one that separates the species and gives similar results. It is the method preferred by Singh et al. (1975) after testing 13 variants of the harvest method. This is the method that will be used in this exercise if it is done in the spring or summer; a single harvest method will be used if it is done in the fall. Details of the two are given separately below.

Fall Sampling: Single Harvest of Peak Standing Crop. Visualize a cornfield in the spring; the biomass (ignoring weeds) is negligible, consisting of whatever weight the planted seed corn has. Over the course of the summer the corn plants grow, tassel, and produce ears. At the end of the summer or early in the fall, you could go into the field, harvest the corn—ear, roots, stalks, and all—and obtain a very close estimate of net primary production. This is the sort of situation in which a single harvest of peak standing crop works best. We suggest, then, that in applying this method you attempt to use phenologically simple communities. If you can find a complaisant farmer, a crop field such as corn or soybeans works very well. It also dovetails with Exercise 23 on the trophic ecology of humans. The first year's growth in an old field is good and sampling could be done in connection with Exercise 19. The first year's growth on a totally new surface (cut or fill at a construction site or along a new highway) is even better. Communities dominated by perennials are less satisfactory since the assumption of negligible biomass at the beginning of the growing season may be wrong; however, unmowed hayfields composed of brome grass or alfalfa are suitable and in this case the sampling suggested here essentially duplicates that suggested in Exercise 14 on grassland biomass.

1. Randomly establish as many sample plots as can be conveniently sampled (three or four students per plot is best). Depending on the vegetation the plots can be of any size from about 0.2 to 1.0 m^2. If neither time nor the removal of the vegetation is a problem, the larger sizes are preferable in biomass sampling because of the reduction of the area:perimeter ratio. In other words, the percentage of the plants that are on an edge and that may be mistakenly called in the quadrat when they should be out, or the reverse, is reduced in larger plots. For the same reason, compact plots (squares or circles) are preferred over elongated plots in biomass sampling.
2. If you are doing this exercise in conjunction with the one on biomass, Exercise 14, follow the directions given there for both field and laboratory. Otherwise, follow steps 3–6, below.
3. At the borders of the plots, part the standing vegetation so that all of the stems rooted in the plot are inside the boundary markers.
4. Clip all of the standing leaves and stems as close to the ground as possible and put the materials into a paper sack labeled "Standing Vegetation." Include all living material plus any material that has obviously died during the current growing season.

5. Remove the litter from about 60 cm^2 of the soil surface and discard it. Using a sturdy, sharpened bulb planter take six cores (or fewer if the plot is small) evenly spaced over the bare surface. Place the cores, soil and all, into a paper sack marked ''Roots.''

6. At the laboratory, handle the samples as follows:

Standing Vegetation. Open the tops of the sacks and put them in the oven for drying. If oven space is limited, air-dry all the sacks, weigh them, and then oven-dry one (or a portion of one) to obtain a conversion factor for calculating the oven-dry weights of the rest. Oven-drying should be continued until a constant weight is reached. This will take several days if a large amount of green vegetation is dried and one or two days for material that was first air-dried.

Roots. Place the soil cores from the ''Root'' sack into a sieve and wash away the dirt over a bucket. This should be done outside if possible. Put all the plant material retained in the sieve into a paper sack for drying. Also stir the water in the bucket to bring any plant fragments to the surface. Add them to the sack before placing it into the oven for drying. Stir this material occasionally during the several days it will take to dry.

Note: It is difficult to remove all the soil from roots (or litter) and a few grains of sand weigh as much as a great deal of vegetation. An alternative approach to that described above, if a muffle furnace is available and the time is warranted, is to get the sample as clean as possible, oven-dry it, and then ash it in the muffle furnace. Weigh the ash, which will include the mineral content of the vegetation plus any adhering soil, subtract this from the oven-dry weight, and express biomass as ash-free dry weight.

7. After oven-drying (80–105° C), weigh the sacks with vegetation, empty them, and re-weigh. Subtract sack weight from total weight to obtain weight of the vegetation.

Spring or Summer Sampling: Trough-Peak Analysis. In a complex herbaceous community dominated by perennials, there will be at the beginning of a growing season dead plant material in the form of litter and, mixed with the mineral soil, humus. There will be living plant material in the form of roots, rhizomes, crowns, seeds, and usually some green shoots. The several compartments that can be recognized for living and dead plant material are shown in Figure 15–1. Time 0 can be any starting point, although the beginning of the growing season is particularly appropriate.

The second column of the figure shows the same compartments one time unit later and indicates the source of the material in each compartment. Live material at Time 1, for example, is composed of the live material that persisted from time 0 plus new production. The recent dead material consists of the persistent recent dead material from time 0 plus whatever live material at time 0 died in the interval, plus whatever newly produced material suffered a rapid death.

Suppose we could, without disturbing a plot, obtain the weight of the living (above-ground) vegetation and any recently dead vegetation. We then repeat the measurement two weeks later. Study of the diagram (enter some figures if necessary) will show that (1) if no loss of living material (including new production) to the dead compartment occurred, simply subtracting living weight at time 0 from living weight at time 1 would give *ANP* for that period; (2) if some loss of living material (including new production or not) to recent dead does occur but no loss of recent dead to litter occurs, then *ANP* is obtained readily by subtracting $Live_0$ from $Live_1$, subtracting Recent $Dead_0$ from Recent $Dead_1$, and adding the two together. It is probably often a reasonable assumption that little transfer to the litter occurs until the end of the growing season; to the degree that the assumption is false, *ANP* would be underestimated. We could, of course, get around this problem by considering not just living and recent dead compartments but also litter.* The argument against doing this is that sampling litter is hard and the added accuracy is small.

Now we cannot readily make the weight measurements on a quadrat without disturbing it. What we must do, consequently, is take one set of quadrats at time 0, clip them, and take another set at time 1 and clip them. If we have enough samples each time to be representative of the ecosystem, there will be no problem in making the calculations even though the same plots are not used.

*The preferred method of Singh et al. (1975) does not include summation of positive increments of litter but does include summation of positive changes in an additional ''old dead'' compartment between ''recent dead'' and litter. The method described here includes ''old dead'' within the ''recent dead'' compartment. Singh et al. also recommend summing only statistically significant increases in the two dead compartments.

Statistical adequacy is not always easy to achieve, however, because of the considerable spatial variability in biomass even in relatively homogeneous vegetation.

Choose any herbaceous area for this method. If it is possible to begin about the start of the growing season and sample every two weeks or every month until the end of the growing season, that

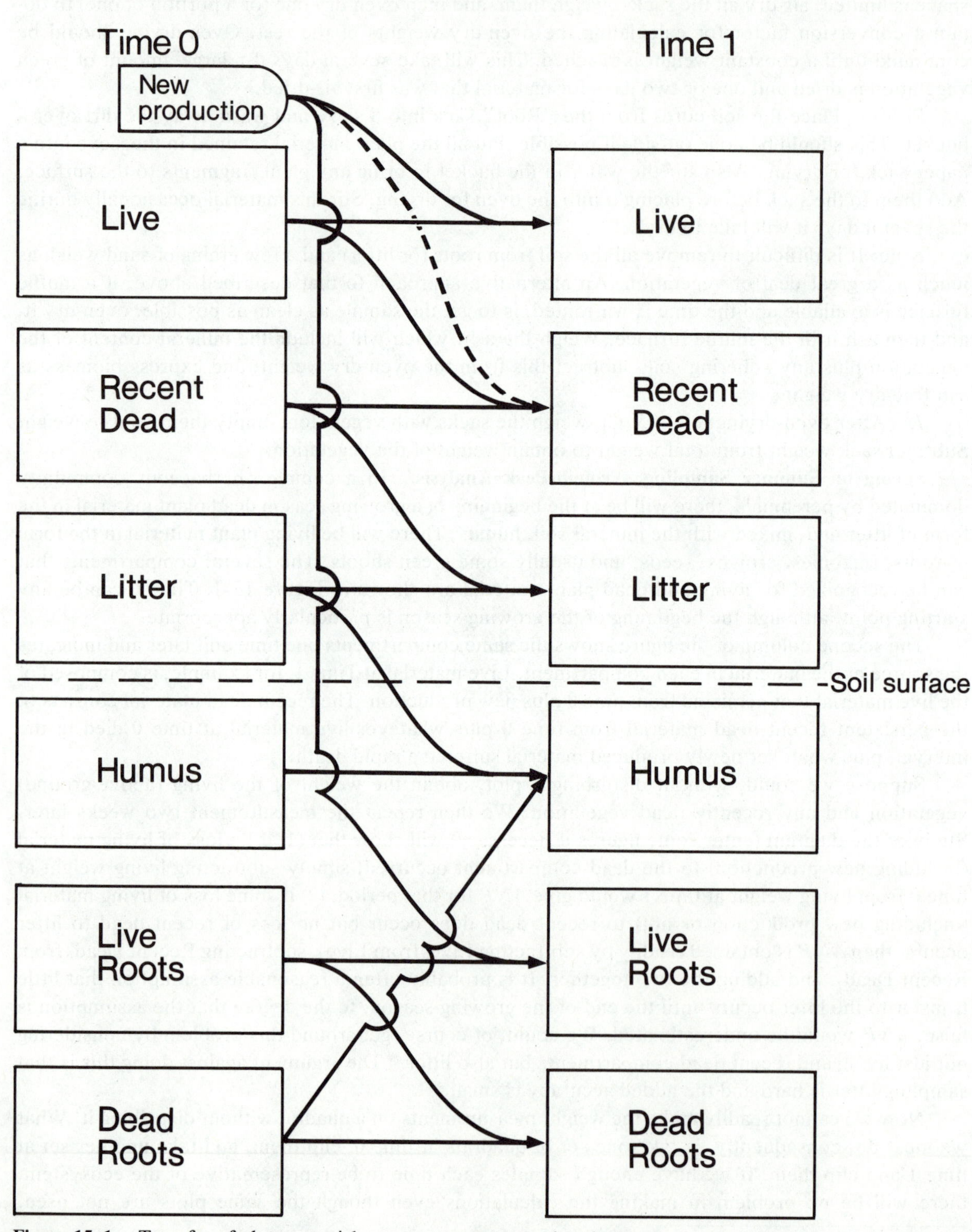

Figure 15–1. Transfer of plant material among compartments in a herbaceous ecosystem in one time interval. Loss by consumption from the various compartments is not diagrammed, nor is loss by physical processes such as solution.

is ideal; however, all that is necessary for the purposes of the exercise is two sampling dates two weeks or a month apart. The specific procedures are as follows:

1. Randomly set up several sampling plots (preferably square or round; see point 1 under Fall Sampling) of 0.2–1.0 m^2. About one plot per 3–4 students works well. At the borders of the plot, part the standing vegetation so that all of the stems (alive plus any that have died during the current growing season) rooted in the plot are inside the boundary markers.

2. From each plot, remove all of the recent dead material (include individual leaves as well as whole plants). This should be the material that has died within the current growing season. On most samplings it will be a small amount; use a small paper bag labeled "Dead."

3. Clip all of the living stems and leaves as close to the ground as possible and put the materials into a paper sack labeled "Live."

4. Remove the litter from about 60 cm^2 of the soil surface but do not keep it. Using a sturdy, sharpened, bulb planter, take six cores (fewer if the plot is less than 1 m^2) in the denuded area. Place the cores, soil and all, into a paper sack marked "Roots."

5. At the laboratory, handle the three kinds of samples as follows:

Live Vegetation and Dead Vegetation. Open the tops of the sacks and put them in the oven for drying. If oven space is limited air-dry all the sacks, weigh them, and then oven-dry one live and one dead to obtain a conversion factor for calculating the oven-dry weights of the rest. Oven drying should be continued until a constant weight is reached. This will take several days if a large amount of green vegetation is dried and one or two days for material that was first air-dried.

Roots. Place the soil cores from the Root sack into a sieve and wash away the dirt over a bucket. This should be done outside if possible. Put all the plant material retained in the sieve into a paper sack for drying. Stir this material occasionally during the several days it will take to dry. (See the note on page 131 for an alternative way to treat root samples.)

6. After oven-drying at 80–105° C, weigh the sacks with vegetation, empty them, and re-weigh. Subtract sack weight from total weight to obtain weight of the vegetation.

7. Repeat the sampling in two to four weeks, following the same procedures. If feasible, repeat throughout the growing season.

Calculation of Above-ground Net Production (*ANP*)

For the fall sampling (single harvest of peak standing crop), simply obtain a mean oven-dry weight of the standing vegetation per sample plot and express this as g/m^2 (henceforth, we will use the notation g m^{-2}). This is your estimate of *ANP* for the growing season (and, in most circumstances, for the year).

For the spring sampling (trough-peak analysis) proceed as follows (use the Trough-Peak Analysis Sheet). Obtain means of live (L) and recent dead (D) vegetation for each sampling period. Your initial sampling will be time 0, the next time 1 (and so on, if more than two samplings were performed). To calculate *ANP* for the period from sampling 0 to 1, subtract L_0 from L_1. If D_1 is greater than D_0, obtain the difference. If $D_1 \leq D_0$, omit the operation.* Add these two remainders ($L_1 - L_0$ and $D_1 - D_0$) together. This is your estimate of *ANP* for the time period 0–1.

If you took additional samplings, do the same thing for period 1–2, 2–3, etc. To obtain growing season (and annual) *ANP*, add the *ANP*'s for all periods together.

Calculation of Below-ground Net Production (*BNP*)

Calculation of root production for ecosystems dominated by annuals is relatively easy. Since you start at zero (or whatever low value the hypocotyl had), all of the biomass of the roots represents net

*If there is no increase or a loss, there are three possibilities: (1) there may have been a large transfer of dead material to litter; (2) no living material was transferred to the dead compartment; or—perhaps most likely—(3) the spatial variation in the field was enough that you sampled spots with a lot of dead material at time 0 and spots with not very much at time 1.

production. For the fall sampling, consequently, simply calculate the area of one bulb planter in cm^2, divide that into 10,000 (the number of square centimeters in a square meter), and multiply the average oven-dry weight of each root core by the resulting conversion factor to obtain root production in $g\ m^{-2}$. Any root production in the square meter below the depth reached by the bulb planter is, of course, missed.

Calculation of root production for herbaceous perennial vegetation is a knotty problem. The roots, unlike the shoots, represent production from more than one growing season (rather as a tree trunk does). Separating older from newer root material (and dead from live) is difficult. The seasonal low point of root biomass is unclear, and just when materials are transported to and taken out of roots is not well known. The upshot of this is that there is really no practical method of estimating root production known to be accurate. We suggest here two possible methods, both easy. The first can be used with either the fall or the spring samples; the second can only be used when samples are taken at intervals, as in the spring method.

The first method consists of assuming that the following proportionality holds:

$$\frac{BNP}{\text{root biomass}} = k \frac{ANP}{\text{above-ground biomass}} .$$

Once above-ground biomass, above-ground production, and root biomass are known, root production is easily estimated if k is known. For the purposes of this exercise, assume that $k = 1$ and estimate root production for the fall sample.

The second method, requiring two or more sampling periods, is analogous to the trough-peak analysis of above-ground parts. The increase in weight from time 0 to 1, time 1 to 2, etc. is taken as an estimate of the net root production. Since no allowance is made for loss of dead material, the estimate will be too low except for the unlikely case that no such loss occurs. For the spring sampling, estimate root production by this method or by both.

DISCUSSION

1. What is the difference between *production* and *productivity?*
2. List in detail the ways in which net primary production is lost from the plants in your samples and which, therefore, cause your *ANP* samples to be too low.
3. List ways in which you might attempt to measure these missed *ANP* elements and thus improve your estimates. Compare the increase in accuracy achieved with the increase in time required.
4. One way in which net primary production is removed from the samples is through consumption, and one way of attempting to correct for this would be to measure *ANP* in exclosures where herbivores were excluded. What problems might there be with figures from exclosures as estimates of *ANP* in grassland ecosystems?
5. If you used the trough-peak analysis for some period shorter than the growing season, you have an estimate of production for only that period of time rather than growing season (or annual) production. Come up with some logical scheme and convert your figure for two weeks or a month or whatever to an estimate of annual productivity for the ecosystem.
6. Convert your annual productivity figures to kilocalories by multiplying by four. (Or, if you can find more accurate figures for the specific plants making up your samples in Cummins and Wuycheck 1971 or elsewhere, use them.) Using either weight or energy content, compare your figures for production with other figures in the literature for a range of ecosystems.
7. What physical factors make an ecosystem highly productive?
8. The emphasis in this exercise has been on productivity at the community level; however, community productivity must be based on processes occurring in the populations and the individual organisms. Describe these underlying processes.

Trough-Peak Analysis Sheet

$L_0 =$ ______ $g\ m^{-2}$	$D_0 =$ ______ $g\ m^{-2}$
$L_1 =$ ______ $g\ m^{-2}$	$D_1 =$ ______ $g\ m^{-2}$
$L_2 =$ ______ $g\ m^{-2}$	$D_2 =$ ______ $g\ m^{-2}$
$L_3 =$ ______ $g\ m^{-2}$	$D_3 =$ ______ $g\ m^{-2}$
$L_4 =$ ______ $g\ m^{-2}$	$D_4 =$ ______ $g\ m^{-2}$

To calculate *ANP* for each period:

Operation	For Time 0–1	For Time 1–2	For Time 2–3	For Time 3–4
1. Subtract L_n from L_{n-1}	$L_1 - L_0 =$ ______ (A)	$L_2 - L_1 =$ ______ (A)	$L_3 - L_2 =$ ______ (A)	$L_4 - L_3 =$ ______ (A)
2. If $D_n > D_{n-1}$, subtract D_{n-1} from D_n	$D_1 - D_0 =$ ______ (B)	$D_2 - D_1 =$ ______ (B)	$D_3 - D_2 =$ ______ (B)	$D_4 - D_3 =$ ______ (B)
3. Add (A) and (B)	$ANP_1 =$ ______	$ANP_2 =$ ______	$ANP_3 =$ ______	$ANP_4 =$ ______

To calculate *ANP* for the period 0–4, sum *ANP* for the separate periods, i.e., $\sum_{i=1}^{4} (ANP)$

9. A transparent and an opaque bottle each containing a sample of lake water are suspended for two hours in a lake. The initial oxygen concentration of the lake water was 6.00 mg O_2 1^{-1}. After the two hour incubation period, the dark bottle contained 5.50 mg O_2 1^{-1} and the light bottle contained 7.00 mg O_2 1^{-1}. Calculate gross productivity, net productivity, and respiration expressed as mg O_2 $1^{-1}hr^{-1}$. Limnologists generally express production as carbon fixed rather than as oxygen liberated. The exact value of the conversion factor depends on what materials besides carbohydrates are being synthesized. If only carbohydrate is being produced, the conversion factor is 1.0, so that 1 mg O_2 liberated = 1 mg C fixed. Usually some other materials are being synthesized and the value of the conversion factor is altered; 1.0 mg carbon = 1.2 mg O_2 is given by Wetzel and Likens (1979) as a "normal" value. Also, results are generally expressed per cubic meter rather than per liter; to make this conversion multiply by 1,000. To summarize, to express either gross or net productivity as carbon:

$$\text{mg C m}^{-3}\text{ hr}^{-1} = \frac{(\text{mg O}_2)\ 1{,}000}{1.2}.$$

10. Tell everything you would have to measure to determine the annual net production of a tree.

11. What additional information would you have to have to be able to state *gross* primary productivity for the site you studied?

12. What are the major consumers of the living plants that you studied? Does more or less of the net production go to grazing or to detritus-based food chains? What are some of the detritivores?

Teaching Notes

A three-hour period is enough for the field work and preparation of materials for drying. Provision will have to be made to weigh material later.

BIBLIOGRAPHY

Cartledge, O., and D. J. Conner. 1972. Field chamber measurements of community photosynthesis. *Photosynthetica* 6: 310–316.

Cummins, K. W., and J. C. Wuycheck. 1971. Caloric equivalents for investigations in ecological energetics. *Mitt. int. Ver. Limnol.* no. 18: 1–158.

Dyer, M. I. 1980. Mammalian epidermal growth factor promotes plant growth. *Proc. Natl. Acad. Sci.* 77: 4836–4837.

Lieth, H., and R. H. Whittaker, eds. 1975. *Primary Productivity of the Biosphere*. Springer-Verlag, New York.

Milner, C., and R. E. Hughes. 1968. *Methods for the Measurement of Primary Production of Grassland*. IBP Handbook no. 6. Blackwell Sci. Publ., Oxford.

Newbould, P. J. 1967. *Methods of Estimating the Primary Production of Forests*. IBP Handbook no. 2. Blackwell Sci. Publ., Oxford.

Odum, E. P. 1960. Organic production and turnover in old field succession. *Ecology* 41: 34–39.

Schwoerbel, J. 1970. *Methods of Hydrobiology (Freshwater Biology)*. Pergamon Press, Oxford.

Singh, J. S., J. S. Lauenroth, and R. K. Steinhorst. 1975. Review and assessment of various techniques for estimating net aerial production in grasslands from harvest data. *Bot. Rev.* 41: 181–232.

Vollenweider, R. A. 1969. *A Manual on Methods for Measuring Primary Production in Aquatic Environments*. IBP Handbook no. 12. Blackwell Sci. Publ., Oxford.

Wetzel, R. G., and G. E. Likens. 1979. *Limnological Analyses*. Saunders, Philadelphia.

Wiegert, R., and F. C. Evans. 1964. Primary production and the disappearance of dead vegetation on an old-field. *Ecology* 45: 49–63.

Exercise 16

FOREST COMPOSITION

Key to Textbooks: Brewer 175–200; Barbour et al. 62–70, 130–134, 170, 177–179, 494–518; Benton and Werner 69–71, 193–223; Colinvaux 71–90; Kendeigh 20–23; Krebs 386–390, 439–442; McNaughton and Wolf 566–573; Odum 380; Richardson 93, 110–111; Ricklefs 589–597; Ricklefs (short) 388–395; Smith 95, 312–314; Smith (short) 432–434; Whittaker 179–183.

OBJECTIVES

Here we compare two forests, using data obtained by both plot and plotless sampling. We relate the composition of forests to features of the sites they occupy and to the processes occurring within them.

INTRODUCTION

Forests are composed of large trees forming a canopy. Beneath the large trees are smaller ones, some of which will contribute to the canopy at a later time. The composition of a forest is described in a *stand table* which tells density of trees by species and size. The table, then, gives the same sort of data as an age pyramid except that sizes are used. Size is not a good measure of age for trees, but since the size of a tree largely determines its role in the forest, size distributions can often be used like age distributions.

The composition of a forest can be influenced by many factors including successional status, soil, microclimate, and historical factors such as a fire a hundred years earlier or a windstorm a week ago. This exercise compares two forests on different sites and speculates on the causes of differences found. We recommend comparing north vs south (or east or west) slopes within a forest stand of uniform history (physiographers refer to the direction a slope faces as its *aspect;* however, this term is more commonly used in ecology as a synonym for season). Doing this at the same site where microclimate readings were made (Exercise 1) makes a good combination. Other possibilities are comparing forests on two different soil types (in which case this exercise could be combined with Exercise 2) or grazed and ungrazed parts of a forest.

Key Words: size distribution, aspect (slope), diameter at breast height (*DBH*), plot and plotless sampling, point-quarter method, mean area, stand table, basal area, importance value (relative frequency, relative dominance, relative density), gap-phase replacement.

MATERIALS

Have several sets of equipment so everyone can participate at the same time.

quadrat frames (see Figure 14–1) and pins (surveyor's arrows)

meter sticks

stakes or surveyor's arrows with a string or tape marked at 178.4 cm (The string revolves around the stake to make a circle of 10 m^2.)

diameter tapes, preferably metric (Calipers can be used, but they are bulky to carry.)

long (> 10 m) metric tapes

compasses

stakes

random numbers table (inside front cover)

tree identification manuals (see Bibliography)

clinometer or Abney level

PROCEDURE

Determination of Slope

If you are comparing two slopes, obtain the angle of each from the horizontal. A simple way to do this is to match two people by height and have one stand 5–20 m upslope of the other. The downslope person should focus on the eyes of the upslope person, using a clinometer (preferably) or an Abney level. With either, the angle from horizontal can be read directly (make sure that the degree scale and not percent grade or topographic units is used).

Determination of Tree Densities

Trees will be sampled in three ways. The object is to determine the density of trees in several size classes for each species. Do not count shrubs or dead trees. To obtain large enough samples it may be desirable to accumulate data from several classes.

Follow the steps below for both of the sites studied.

1. **Identify the Common Trees.** Be sure to identify the seedlings (see U.S. Forest Service 1948), which for some species do not look like the adults. The class should also learn how to measure tree diameters. This is conventionally done at 1.5 m above the ground (''breast height'') and the measurement is called diameter at breast height (*DBH*). Diameters are readily measured with diameter tapes, which are scaled in centimeters (or inches) multiplied by π. The tape around the circumference of the more or less circular trunk then shows the diameter. Everyone should find out about how high 1.5 m is on his or her body so it doesn't need to be measured on each tree. On slopes, most persons measure from average ground level but the National Forest Service measures from the upslope side.

At this point, divide into groups of two or three; each group should do all types of sampling.

2. **One Square Meter Quadrats.** Sample trees which are under 0.5 m tall (''very small'') in 1 m^2 quadrats. Locate the quadrats randomly (see Exercise 5 for ways of choosing random points). Identify and count all the very small trees rooted within the quadrat; ignore trees $\geq$ 0.5 m tall. Check carefully for very small seedlings. Record on Data Sheet 1.

3. **Circle Plots.** Sample trees 0.5 m and taller, but less than 2.5 cm *DBH* (''small'') in circle plots with an area of 10 m^2. To set up the round plots, put the stake at a randomly-chosen point and

Figure 16–1. The version of the point-centered quarter method used here obtains two sets of four trees at each point, one set medium (2.5 – 12.99 cm *DBH*) and one set large (13.00 cm or greater *DBH*). The nearest tree in each of the two size classes in each quarter is chosen. Dotted lines show measurement to medium trees, dashed lines to large trees.

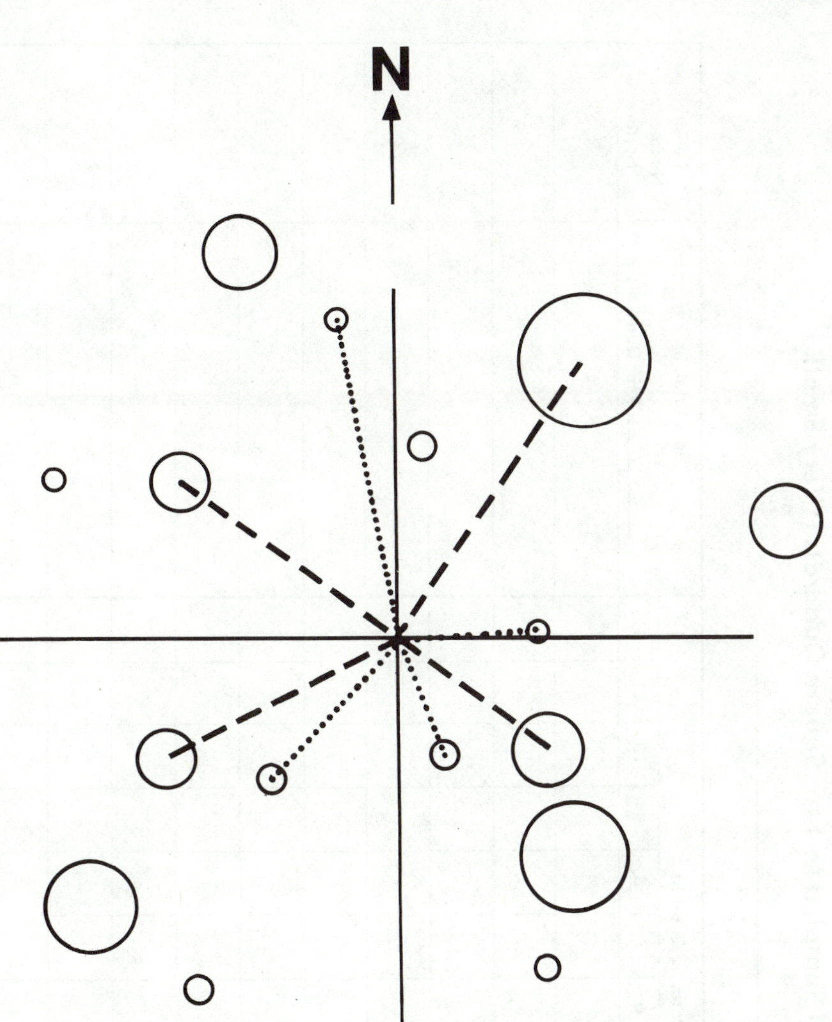

describe a circle with a radius of 178.4 cm, as marked on the tape or string. Tally each small tree rooted within the circle on Data Sheet 2.

4. **Point-Centered Quarter (or Simply Point-Quarter) Sampling.** The point-quarter method was developed by Cottam and Curtis (1956) as a speedier alternative to using large quadrats (e.g., 100 m^2) for sampling large trees. Its approach is to use point to plant distances to estimate the area occupied by an average tree. This value (expressed, for example, as square meters per tree) is the inverse of density (trees per square meter). That is, density = 1/mean area, where the 1 is the unit in which density is specified, generally hectares or acres. If one wishes density in hectares and the distance measurements are in meters, the 1 translates to 10,000 m^2 (= 1 ha).

Mean area is related to the average distance ($\bar{D}$) from point to tree, as determined by the point-quarter method as

$$\text{mean area} = \bar{D}^2.$$

There is no intuitive way of seeing why this satisfyingly simple relationship holds; Cottam and Curtis (1956) and Morisita (1954) give empirical and theoretical derivations.

Use this method to sample trees 2.5 cm *DBH* and larger. First, choose random points (a suitable way is to choose a random direction and walk a random number of paces in that direction and sample from a stake set at that point). Mentally divide the space around the stake into quarters by cardinal points of the compass. You will, then, have northeast, southeast, southwest, and northwest quarters. In each quarter, find the tree that is closest to the stake in two size classes: 2.5 cm to 12.99 cm (''medium'') and 13.00 cm and over (''large''). You will, thus, choose two trees in each quarter, or eight

Data Sheet 1. Trees under 0.5 m Tall Sampled in 1 m² Square Quadrats ("Very Small")

Site/Date ______

Species	Your Data, by Quadrat												Class Total, in ___ 1 m² Quadrats	Density, Number per Hectare
	1	2	3	4	5	6	7	8	9	10	11	12		
												Total		

Data Sheet 2. Trees at Least 0.5 m Tall and Less Than 2.5 cm *DBH* Sampled in 10 m^2 Circle Plots ("Small")

Site/Date ______________

Species	Your Data, by Quadrat												Class Total, in ______ 1 m^2 Quadrats	Density, Number per Hectare
	1	2	3	4	5	6	7	8	9	10	11	12		
Total														

Data Sheet 3. Point-Centered Quarter Sampling*

Site/Date ____________________

	Sample Point 1		Sample Point 2		Sample Point 3		Sample Point 4	
	Medium	Large	Medium	Large	Medium	Large	Medium	Large
Species								
DBH								
Distance								
Species								
DBH								
Distance								
Species								
DBH								
Distance								
Species								
DBH								
Distance								
	Sample Point 5		Sample Point 6		Sample Point 7		Sample Point 8	
Species								
DBH								
Distance								
Species								
DBH								
Distance								
Species								
DBH								
Distance								
Species								
DBH								
Distance								

*Medium—2.5–12.99 cm *DBH*; large—13.0 cm *DBH* and larger; distance in m, *DBH* in cm.

trees (four medium and four large) for each point. For each tree record distance* from the stake to the tree (measure to the middle of the tree), species, and *DBH* (Fig. 16–1). Record on Data Sheet 3.

For trees on the line between quarters, try by careful sighting with a compass to determine if they are more in one quarter than the other. If they seem half and half, flip a coin to decide which quarter they go in. It is permissible to include the same tree in the samples for two different points; however, the spacing between the points should be great enough to make this a rare occurrence.

You will need to adopt some logical convention for dealing with multiple-trunked trees. One possibility is to convert the *DBH*'s of the individual trunks to basal area, sum the basal areas, and then consider the tree to be the equivalent of a single-trunked tree of the *DBH* corresponding to the calculated basal area. Basal area is the cross-sectional area of a trunk at a standard height, usually breast height (see Table 16–1).

RESULTS

Stand Table

The present composition of the forest will be described in a *stand table*. This tabulates the densities of the various tree species in several diameter classes.

1. **Quadrat Data.** For the very small and small trees, which were sampled in 1 m^2 and 10 m^2 plots, convert the raw data to the number of trees per hectare (1 ha = 10,000 m^2) as

$$\frac{\text{number of trees}}{\text{area sampled, in m}^2} \times 10{,}000 = \text{trees/ha.}$$

For example, if the class sampled 50 circle plots the area sampled is $50 \times 10\ m^2 = 500\ m^2$. If that included 16 sassafras trees (< 2.5 cm *DBH* and ≥ 0.5 m tall) the number per hectare is

$$\frac{16 \text{ trees}}{500 \text{ m}^2} \times 10{,}000 = 320 \text{ sassafras (in that size class).}$$

Figure the number of trees in both size classes (very small and small) for each species and record in the appropriate columns of the stand table.

2. **Point-Centered Quarter Data.** For the stand table, we are interested in tree density. The procedure is, first, to calculate overall density of the trees (separately for the medium and large trees) and, then, to apportion this overall density among the various species. For the large trees, further calculations are done to make use of additional size classes and to compute importance values. Note that the same operations could be performed on the data for medium trees if desired.

Medium Trees. Follow the procedures outlined on the Analysis Sheet for Medium Trees. Enter the figures for trees per hectare for each species in the stand table.

Large Trees. Follow the procedures outlined on the Analysis Sheet for Large Trees. Enter the figures trees per hectare in each species' size category in the stand table. This completes the stand table.

Importance Values

In some circumstances, detailed information on density by size class may be less useful than some other approach to describing the structure of a forest. The approach described below is

---

*Optical rangefinders may speed up obtaining distances in some circumstances but not in others (finding and focusing on the right tree may be difficult if stem density is high or light poor). If rangefinders are used, they should be carefully calibrated and should be accurate to 0.1 m over the whole range of distances encountered.

TABLE 16–1. Basal Area in m² by *DBH* from 13.0–62.9 cm*

DBH	Basal Area	*DBH*	Basal Area	*DBH*	Basal Area	*DBH*	Basal Area	*DBH*	Basal Area
13.0	0.0133	18.0	0.0254	23.0	0.0415	28.0	0.0616	33.0	0.0855
13.1	0.0135	18.1	0.0257	23.1	0.0419	28.1	0.0620	33.1	0.0860
13.2	0.0137	18.2	0.0260	23.2	0.0423	28.2	0.0625	33.2	0.0866
13.3	0.0139	18.3	0.0263	23.3	0.0426	28.3	0.0629	33.3	0.0871
13.4	0.0141	18.4	0.0266	23.4	0.0430	28.4	0.0633	33.4	0.0876
13.5	0.0143	18.5	0.0269	23.5	0.0434	28.5	0.0638	33.5	0.0881
13.6	0.0145	18.6	0.0272	23.6	0.0437	28.6	0.0642	33.6	0.0887
13.7	0.0147	18.7	0.0275	23.7	0.0441	28.7	0.0647	33.7	0.0892
13.8	0.0150	18.8	0.0278	23.8	0.0445	28.8	0.0651	33.8	0.0897
13.9	0.0152	18.9	0.0281	23.9	0.0449	28.9	0.0656	33.9	0.0903
14.0	0.0154	19.0	0.0284	24.0	0.0452	29.0	0.0661	34.0	0.0908
14.1	0.0156	19.1	0.0287	24.1	0.0456	29.1	0.0665	34.1	0.0913
14.2	0.0158	19.2	0.0290	24.2	0.0460	29.2	0.0670	34.2	0.0919
14.3	0.0161	19.3	0.0293	24.3	0.0464	29.3	0.0674	34.3	0.0924
14.4	0.0163	19.4	0.0296	24.4	0.0468	29.4	0.0679	34.4	0.0929
14.5	0.0165	19.5	0.0299	24.5	0.0471	29.5	0.0683	34.5	0.0935
14.6	0.0167	19.6	0.0302	24.6	0.0475	29.6	0.0688	34.6	0.0940
14.7	0.0170	19.7	0.0305	24.7	0.0479	29.7	0.0693	34.7	0.0946
14.8	0.0172	19.8	0.0308	24.8	0.0483	29.8	0.0697	34.8	0.0951
14.9	0.0174	19.9	0.0311	24.9	0.0487	29.9	0.0702	34.9	0.0957
15.0	0.0177	20.0	0.0314	25.0	0.0491	30.0	0.0707	35.0	0.0962
15.1	0.0179	20.1	0.0317	25.1	0.0495	30.1	0.0712	35.1	0.0968
15.2	0.0181	20.2	0.0320	25.2	0.0499	30.2	0.0716	35.2	0.0973
15.3	0.0184	20.3	0.0324	25.3	0.0503	30.3	0.0721	35.3	0.0979
15.4	0.0186	20.4	0.0327	25.4	0.0507	30.4	0.0726	35.4	0.0984
15.5	0.0189	20.5	0.0330	25.5	0.0511	30.5	0.0731	35.5	0.0990
15.6	0.0191	20.6	0.0333	25.6	0.0515	30.6	0.0735	35.6	0.0995
15.7	0.0194	20.7	0.0337	25.7	0.0519	30.7	0.0740	35.7	0.1001
15.8	0.0196	20.8	0.0340	25.8	0.0523	30.8	0.0745	35.8	0.1007
15.9	0.0199	20.9	0.0343	25.9	0.0527	30.9	0.0750	35.9	0.1012
16.0	0.0201	21.0	0.0346	26.0	0.0531	31.0	0.0755	36.0	0.1018
16.1	0.0204	21.1	0.0350	26.1	0.0535	31.1	0.0760	36.1	0.1024
16.2	0.0206	21.2	0.0353	26.2	0.0539	31.2	0.0765	36.2	0.1029
16.3	0.0209	21.3	0.0356	26.3	0.0543	31.3	0.0769	36.3	0.1035
16.4	0.0211	21.4	0.0360	26.4	0.0547	31.4	0.0774	36.4	0.1041
16.5	0.0214	21.5	0.0363	26.5	0.0552	31.5	0.0779	36.5	0.1046
16.6	0.0216	21.6	0.0366	26.6	0.0556	31.6	0.0784	36.6	0.1052
16.7	0.0219	21.7	0.0370	26.7	0.0560	31.7	0.0789	36.7	0.1058
16.8	0.0222	21.8	0.0373	26.8	0.0564	31.8	0.0794	36.8	0.1064
16.9	0.0224	21.9	0.0377	26.9	0.0568	31.9	0.0799	36.9	0.1069
17.0	0.0227	22.0	0.0380	27.0	0.0573	32.0	0.0804	37.0	0.1075
17.1	0.0230	22.1	0.0384	27.1	0.0577	32.1	0.0809	37.1	0.1081
17.2	0.0232	22.2	0.0387	27.2	0.0581	32.2	0.0814	37.2	0.1087
17.3	0.0235	22.3	0.0391	27.3	0.0585	32.3	0.0819	37.3	0.1093
17.4	0.0238	22.4	0.0394	27.4	0.0590	32.4	0.0824	37.4	0.1099
17.5	0.0241	22.5	0.0398	27.5	0.0594	32.5	0.0830	37.5	0.1104
17.6	0.0243	22.6	0.0401	27.6	0.0598	32.6	0.0835	37.6	0.1110
17.7	0.0246	22.7	0.0405	27.7	0.0603	32.7	0.0840	37.7	0.1116
17.8	0.0249	22.8	0.0408	27.8	0.0607	32.8	0.0845	37.8	0.1122
17.9	0.0252	22.9	0.0412	27.9	0.0611	32.9	0.0850	37.9	0.1128

*For trees larger or smaller, use the formula for the area of a circle ($A = \pi r^2$) or the equivalent but more convenient form, basal area = 0.7854 $(DBH)^2$.

TABLE 16–1 continued

DBH	Basal Area	DBH	Basal Area	DBH	Basal Area	DBH	Basal Area	DBH	Basal Area
38.0	0.1134	43.0	0.1452	48.0	0.1810	53.0	0.2206	58.0	0.2642
38.1	0.1140	43.1	0.1459	48.1	0.1817	53.1	0.2215	58.1	0.2651
38.2	0.1146	43.2	0.1466	48.2	0.1825	53.2	0.2223	58.2	0.2660
38.3	0.1152	43.3	0.1473	48.3	0.1832	53.3	0.2231	58.3	0.2669
38.4	0.1158	43.4	0.1479	48.4	0.1840	53.4	0.2240	58.4	0.2679
38.5	0.1164	43.5	0.1486	48.5	0.1847	53.5	0.2248	58.5	0.2688
38.6	0.1170	43.6	0.1493	48.6	0.1855	53.6	0.2256	58.6	0.2697
38.7	0.1176	43.7	0.1500	48.7	0.1863	53.7	0.2265	58.7	0.2706
38.8	0.1182	43.8	0.1507	48.8	0.1870	53.8	0.2273	58.8	0.2715
38.9	0.1188	43.9	0.1514	48.9	0.1878	53.9	0.2282	58.9	0.2725
39.0	0.1195	44.0	0.1521	49.0	0.1886	54.0	0.2290	59.0	0.2734
39.1	0.1201	44.1	0.1527	49.1	0.1893	54.1	0.2299	59.1	0.2743
39.2	0.1207	44.2	0.1534	49.2	0.1901	54.2	0.2307	59.2	0.2753
39.3	0.1213	44.3	0.1541	49.3	0.1909	54.3	0.2316	59.3	0.2762
39.4	0.1219	44.4	0.1548	49.4	0.1917	54.4	0.2324	59.4	0.2771
39.5	0.1225	44.5	0.1555	49.5	0.1924	54.5	0.2333	59.5	0.2781
39.6	0.1232	44.6	0.1562	49.6	0.1932	54.6	0.2341	59.6	0.2790
39.7	0.1238	44.7	0.1569	49.7	0.1940	54.7	0.2350	59.7	0.2799
39.8	0.1244	44.8	0.1576	49.8	0.1948	54.8	0.2359	59.8	0.2809
39.9	0.1250	44.9	0.1583	49.9	0.1956	54.9	0.2367	59.9	0.2818
40.0	0.1257	45.0	0.1590	50.0	0.1963	55.0	0.2376	60.0	0.2827
40.1	0.1263	45.1	0.1598	50.1	0.1971	55.1	0.2384	60.1	0.2837
40.2	0.1269	45.2	0.1605	50.2	0.1979	55.2	0.2393	60.2	0.2846
40.3	0.1276	45.3	0.1612	50.3	0.1987	55.3	0.2402	60.3	0.2856
40.4	0.1282	45.4	0.1619	50.4	0.1995	55.4	0.2411	60.4	0.2865
40.5	0.1288	45.5	0.1626	50.5	0.2003	55.5	0.2419	60.5	0.2875
40.6	0.1295	45.6	0.1633	50.6	0.2011	55.6	0.2428	60.6	0.2884
40.7	0.1301	45.7	0.1640	50.7	0.2019	55.7	0.2437	60.7	0.2894
40.8	0.1307	45.8	0.1647	50.8	0.2027	55.8	0.2445	60.8	0.2903
40.9	0.1314	45.9	0.1655	50.9	0.2035	55.9	0.2454	60.9	0.2913
41.0	0.1320	46.0	0.1662	51.0	0.2043	56.0	0.2463	61.0	0.2922
41.1	0.1327	46.1	0.1669	51.1	0.2051	56.1	0.2472	61.1	0.2932
41.2	0.1333	46.2	0.1676	51.2	0.2059	56.2	0.2481	61.2	0.2942
41.3	0.1340	46.3	0.1684	51.3	0.2067	56.3	0.2489	61.3	0.2951
41.4	0.1346	46.4	0.1691	51.4	0.2075	56.4	0.2498	61.4	0.2961
41.5	0.1353	46.5	0.1698	51.5	0.2083	56.5	0.2507	61.5	0.2971
41.6	0.1359	46.6	0.1706	51.6	0.2091	56.6	0.2516	61.6	0.2980
41.7	0.1366	46.7	0.1713	51.7	0.2099	56.7	0.2525	61.7	0.2990
41.8	0.1372	46.8	0.1720	51.8	0.2107	56.8	0.2534	61.8	0.3000
41.9	0.1379	46.9	0.1728	51.9	0.2116	56.9	0.2543	61.9	0.3009
42.0	0.1385	47.0	0.1735	52.0	0.2124	57.0	0.2552	62.0	0.3019
42.1	0.1392	47.1	0.1742	52.1	0.2132	57.1	0.2561	62.1	0.3029
42.2	0.1399	47.2	0.1750	52.2	0.2140	57.2	0.2570	62.2	0.3039
42.3	0.1405	47.3	0.1757	52.3	0.2148	57.3	0.2579	62.3	0.3048
42.4	0.1412	47.4	0.1765	52.4	0.2157	57.4	0.2588	62.4	0.3058
42.5	0.1419	47.5	0.1772	52.5	0.2165	57.5	0.2597	62.5	0.3068
42.6	0.1425	47.6	0.1780	52.6	0.2173	57.6	0.2606	62.6	0.3078
42.7	0.1432	47.7	0.1787	52.7	0.2181	57.7	0.2615	62.7	0.3088
42.8	0.1439	47.8	0.1795	52.8	0.2190	57.8	0.2624	62.8	0.3097
42.9	0.1445	47.9	0.1802	52.9	0.2198	57.9	0.2633	62.9	0.3107

Stand Table for ______________________________*

(Site) (Date)

Species	Very Small (<0.5 m Tall)	Small (≥0.5 m Tall, <2.5 cm *DBH*)	Medium (2.5–12.99 cm *DBH*)	Large (cm *DBH*)				Totals
				13.0–22.99	23.00–32.99	33.00–42.99	43.0 and Larger	
Totals								

*Numbers are trunks per hectare.

Point-Centered Quarter Analysis Sheet for Medium Trees

Site/Date ______________________

Number of distance measurements
(This will be 4 times the number of points sampled) N = ________

Sum of the distance measurements ΣD = ________ m

Mean point-to-plant distance ($\Sigma D \div N$) $\bar{D}$ = ________ m

The mean area occupied by an average tree is given by squaring $\bar{D}$ $\bar{D}^2$ = ________ m^2

Density in trees per square meter is ($1 \div \bar{D}^2$) Trees/m^2 = ________

Density in trees per hectare (Trees/m^2 × 10,000) Trees/ha = ________

For each species fill in the columns below. Relative density of each species is given by dividing the number of trees of that species by the total number of trees in the sample (N). To calculate trees per hectare for each species multiply its relative density by trees/ha for all species (calculated in the box above).

Species	Number of Trees in Sample	Relative Density	Trees/Hectare

The sum of trees/hectare for all species should be equal to the total density (trees/ha) calculated in the box at the top of the page.

Point-Centered Quarter Analysis Sheet for Large Trees

Site/Date ____________________

Number of distance measurements
(This will be 4 times the number of points sampled) $N =$ ________

Sum of the distance measurements $\Sigma D =$ ________ m

Mean point-to-plant distance ($\Sigma D \div N$) $\bar{D} =$ ________ m

The mean area occupied by an average tree is given by squaring $\bar{D}$ $\bar{D}^2 =$ ________ m^2

Density in trees per square meter is ($1 \div \bar{D}^2$) Trees/m^2 = ________

Density in trees per hectare (Trees/m^2 × 10,000) Trees/ha = ________

For each species, fill in the raw numbers in the appropriate cell below; then divide the number in each cell by N to obtain the proportion of trees in each species:size class; enter this value in the appropriate cell using a different-colored ink or circling it to make it distinct from the raw numbers.

Species	13.0–22.99	23.00–32.99	33.0–42.99	43.0 and up

Finally, multiply each proportional value by trees per hectare (calculated above, in the box). These numbers are trees per hectare for each species:size category, to be entered in the stand table. The sum of all the trees per hectare for each species in each size class should be equal to total density (trees/ha) calculated in the box at the top of the page.

Work Sheet for Importance Values from Point-Centered Data

Site/Date ______________________

Species	P No. of Points of Occurrence	Q Number of Trees	Basal Area	Frequency	RF Relative Frequency	RB Relative Dominance	RD Relative Density	IV Importance Value
Sums		$= N$			100	100	100	300

Enter P and Q from the data. Obtain basal areas of each individual tree (see Table 16–1); then add the individual basal areas together and enter the sum as the basal area for that species. Formulas for the remaining values for species i are as follows:

$$\text{Frequency} = \frac{\text{points of occurrence for species } i}{\text{total number of points sampled}}$$

$$\text{Relative frequency} = \frac{\text{frequency of species } i}{\text{sum of frequencies for all species}} \times 100$$

$$\text{Relative dominance} = \frac{\text{basal area for species } i}{\text{sum of basal areas for all species}} \times 100$$

$$\text{Relative density} = \frac{\text{number of trees of species } i}{\text{total number of trees of all species}} \times 100$$

$$\text{Importance value} = RF + RB + RD$$

especially useful when information is needed only for the larger trees. Point-quarter sampling can be restricted to trees of 10 or 15 cm *DBH* (alternatively, large quadrats could be used). Generally, measures of density, frequency, and size are calculated. The reasoning is that the influence of a tree species in a stand is dependent on its numbers (density) but also on how generally it is distributed (measured by frequency) and on the size or ''dominance'' of the individuals. The last is usually measured by basal area, the cross-sectional area of a tree at a standard height, generally breast height. Biomass can be used; sometimes, when the same approach is applied to non-forest vegetation, a measure of cover is substituted.

There is a persistent fascination in trying to express the ecological importance of a species in a community as a single number. J. T. Curtis and his co-workers, who developed the point-quarter method, combined density, frequency and dominance to produce an index number that they called the *importance value* of each species. The procedure they followed was to relativize the three traits so that each made an equal contribution and then sum them. Follow the Importance Value Work Sheet to make these calculations for your data for trees 13.0 cm and larger. Note that the highest value that an importance value can take is 300, when a forest consists of a single species. Similarly, in a forest composed of several species, the importance values for all species combined sums to 300.

DISCUSSION

1. Give names to the forests you studied using one to three of the tree species.
2. Summarize the major differences in the two forests. Include tree sizes, densities, species composition, and diversity (as measured by number of species; if desired, other measures of diversity given in Exercise 22 can be used).
3. If one of your sites was a generally south-facing slope, construct a diagram showing the increased solar radiation on that slope compared with a horizontal surface. Use graph paper and a protractor and compare the slope and a horizontal surface (profile view) with the sun at noon on June 22 and December 22 (obtain the angle of incoming sunlight from a sun path diagram, such as Figure 1–2, in Exercise 1). Remember that you are comparing areas, so you should square the linear measurements for the comparison. Is the difference between the two sites larger in summer or winter? What are some probable microclimatic differences as a result of the differences in solar radiation?
4. How do the two sites differ in terms of microclimate, soil, or history? Suggest which specific climatic, edaphic, or historical factors may be responsible for the differences in vegetational composition given in your answer to question 2. Suggest observations or experiments that could be done to test whether your suggested explanations are correct.
5. Categorize the species you recorded by shade tolerance using Table 2–3 in *Principles of Ecology* by Brewer (p. 33) or some other source.
6. Did you find any very large trees of relatively shade-intolerant species? How do you account for their presence?
7. List any subcanopy tree species you found (that is, species for which few or no individuals ever grow tall enough to form part of the canopy).
8. Were the same species represented in the large trees as in the medium ones? Were the same species represented in the large trees as in the very small ones? Can you get clues from such data as to whether the forest is stable or whether some species are increasing or decreasing?
9. Does the presence of a large number of seedlings necessarily indicate a species that is increasing? Why or why not? What is *gap-phase replacement* in forest? Do some tree species seem especially well suited for filling this role? What physical and life history traits would be advantageous for gap phase species?
10. Draw a size pyramid for species that are stable in numbers in an undisturbed forest.
11. What, if any, of the differences in forest composition that you detected between the two sites would disappear in time through successional processes?

12. If the forests you studied seem relatively stable, discuss the differences between them in terms of the monoclimax, polyclimax, and climax pattern views of vegetation.

13. The point-quarter method gives erroneous densities if the trees (all species together) are not randomly spaced. Specifically, it overestimates average point-to-plant distances (and thus underestimates density) if the distribution is clumped and underestimates point-to-plant distances (and thus overestimates density) in an evenly spaced distribution. Show by diagrams why this would be so.

14. Do you suspect that the tree distribution (all species together) of the forests you sampled was non-random? How could you test whether your suspicions were correct? If your calculated densities were off by 10 or 20% would it affect any of the conclusions you have drawn in answers to previous questions?

Teaching Notes

Even with minimum travel time it will be difficult to obtain an adequate sample from more than one site in a three-hour period. Using two periods or accumulating data over the years are solutions; so also are sampling only a single stand, thus eliminating the comparative element, or sampling only medium and/or large trees. Forestry Suppliers, Inc., 205 West Rankin St., P.O. Box 8397, Jackson, Mississippi 39204, carries most forest sampling equipment including fiberglass diameter tapes, metric tapes, plastic flagging, surveyor's arrows, Suunto clinometers and compasses.

If it is not feasible to do this as a field exercise, the point-quarter method can be applied to the artificial plant population map given with Exercise 5. If 1 cm on the map is assumed equal to 1 m on the ground, reasonably realistic densities are obtained. One approach is to use the different symbols as different species. Another approach that allows an opportunity to test the method on populations of known spacing is to sample three times, on squares, stars, and suns, considering each of these in turn to represent the total tree population. The transparency master (next to the page of aerial photographs in Exercise 18) has crossed lines that can be used to define the quadrants for point-quarter sampling on the artificial plant population.

BIBLIOGRAPHY

Bormann, F. H., and G. E. Likens. 1979. *Pattern and Process in a Forest Ecosystem: Disturbance, Development, and the Steady State Based on the Hubbard Brook Ecosystem Study*. Springer-Verlag, New York.

Bray, J. R. 1956. Gap phase replacement in a maple-basswood forest. *Ecology* 37: 598–600.

Brewer, R., and P. G. Merritt. 1978. Wind throw and tree replacement in a climax beech-maple forest. *Oikos* 30: 149–152.

Cantlon, J. E. 1953. Vegetation and microclimates on north and south slopes of Cushetunk Mountain, New Jersey. *Ecol. Monogr.* 23: 241–270.

Cottam, G., and J. T. Curtis. 1956. The use of distance measures in phytosociological sampling. *Ecology* 37: 451–460.

Curtis, J. T. 1959. *The Vegetation of Wisconsin*. Univ. Wisconsin Press, Madison.

Daubenmire, R. 1968. *Plant Communities: A Textbook of Plant Synecology*. Harper & Row, New York.

Forbes, R. 1961. *Forestry Handbook*. Ronald Press, New York.

Harper, J. L. 1977. *Population Biology of Plants*. Academic Press, New York.

Morisita, M. 1954. Estimation of population density by spacing method. *Kyushu Univ., Mem. Fac. Sci. Ser. E* 1: 187–197.

Mueller-Dombois, D., and H. Ellenberg. 1974. *Aims and Methods of Vegetation Ecology*. Wiley, New York.

Parker, J. 1952. Environment and forest distribution of the Palouse Range in northern Idaho. *Ecology* 33: 451–461.

Pielou, E. C. 1974. *Population and Community Ecology: Principles and Methods*. Gordon and Breach, New York.

Potzger, J. E. 1939. Microclimate and a notable case of its influence on a ridge in central Indiana. *Ecology* 20: 29–37.

Schneider, G. 1966. *A 20-year Investigation in a Sugar Maple-Beech Stand in Southern Michigan*. Agric. Exp. Sta., Michigan State Univ. Research Bull. 15, East Lansing.

Williamson, G. B. 1975. Pattern and seral composition in an old-growth beech-maple forest. *Ecology* 56: 727–731.

Tree Identification and Life History Sources

Baker, F. S. 1949. A revised shade tolerance table. *J. Forestry* 47: 179–181.

Fowells, H. A., comp. 1965. *Silvics of Forest Trees of the United States*. U.S. Division of Timber Management Research, Forest Service, Agriculture Handbook no. 271. U.S. Dept. Agriculture, Washington, D.C.

Grimm, W. C. 1967. *Familiar Trees of America*. Harper & Row, New York.

Harrar, E. S., and J. G. Harrar. 1946. *Guide to Southern Trees*. McGraw-Hill, New York (reprinted 1961, Dover, New York).

Muenscher, W. C. 1950. *Keys to Woody Plants*. 6th ed., rev. Comstock Publ., Ithaca, New York.

Peattie, D. C. 1953. *A Natural History of Western Trees*. Houghton Mifflin, Boston.

———. 1966. *A Natural History of Trees of Eastern and Central North America*. 2nd ed. Houghton Mifflin, Boston.

Preston, R. J., Jr. 1976. *North American Trees (Exclusive of Mexico and Tropical United States): A Handbook Designed for Field Use, with Plates and Distribution Maps*. 3rd ed. Iowa State Univ. Press, Ames.

U.S. Forest Service. 1948. *Woody-plant Seed Manual*. U.S. Dept. Agriculture Misc. Publ. No. 654. U.S. Gov't. Printing Office, Washington, D.C.

Watts, M. T. 1963. *Master Tree Finder*. Nature Study Guild, Berkeley, California.

Exercise 17

ISLAND BIOGEOGRAPHY

Key to Textbooks: Brewer 169–172; Emlen 368–370, 415–421; Kendeigh 285–286; Krebs 478–484; McNaughton and Wolf 460–466; Pianka 308–330; Richardson 76–78, 221–222, 384–386; Ricklefs 709–712; Ricklefs (short) 342–345; Smith 603–605; Smith (short) 163–165; Whittaker 98–99.

OBJECTIVES

We look at species number on habitat islands, specifically artificial substrates in a stream. The approach is that of MacArthur and Wilson's island biogeography model, to which we attempt to put actual figures for immigration and extinction rates.

INTRODUCTION

How many species will there be on an island? There are some intuitive answers to that question—it depends, perhaps, on such qualities as the size of the island and how close it is to the mainland (where the invading organisms come from). At any one time, the number of species on an island is surely the number which got there minus the ones which have left (by becoming locally extinct). At the beginning of the settlement of the island, each species which arrives is new to the island, so the immigration rate is high. As time passes, the immigration rate would decrease since the organisms which arrive are of species which are already there; only new species are counted in an immigration rate. Counter to the immigration rate is the extinction rate. At first, the extinction rate is low: there are few species to become extinct, and pressures such as predation and competition are low on the uncrowded island. Later, as more species and more individuals settle on the island, extinction becomes more likely.

These sorts of trends were included in a model by MacArthur and Wilson (1963, 1967). In general, as the number of species present on an island increases (the maximum is the mainland species pool), the immigration rate decreases and the extinction rate increases (Fig. 17–1). At the intersection of the curves, where the extinction rate equals the immigration rate, the number of species is at an equilibrium called $\hat{S}$ (read as ''S hat'').

This equilibrium number of species can be approached from the other direction, too; an island could have, temporarily, more species than $\hat{S}$. This can occur when a large island becomes smaller (caused, perhaps, by a higher sea level or by the construction of a parking lot). The species number is said to ''relax'' to equilibrium as species become extinct there.

Key Words: species diversity, equilibrium number of species ($\hat{S}$), immigration rate, colonization and extinction curves, Hester-Dendy sampler, artificial substrate, habitat island, island biogeography model.

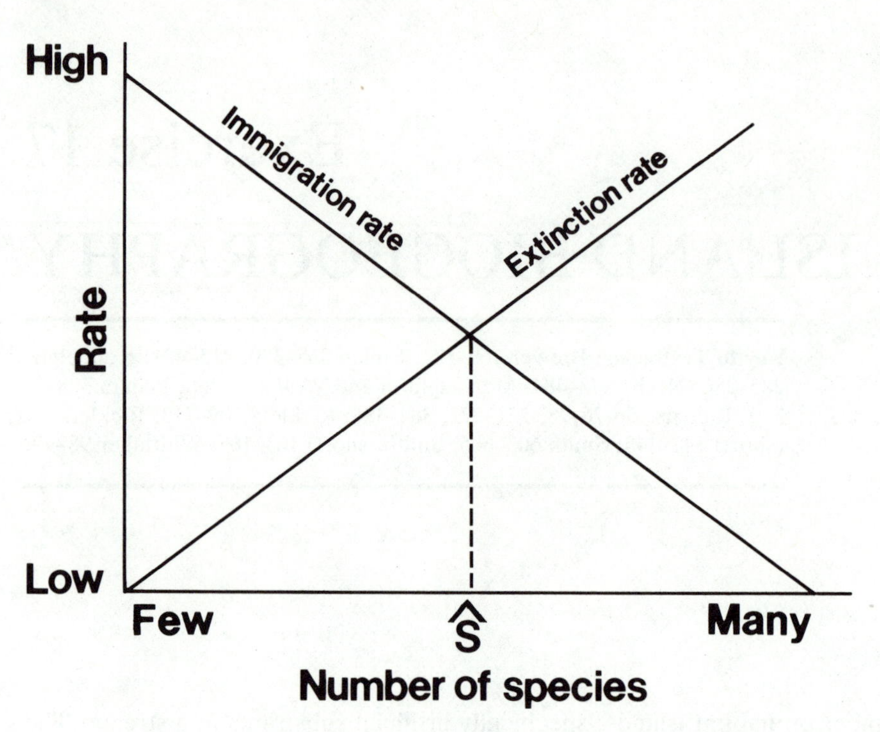

Figure 17–1. The number of species on an island as determined by immigration and extinction rates. Equilibrium species number is at the point where the curves intersect.

$\hat{S}$ can be estimated from extinction and immigration rates from either direction. Another approach is by a colonization curve. This is a plot of the number of species present over time. The colonization curve is expected to level off as $\hat{S}$ is approached.

All of these concepts can be applied to islands of habitat as well as ordinary islands surrounded by water. A vacant lot surrounded by city cement, a decaying log in a forest, and a parasite host can all be studied as islands. The islands used in this exercise are Hester-Dendy samplers placed in streams. These artificial substrates provide surfaces for stream organisms which ordinarily cling to rocks. Immigration and extinction rates and a colonization curve are plotted from the tally of species on the samplers at various times.

MATERIALS

30 Hester-Dendy samplers (Fig. 17–2)
30 wire stakes about 30 cm long
plastic flagging tape, at least 4 colors
30 strong plastic bags large enough to hold a Hester-Dendy sampler
30 twist ties
30 sorting trays
forceps
medicine droppers
dissecting microscopes
at least 90 small fingerbowls or jars
wax pencils or labeling pens
at least 30 vials and a supply of neutral formalin* (if specimens are to be kept more than a day)

*Make this by saturating 37% formaldehyde with magnesium carbonate.

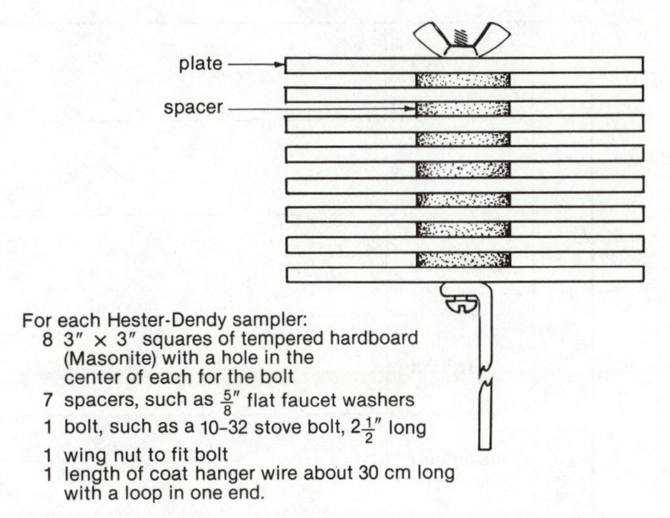

Figure 17–2. An assembled Hester-Dendy sampler. It consists of eight 3 inch by 3 inch plates separated by ⅛ inch thick spacers held together by a bolt. The wire stake is pushed into the stream bottom.

PROCEDURE

Thirty Hester-Dendy samplers which have been in a stream one to five weeks are the islands under consideration. All 30 samplers can be put out the same day, then taken in six at a time at weekly intervals for five weeks, or six samplers can be put out each week for the five weeks and all collected for the class day. In either case, be sure to keep samples of different ages separate. Tie a different color of flagging tape on the samplers for each week.

Place them in similar situations in the swiftly-flowing part of a stream at least 3 m apart. Push the wire stake into the stream bottom to hold the sampler about 3 cm from the bottom. The plates of the sampler should be horizontal. Record landmarks so you can find them again.

Collect each sampler with its associated organisms in a plastic bag. Do this gently, underwater, so that any dislodged organisms are caught in the bag. Close the bag with sampler and water inside and the wire stake sticking out at the closure. Transport them back to the laboratory.

Inside, give each of six student groups a sampler of each age, one to five weeks. The groups can work independently, using their own data for determining colonization, immigration, and extinction rates. Data from the whole class will then be used to plot colonization, immigration, and extinction curves.

Dump the contents of each bag into a sorting tray, keeping track of the ages of the samples. At each tray, disassemble the Hester-Dendy sampler. Use forceps and medicine droppers to remove macroscopic organisms and to sort them into small fingerbowls, a set for each sampler age.

The aim is to sort the organisms to species; it is not, however, necessary to identify every species in the sense of knowing its scientific name. All that is necessary is to know how many species were present for each time period and how many of these were also present in the preceding time period. Probably the best way to proceed is to identify all organisms to order, family, or genus, depending on the difficulty of the group and the experience of the class. Then, within each taxon (Ephemeroptera or Odonata, for example), separate the different species. Make your decision on traits that are likely to separate one species from another, such as arrangement of setae, shapes of segments, etc., rather than on traits like size that (for immature stages) will be variable within a species. Use a dissecting microscope as necessary.

When all the organisms for each week have been sorted to species, give each species an identifying code name (for example, Trichoptera species A) and make inter-week comparisons. Record on the Data Sheet and tally as:

continuing species, present last week as well as this week;

immigrant species, not present last week but present this week; or

extinct species, present last week but not present this week.

Data Sheet for Hester-Dendy "Islands"

Make checks in the appropriate columns.

Actual Sample Intervals	1 Week			2 Weeks			3 Weeks			4 Weeks			5 Weeks		
Species	Continuing	Immigrant	Extinct	Continuing	Immigrant	Extinct	Continuing	Immigrant	Extinct	Continuing	Immigrant	Extinct	Continuing	Immigrant	Extinct
Totals	X														
Species Present (Continuing + Immigrants)			X			X			X			X			X
Immigration and Extinction Rates	X			X			X			X			X		

The number of species present for a given week is, then, the continuing species + the immigrant species. Make comparisons only between the last week and the present week; a species may immigrate and become extinct more than once in the series. For the first week, all the species are immigrants and there are no extinct species.

If specimens are to be kept more than a day, preserve them in vials of 4–5% neutral formalin.

RESULTS AND DISCUSSION

1. Use your data to make a colonization curve by plotting the number of species present against time (on the horizontal axis). Do the same using the class means for each sample date. Can you make an estimate of the equilibrium number of species from the curve(s)? If so, what is $\hat{S}$?
2. Figure immigration and extinction rates for each interval. If the samples were one week apart, the rates will be simply the numbers immigrating and going extinct per week.
3. Plot immigration and extinction rates on the same graph (see Fig. 17–1). For the horizontal axis, use the number of species present at the beginning of the week (that is, the number at the end of the preceding week). For example, plot rates for week 2 against species present for week 1. For week 1, species present will be 0, the number going extinct will be 0, and the number immigrating will be however many species were found. Use data from the entire class, for 30 data points per line. Draw a best-fit straight line for immigration and one for extinction.
4. Do the lines cross? (Extrapolate, if necessary). Estimate $\hat{S}$.
5. Although you plotted immigration and extinction curves as straight lines, do you think some other shape might be more likely biologically? Why?
6. Suppose a Hester-Dendy sampler was left out ten rather than five weeks. What extinction and immigration rates would you expect? Would there be twice as many species?
7. What sorts of animals and plants are likely to become extinct on small islands? Why?
8. List behaviors or structures of plants and animals which are likely to be early immigrants to oceanic islands.
9. Which would have more species, a small or a large island? Is the effect due to size alone? Would a doubling of size result in a doubling of species number?
10. Which would have more species, an island close to the mainland or one farther away? Would this effect change with time? If so, how?
11. Species which are found only on species-poor islands are called "supertramps" by J. M. Diamond. Explain why such supertramps are absent from species-rich islands.
12. List four kinds of habitat islands not mentioned in the Introduction. What makes them islands? How are habitat islands different from real islands?
13. Suppose you are responsible for buying wilderness land for a wildlife preserve. (a) Your choice is between one large parcel and several small parcels which combined have the same area and similar habitats as the large parcel. The land around the preserve(s) will be urban. Which should you choose? Why? (b) Suppose you can only buy several small parcels. What can be done to the area between the preserves to make the system a better preserve (that is, to maintain more species)? (c) Your choice is between a long, narrow parcel and a square parcel; everything else is similar. Which should you choose? Why?
14. What biological processes might lead, over the course of time, to a rise in the equilibrium species number on a remote island?

Teaching Notes

The Hester-Dendy samplers need to be put in a stream five weeks to one week ahead. Within a three-hour lab period, each group can identify and count the organisms on their samplers and post their data for class use. A reference collection will hasten identification. As an alternative exercise, diatoms on slides may be studied. Both Hester-Dendy samplers and Wildco Periphyton Samplers (which hold slides) are available from Wildlife Supply Company, 301 Cass Street, Saginaw, Michigan 48602; however, both are easily fabricated from locally available materials.

BIBLIOGRAPHY*

Cairns, J., Jr., M. L. Dahlberg, K. L. Dickson, N. Smith, and W. T. Waller. 1969. The relationship of fresh-water protozoan communities to the MacArthur-Wilson equilibrium model. *Am. Nat.* 103: 439–454.

Diamond, J. M. 1975. Assembly of species communities. Pp. 342–444 in M. L. Cody and J. M. Diamond (eds.). *Ecology and Evolution of Communities*. Harvard Univ. Press, Cambridge.

———, and R. M. May. 1976. Island biogeography and the design of natural reserves. Pp. 163–186 in R. M. May (ed.). *Theoretical Ecology: Principles and Applications*. Saunders, Philadelphia.

Gilpin, M. E., and J. M. Diamond. 1976. Calculation of immigration and extinction curves from the species-area-distance relation. *Proc. Natl. Acad. Sci. U.S.* 73: 4130–4134.

Hester, F. E., and J. S. Dendy. 1962. A multiple-plate sampler for aquatic macroinvertebrates. *Trans. Am. Fish. Soc.* 91: 420–421.

MacArthur, R. H., and E. O. Wilson. 1963. An equilibrium theory of insular zoogeography. *Evolution* 17: 373–387.

———. 1967. The theory of island biogeography. Princeton Univ. Press, Princeton.

Simberloff, D. S., and L. G. Abele. 1976. Island biogeography theory and conservation practice. *Science* 191: 285–286.

———, and E. O. Wilson. 1969. Experimental zoogeography of islands: the colonization of empty islands. *Ecology* 50: 278–296.

Wilcox, B. A. 1978. Supersaturated island faunas: a species-age relationship for lizards on post-Pleistocene land-bridge islands. *Science* 199: 996–998.

*Books for identification of aquatic invertebrates are listed in the Bibliography of Exercise 21.

Exercise 18

ECOLOGICAL USE OF REMOTE SENSING

Key to Textbooks: Barbour et al. 195–201; Odum 468–483; Smith 725, 728.

OBJECTIVE

To explore some applications of remote sensing, especially aerial photography, to ecology.

INTRODUCTION

Air photos can be useful to an ecologist. A study area, for instance, can readily be mapped and different vegetation types delineated. Photos from different years can be compared to show invasion of woody plants into a field or the spread of suburbia or the increase in area devoted to parking lots. Photos at large scales can be used to census deer and other large animals, or to estimate timber volume. Because many air photos can be viewed stereoscopically, landforms and changes in topography can be seen.

There are several different kinds of air photos, but this exercise will use only the readily available government photographs taken from airplanes. The basic kind of air photo is a black and white 9½ × 9½ inch contact print available from the Agricultural Stabilization and Conservation Services (ASCS).* They photograph all parts of the United States periodically. The photos are taken along flight lines with about 65% overlap between adjacent photos and 30% overlap between flight lines. The overlap areas on adjacent photos show the same place from different points in the air, allowing stereoscopic viewing.

Enlargements large enough to give a scale of about 1 inch to 200 feet are available through the ASCS. A useful map of a study area can be traced directly from such an enlargement, but it will not be planimetric; that is, the scale will not be quite the same all over, since the land surface is not flat and parallel to the film. If a planimetric map is needed, it is simplest to transfer photo details to a planimetric base map of a suitable scale.

The best way to get an ASCS photo of a particular area is first to order the photo index for your county. The index is a mosaic of the many contact prints which cover the county. Each of these has a number on it. Find the desired area on the photo index (use a map as a guide, if necessary), find the

Key Words: aerial photographs, mapping, quadrangle (topographic) maps, ground-truthing, stereo pair, representative fraction (*RF*), planimeters, linear and areal scales, Congressional Land Survey system (legal description).

*Address: Aerial Photography Field Office, Administrative Services Division, ASCS–USDA, 2505 Parley's Way, Salt Lake City, Utah 84109.

numbers of the two or more contact prints which include your area where they overlap, and order those contact prints by number.

There are other ways of getting photographs of the area you want: consult your local ASCS office, or write to the ASCS with the legal description of the area and request stereo coverage.

Other types of air photos may be available for your area; local agencies such as planning boards are likely to know about them. Conventional color photographs are expensive and give only a little more useful information than black and white. Infrared photography, on the other hand, provides much information for the ecologist. Infrared films record the degree of infrared reflectiveness, not the visible light reflectiveness, of objects. On black and white infrared photos, coniferous trees are dark and broad-leaved trees are light. With "false-color" infrared film, live, healthy, broad-leaved trees are shown in a deep pink, healthy evergreens are bluish, and dead or unhealthy leaves are green. In either type of infrared film, waterways are very dark. This helps in mapping small streams and can also show, for example, a marshy place nearly covered by vegetation.

Other types of remotely-sensed information include radar, thermography, and multiband spectral imagery. Radar images are not very useful to the plant ecologist, as the signals penetrate vegetation to show what is beneath. Thermography records heat emission, which is useful in detecting wet areas which are likely to be cooler than the surrounding dry land. Multiband imagery records several qualities at once—infrared, visible light, thermal, whatever—and is generally done from satellites (such as LANDSAT) at a scale most useful for general, regional projects.

Regardless of the type of photos or images used, success in interpretation depends on the experience of the viewer. Ground-truthing, or viewing the area on the ground, is important especially when the interpreter is not familiar with the objects and vegetation photographed. Experienced forest interpreters can tell the species of trees in a stand and estimate board feet from air photos.

MATERIALS

for every 1–2 students:

- lens stereoscope
- ruler
- magnifying glass
- plastic overlay made from page 161

for the class:

- local air photos
- USGS topographic maps (quadrangle maps) of the same area*
- scissors

*Available at local business supply or camping equipment stores or directly from the USGS. For maps west of the Mississippi, write to the Branch of Distribution, Central Region, U.S. Geological Survey, Box 25286, Denver Federal Center, Denver, Colorado 80225. For maps of areas east of the Mississippi (including Minnesota), write to the Branch of Distribution, Eastern Region, U.S. Geological Survey, 1200 South Eads St., Arlington, Virginia 22202.

The opposite page includes several items that need to be on transparent plastic. To avoid the substantial addition to the price of this book that printing and binding such a page would have necessitated, we include instead a page that can be used as a master for a plastic transparency produced on a photocopy machine for a few cents.

The diagrams on this page are as follows:

Upper left—Overlay for Figure 18–1. The overlay fits the upper left of that page. The cross in E indicates the road intersection in the left middle photo. Several of the outlines fit recognizable patches on the photos, so alignment should be easy.

Upper right—The quadrat 5 cm on a side is for use (Exercise 5) with the artificial plant population map. The two crossed lines are for applying the point-quarter method (Exercise 16) to the artificial plant population map.

Lower—The bottom of the page is a random dot planimeter for measuring areas on aerial photos (Exercise 18) and in the territoriality study (Exercise 8).

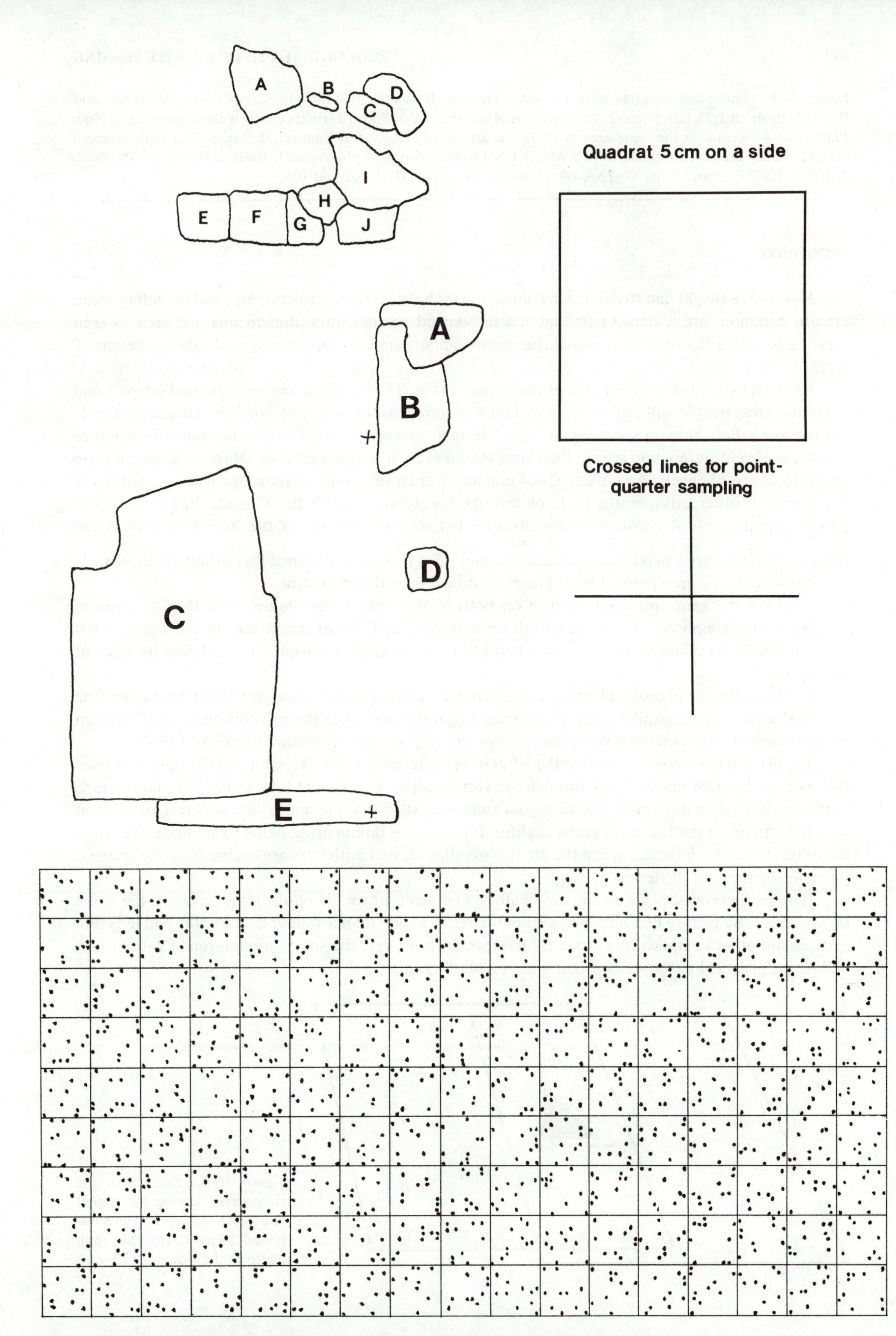
A
B
D
C
I
E
F
H
G
J
Quadrat 5 cm on a side
A
B
Crossed lines for point-quarter sampling
D
C
E

Figure 18–1. Photo pair 1 (top) is set up for stereo viewing. It was taken in 1967, in Barry County, Michigan, and the scale is about 1:21,120. Photo 2 (left center) was taken in 1950. Photo 3 (right center) is the same area in 1960. Photo pair 4 (bottom) is the same area in 1967. The area is in Kalamazoo County, Michigan. The scale is about 1:21,120. They are parts of the following ASCS–USDA contact prints: photo pair 1, BDB–2HH 44 and 45; photo 2, BDW–2G–200; photo 3, BDW–2AA–60; photo pair 4, BDW–3HH–102 and 103.

PROCEDURE

How to See Height and Relief in Air Photos. Air photos taken consecutively, and including some area in common, are a *stereo pair* and can be viewed so that three dimensions are seen. Stereo viewing provides information that a single photo cannot, such as topography and relative height of trees.

Read steps 1–3 below, then look at photo pair 1 (Fig. 18–1), which has been aligned (steps 1 and 2) so that most people will see the trees and hills in stereo. Look at it, following the advice in step 3. Do you see relief? Do you see trees and hills? Heights are exaggerated; if you are not sure whether you are seeing in stereo, you aren't. Rest your eyes and try a simple exercise. Draw an X and an O on separate scraps of paper. Place them 62–64 mm apart, then set up the stereoscope with one lens over the X and the other lens over the O. Look through the stereoscope. Is the X inside the O? Move the papers slightly, until it is. Measure the distance between the centers of the X and O. This is the distance between your pupils:____________ mm. Use this same distance for setting up photos.

Now, look at stereo pairs of local places as directed by the instructor.

1. Find the spot you want to view on both photos. Stack the photos such that the area of overlap is superimposed, the shadows fall toward you, and the numbers are on the right or left (Fig. 18–2). Either photo may be on top; it helps to have the spot you want to look at near the edge of the top photo.

2. Slide the top photograph to the side so that the corresponding spots are about 62–64 mm (the distance between your pupils) apart. If, in order to get the two spots the right distance apart, the top photo covers the spot on the lower photo, curve the edge of the top photo as in Figure 18–3.

3. Place the stereoscope so that the left lens is centered over the left spot and the right lens over the spot on the right photo. Look through the stereoscope. You should look at the left photo image with your left eye and the right one with your right eye, such that the images are superimposed. You may have to move the top photograph slightly. If one eye is dominant, it helps to shut one eye, then the other, to see each image. When the photos are aligned, and both eyes are seeing separate images, you will see the spot in stereo.

How to Determine Scale on an Air Photo. The scale of an air photo is not exactly the same throughout. The picture of a hilltop has more inches to a mile than a valley; camera tilts and lens distortions are other complications. In fairly flat terrain, however, the scale you determine with a ruler will be for practical purposes constant within a photograph.

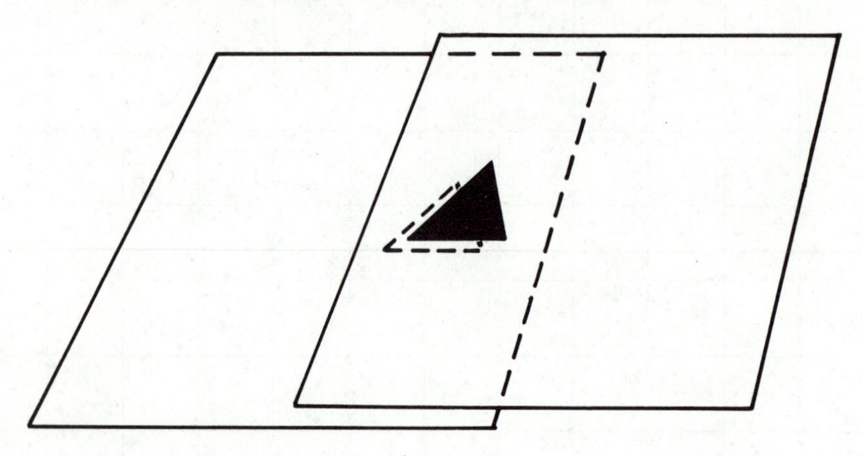

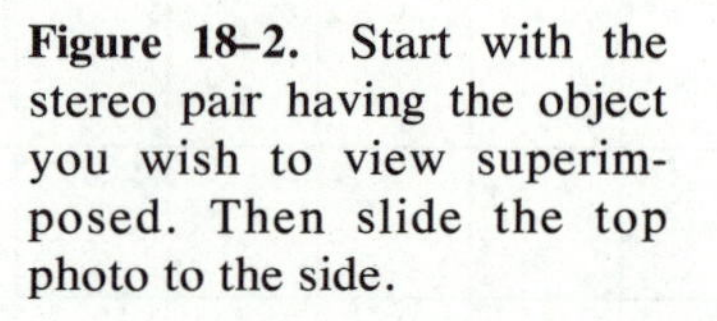

Figure 18–2. Start with the stereo pair having the object you wish to view superimposed. Then slide the top photo to the side.

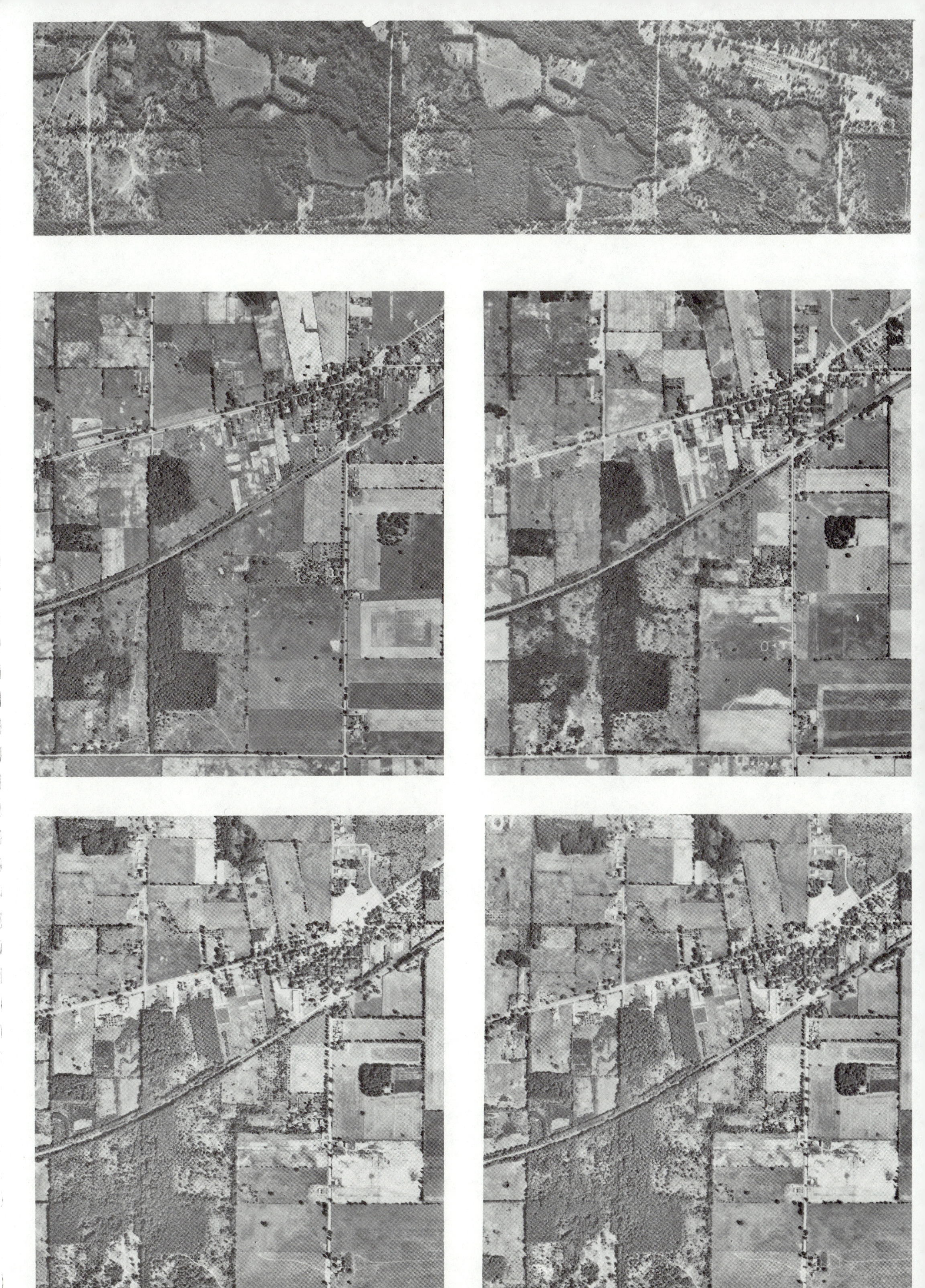

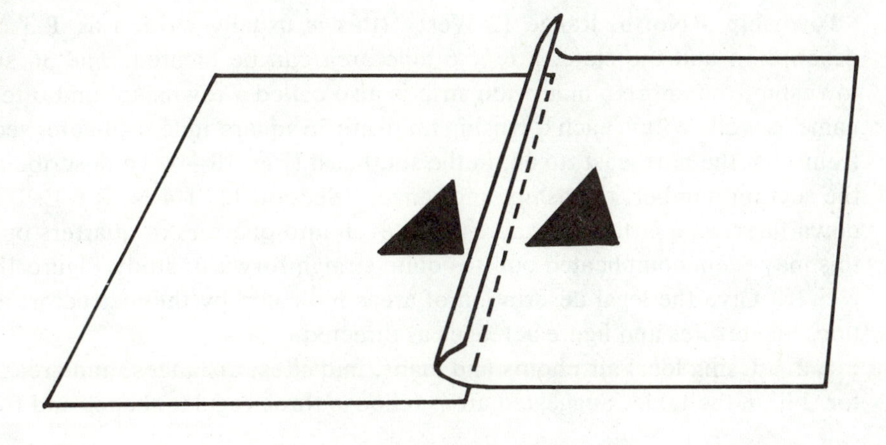

Figure 18–3. If the top photo covers part of what you want to see, curl the edge of the top photo up.

To determine the scale, you need a known ground distance in the photograph. It is best for this to be long, straight, and through the center of the photo. Country roads are ideal. A corresponding map (such as a USGS quadrangle map) will provide the ground distance.

For an example, use photo 2. The distance between the crosses (on the overlay) is 3 inches on the photo and 1 mile on the ground. The linear scale is then 1 mile = 3 inches or 1 inch = ⅓ mile. The representative fraction (*RF*) is then

$$1: \frac{63{,}360 \text{ inches per mile on the ground}}{3 \text{ inches per mile on the photo}} \text{ or } 1{:}21{,}120.$$

The *RF* has no units. It means that 1 of any linear unit on the photo shows 21,120 of that unit on the ground.

Most of us are less familiar with areal scales, so calculating areas from aerial photographs requires a little more thought. An *RF* of 1:21,120 does *not* mean that 1 square inch on the photo represents 21,120 square inches on the ground, but instead $21{,}120^2$ or 446,054,400 square inches on the ground (446,054,400 square inches = 1/9 square mile or 71.1 acres). When the area you want to measure is regular in shape, simple geometry will tell you the area on the photo, which can then be converted to the area on the ground. For irregular areas, such as lakes, use a random dot planimeter such as the one included on the plastic overlay.* To use the random dot planimeter, place it over the area to be measured and count the dots within the area. Dots on the border count ½ (each square has 10 dots). Total the number of dots in the area and divide by 10 to give the number of square centimeters in the area.

For example, within the outline of woodlot C of photo 4 there are about 235 dots, or 23.5 square centimeters on the photo. How many square miles is that? ______________ How many acres? (640 acres = 1 square mile) ______________

Legal Descriptions. The Ordinance of 1785 included a provision for surveying public land using the township and range system. For U.S. lands acquired since then, this is the preferred way of describing the location of a study site in scientific manuscripts; in effect, it gives the legal description of the study site. In the Congressional Land Survey system strips of land 6 miles wide running east-west are called townships. They are numbered with reference to a baseline (there are many baselines in the U.S.) as Township 1 South, Township 2 South, Township 1 North, Township 2 North, and so on. In a similar manner, 6-mile wide strips called ranges, referenced to various meridians, run north-south. A square of land 6 miles on a side is described by a township and a range number, such as

*There are other, generally less satisfactory, ways of determining such areas. Instruments called *compensating polarizing planimeters* are often used but tend in practice to be less accurate than random dot planimeters.

"Township 3 North, Range 12 West" (this is usually written as T 3 N, R 12 W). Given such a description and the state, a 6 × 6 mile area can be located. The 36 square-mile area is called a township (the entire 6 mile wide strip is also called a township) and often has a local governmental name as well. Within each township there are 36 square-mile (640-acre) sections. These are numbered from 1, in the northeast, to 36, in the southeast (Fig. 18–4). To describe a 1 square-mile parcel, give the section number, township, and range: "Section 12, T 4 N, R 6 E." The section is often broken down finer, as quarters (160 acres) and even into quarters of quarters or halves of quarters. At first this may seem complicated but it is quite straightforward; study Figure 18–4.

1. Give the legal description of areas indicated by the instructor; and given the legal description, locate sites and figure acreages as directed.

2. Using local air photos and maps, find sites, distances, and areas as directed by the instructor. Fill in the table. Suggested areas (choose three regular shapes and two irregular shapes):

highway interchange	golf course
orchard	field
woods	lake
shopping mall	a study area
campus	a new subdivision

Photograph or Map	Site	Photo or Map Distance (inches)	Ground Distance (miles)	*RF*	Photo or Map Area (cm^2)	Ground Area (acres)

Identification of Vegetation. Study, in stereo, photo pair 1. Look particularly at the shadows of trees, and at the lay of the land. The *RF* is 1:21,120. Use the plastic overlay to identify the areas mentioned below by letter.

1. Which areas have a closed-canopy forest? ______________ Do they all have the same type of trees? ____________ Which forested area looks the least like the others? ____________ In which area are the trees more dense, D or F? ____________ Less dense, F or H? ____________

2. Which area has the tallest trees, E or F? ____________ Are there difficulties in comparing forests on rolling terrain? ____________ Is that a problem in this photo? ____________ Why or why not? ____

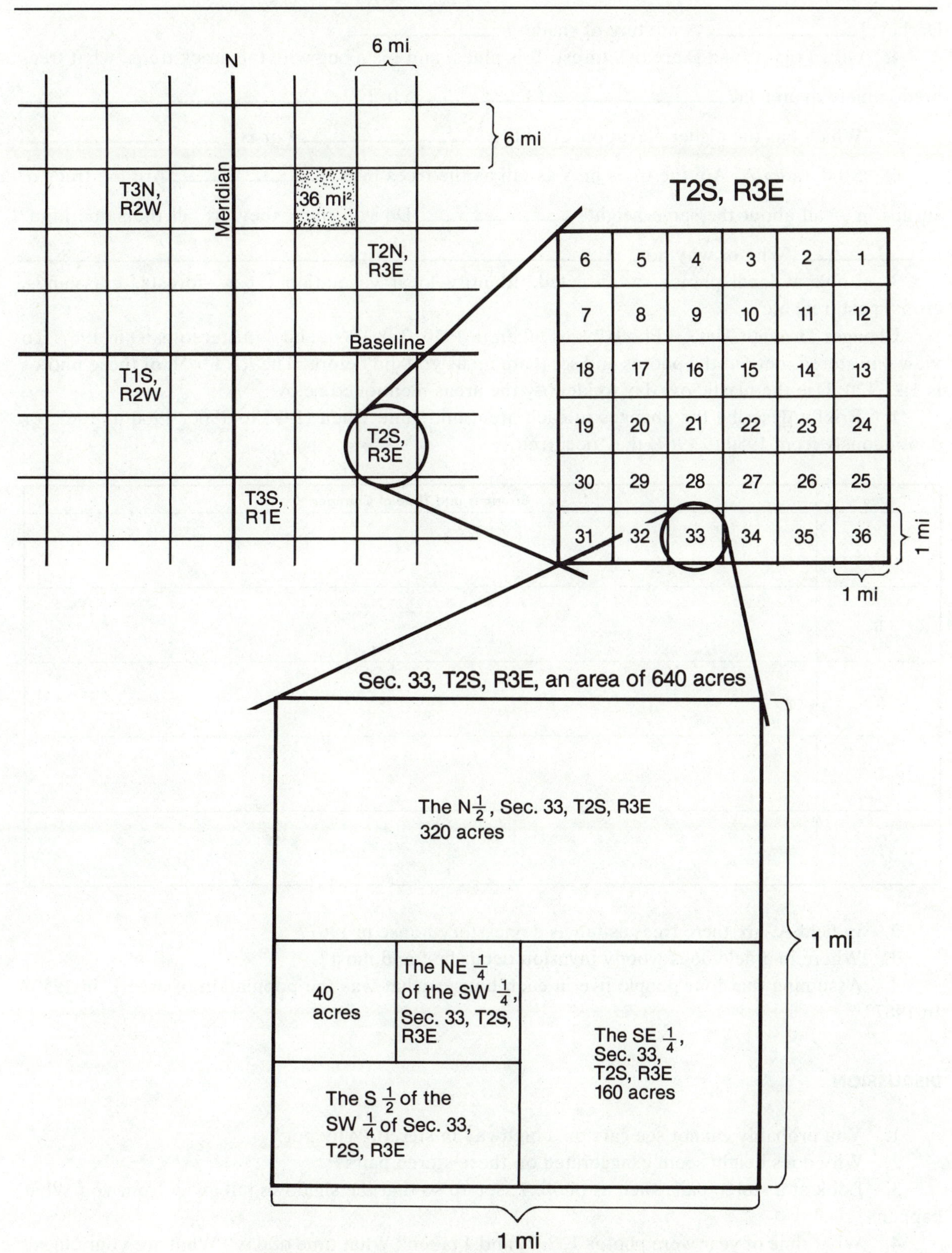

Figure 18–4. The Congressional Land Survey system.

In which area(s) is the height of the trees quite uniform? ____________ What do you suppose caused that? ____________

3. Look at the crowns of the trees, both as shadows and as treetops. In which area(s) are the crowns largest? ________ Smallest? ________ Most pointed? ________ Darkest? ________ A mixture of shades? ________

4. Given that C and D are oak forest, B is pines, and I is a bog with tamarack trees, what trees predominate in area G? ____________ In F? ____________

5. Which has the higher elevation, A or I? ____________ J or H? ____________

6. Study area A. Are the trees in A as tall as the trees in B? ________ Are the trees or shrubs in A all about the same height? ________ Do you think they are all the same kind? ________ Why or why not? ____________

7. Look at local photos, as directed. Identify local vegetation types—forests, grasslands, crops, pastures, etc.

Changes Through Time. Photo 2 was taken in 1950, 3 in 1960, and 4 (a stereo pair) in 1967. To view 4 in stereo, cut out the photos and set them up as you did before. The *RF* for all of these photos is 1:21,120. Use the plastic overlay to identify the areas mentioned below.

1. Briefly describe the changes in each area, and state when (1950 to 1960, 1960 to 1967, or continuously from 1950 to 1967) they occurred.

Area	Changes and Time of Changes
A	
B	
C	
D	
E	

2. In field A, are there trees as tall as a one-story house in 1967?
3. Where in a field does woody invasion occur first and most?
4. Assuming that four people live in each house, what was the population of area E in 1950? In 1967?

DISCUSSION

1. You probably cannot see cars on a highway in stereo. Why not?
2. Why does height seem exaggerated on these stereo pairs?
3. Look at a stereo pair, such as photo 1, set up so that the shadows fall away from you. What happens?
4. What time of year were photos 1, 2, 3, and 4 taken? What time of day? What are your clues?
5. When figuring distances and areas on these photos, why is it easier to use inches, miles, and acres instead of metric units? (It's not just the photography; there is something about the land.)

Teaching Notes

If enough copies of all photos are available to prevent bottlenecks, this exercise fits easily into three hours. Allow six to eight weeks to get air photos from ASCS. The inexpensive student stereoscope (2.2×) distributed by Hubbard Scientific Co., P.O. Box 105, Northbrook, Illinois 60062, and available from many scientific supply houses is satisfactory. Random dot planimeters in three degrees of precision and in sizes up to 11 by 15½ inches on Mylar plastic are commercially available as Bruning Areagraphs from cartographic supply houses. Done early in the term with well-chosen maps and photos this exercise can give students a preview of field trip sites and help familiarize them with local vegetation.

BIBLIOGRAPHY

Aldrich, R. C. 1979. *Remote Sensing of Wildland Resources: A State-of-the-Art Review*. Rocky Mountain Forest and Range Expt. Sta., U.S. Dept. Agriculture Forest Service, General Tech. Rept. RM-71: 1–56.

Avery, T. E. 1977. *Interpretation of Aerial Photographs*. 3rd ed. Burgess, Minneapolis.

Colwell, R.N. 1967. Remote sensing as a means of determining ecological conditions. *BioScience* 17: 444–449 and cover.

Howard, J. A. 1970. *Aerial Photo-ecology*. American Elsevier, New York.

Sayn-Wittgenstein, L. 1960. *Recognition of Tree Species on Air Photographs by Crown Characteristics*. Canada Dept. Forestry, Forest Research Div., Technical Note no. 95.

Zsilinszky, V. G. 1964. The practice of photo interpretation for a forest inventory. *Photogrammetria* 19: 192–208.

Exercise 19

SUCCESSION ON ABANDONED FIELDS

Key to Textbooks: Brewer 178–190; Barbour et al. 69–76, 94–102, 202–236; Benton and Werner 113–114, 193–218, 365; Colinvaux 71–90; 549–572; Emlen 351–362, 368–370; Kendeigh 24–26, 265; Krebs 432–442; McNaughton and Wolf 410–448; Odum 251–265; Pianka 60–64, 122; Richardson 114–117, 384–386; Ricklefs 751–759; Ricklefs (short) 380–390, 397–399; Smith 515–516, 612–628; Smith (short) 176–177, 180–193; Whittaker 171–185.

OBJECTIVES

In this exercise we will study the plants in abandoned fields of various ages for insight into the workings of succession. Life history strategies are emphasized.

INTRODUCTION

Succession on abandoned cropland—''old field succession''—is a new phenomenon in geological history but does follow a predictable pattern and is often used to demonstrate succession. The actual species involved depend on the geographic location of the field, as does succession on ''natural'' sites such as sand-bars or lava flows. The previous treatment of the field, such as whether the last plowing was done in the fall or the spring, makes a difference in the species which are found there in the first year after abandonment, but the effects are much less in later years. Thus, it is reasonable to compare a series of fields abandoned at various times even though the detailed history of each field is not known.

MATERIALS

several long metric tapes; 1 or 2 at least 25 m long plus 1 to 4 at least 10 m long will save time, but 1 tape and several strings and meter sticks can be used

stakes

sighting compass(es) or optical surveying square(s)

meter sticks

plant identification books (see Bibliography)

Key Words: old field, primary and secondary succession, transect (line intercept) and quadrat sampling, increment boring, seres, annual, biennial, and perennial life cycles, pioneer species, r- and K-selection, dispersal, reactions and coactions, allelopathy.

Table 19–1. Characteristics of Dominant Old Field Plants

Species	Age of Field	Annual, Biennial, or Perennial?	Herb, Shrub, or Tree?	Seed Dispersed by Wind, Animals, or Other?	Many or Few Seeds per Plant?	Other Notable Characteristics—Winter Annual? Allelopathy?

PROCEDURE IN THE FIELD

Study sites should be cropland abandoned one, three to five, and ≥10 years ago. It is not necessary to know the exact history of each site. If all three cannot be visited, collect data on two (preferably the younger sites) and supplement with data from the literature (see Bibliography) or from previous classes.

At each site, sample herbs and woody plants less than a meter tall along a line transect and woody plants a meter tall or taller in quadrats. As you sample, look at the sites and at the plants. Observe the robustness of the plants; particularly compare the same species at different sites. Do some species tend to grow in patches? Do they seem to be in a patch because of vegetative reproduction or because of an unusual habitat patch such as a moist depression or something else? What plants are around the edge of the field? How are they dispersed? Make other observations to fill in, as much as practical, Table 19–1; later you will complete it using literature sources.

Transects. A *transect* is an elongated quadrat. It is particularly well suited to describing vegetational transitions. For example, a transect run along a radius from the middle of a pond to dry land could be analyzed to show the spatial changes in plant distribution called *zonation*. Reducing the width to a line produces a *line transect* (Clements 1907) generally studied by some variety of the *line intercept method* (Canfield 1941). We use line transects here not to study spatial change but instead to take advantage of the fact that elongated plots sample certain community traits more efficiently than compact plots (circles and squares). Although one of the strengths of the line intercept method is estimating cover, especially in communities such as grassland and desert, we use it here primarily to provide an estimate of frequency for each species. In using any transect method, it should be remembered that the various sections of the line or strip are not statistically independent so that for many statistical analyses each transect must be considered a single sample.

A small group should set up the lines before the class tramples the area. Once they are set up, stay away from the lines unless you are sampling there.

1. To set up the lines, randomly choose a starting point (see Exercise 5) and run the long tape a random direction from that point. You might do this by going a random number of paces in a random direction (say, 1=south, 2=west, etc.) to a starting point, then choosing another compass bearing. (Stay at least 5 m from the edge of the study area.) Stretch the tape in that direction and secure the ends with stakes. It should be just high enough to be straight. At least 50 m should be sampled (it goes quickly); several lines may be used.

2. After the lines are marked, the class should identify the common plants in the study area. Only those with a high frequency (dominants) will be considered later. Although we can suggest likely dominants (Table 19–2), only experience with the sites will indicate which species need to be identified there. So, identify those the instructor suggests, and use "unknown #1" and so forth for the others. Be sure everyone is using the same name for each kind, and take specimens of the unknowns so that they can be identified later.

3. Record data separately from each 1 m section. Look down at the plants which are along one edge of the tape at any level. Each plant with any part, even only a leaf, between the edge of the tape (the line) and the ground is considered present. Tally the species on the Data Sheet (use separate sheets for different fields as necessary). You do not need to count the plants, just record them as present along that 1 m section of line. Include herbs and woody plants up to 1 m tall.

You may, if the instructor directs, measure the distance each plant occupies along the line to obtain a measure of cover. Conventionally, the intercept length measured for grasses and other herbs is near ground level; however, if crown coverage is considered more informative, the projection of the foliage on the line could be measured. The latter method, measuring the length of the line that the projected crown intercepts, is always used when trees and shrubs are included.

Quadrats. At each site, identify and find the density of woody plants ≥1 m tall. If there are only a few, count all of them and measure the area of the study site; the density is the number divided by the area. If there are many plants in this category sample in quadrats.

1. Place a stake at a random point in the field. Use this as the SW (or any unbiased) corner of a square quadrat 10 m on a side. Set up the other three corners, taping the distances and obtaining 90°

Table 19–2. Plants Often Found in Earlier Stages of Succession on Abandoned Cropland

Grasses and Grass-Like Plants
Andropogon gerardii Bluestem; Turkeyfoot
A. scoparius Little Bluestem
A. virginicus Broom Sedge
Aristida oligantha Poverty Grass
Cenchrus longispinus Sandbur
Cynodon dactylon Bermuda Grass; Wire Grass
Dactylis glomerata Orchard Grass
Digitaria sanguinalis Crabgrass
Eragrostis spp. Love Grass
Festuca elatior Meadow Fescue
Juncus tenuis Path Rush
Panicum capillare Witch Grass
P. dichotomum Panic Grass
Poa compressa Canada Bluegrass

Forbs
Abutilon theophrasti Velvet-leaf
Allium vineale Field Garlic
Amaranthus hybridus Prince's Feather
Ambrosia artemisiifolia Common Ragweed
A. psilostachya Western Ragweed
Aster dumosus Wild Aster
A. pilosus Wild Aster
Chenopodium album Lamb's Quarters
Convolvulus sepium Hedge Bindweed
Conyza canadensis Horseweed (This has also been called *Leptilon canadensis* and *Erigeron canadensis*.)
Coreopsis tinctoria Tickseed
Daucus carota Wild Carrot; Queen Anne's Lace
Diodia teres Buttonweed
Erigeron annuus Daisy Fleabane
E. pulchellus Daisy Fleabane
Eupatorium perfoliatum Boneset; Thoroughwort
Fragaria virginiana Strawberry
Gnaphalium obtusifolium Cudweed; Everlasting
Haplopappus divaricatus
H. ciliatus
Helianthus annuus Common Sunflower
Hieracium pratense Hawkweed
Ipomoea hederacea Morning-glory
Lactuca serriola Prickly Lettuce
Linaria vulgaris Butter-and-eggs
Oxalis stricta Wood-sorrel; Sheep-sorrel
Picris hieracioides Bitterweed
Plantago lanceolata English Plantain; Ribgrass
Polygonum pensylvanicum Smartweed
Potentilla simplex Cinquefoil; Five-finger
Solanum carolinense Horse-nettle
Solidago canadensis Canada Goldenrod
S. graminifolia Grass-leaved Goldenrod
S. nemoralis Gray Goldenrod
S. rugosa Rough-stemmed Goldenrod
Strophostyles umbellata Wild Bean
Verbena hastata Vervain

angles using either a sighting compass (such as a Brunton pocket transit) or a prism device (optical surveying square). If your angles are correct, the diagonal should be 14.14 m. You may wish to run twine between the stakes to outline the quadrat but if woody plants are sparse and none is close to the edges, great precision in defining the exact boundary lines is unnecessary.

2. Identify and count the number of individuals of each kind of woody plant ≥1 m tall rooted within the quadrat. Record on the Data Sheet.

3. Each student should participate in the quadrat sampling. Do as many as time permits or as the instructor assigns.

Increment Boring. To determine the age (number of years since abandoned) of the oldest site, count tree rings in cores obtained by increment boring. Choose one or more of the largest trees, preferably of a species that has entered your early fields as seedlings. Use an increment borer longer by several centimeters than the radius of the tree. For the purposes of this exercise, bore as low as feasible, probably about 1 m. Drill straight in on a radius, going slightly past the middle of the tree. In-

Data Sheet for Line Transect Sampling of Herbs and Woody Plants Less Than 1 m Tall

Species	Age of Field	Number of 1 m Line Sections in Which the Species Is Found		Total Number of 1 m Line Sections Sampled (*B*)	% Frequency $\frac{A}{B} \times 100\%$
		Your Data	Class Total (*A*)		

Data Sheet for Quadrat Sampling of Woody Plants 1 m Tall or Taller

Species	Age of Field	Record the Number Found in Each 10 m × 10 m Quadrat	(A) In How Many Quadrats Was It Found?	(B) Number of Quadrats Sampled	Percent Frequency $\frac{A}{B} \times 100$	(C) Total Number of Individuals Found	Density (Number per 100 m^2) $\frac{C}{B}$
		1 2 3 4 5 6 7 8 9 10					

sert the extractor as far as possible, groove down. Back the borer off one-half turn. This will break the core loose from the center of the tree. Pull the extractor (with the core) out and immediately unscrew the borer from the tree.

For hardwoods, lubricating the bit with beeswax prior to boring will greatly increase speed and ease, especially with large trees. For leaning trees and trees on slopes, it is best to avoid the top and bottom and bore from the side. Avoid boring trees with dead heartwood (check for hollows at the base of the tree); you will not be able to count the inner rings and, worse, you may get the borer trapped. Three-threaded borers are best if hardwoods are to be cored.

If the core is to be taken back to the laboratory, protect it in a soda straw or thermometer case. However, it may be feasible to count rings in the field. Remember that the years of growth prior to the time the tree reached the height of the core are not included in your record. If the rings are difficult to discern, you may need to take the core back to the laboratory, wet it (alternatively stain with phloroglucinol), and examine under a dissecting microscope. If there are obvious variations in the width of growth rings, try to get an idea of the years in which rapid and slow growth occurred.

PROCEDURE IN THE LABORATORY

1. Pool the data from the entire class. For the transect data, calculate the frequency of each species. Frequency (p. 49) of species A is

$$\frac{\text{number of 1 m line sections that species A occurred in}}{\text{total number of 1 m segments in the transect}}.$$

If you recorded cover data, calculate coverage as

$$\frac{\text{total distance occupied by species A}}{\text{total distance measured}}.$$

Multiply by 100 to express as percent. Species with frequencies or cover over 50% (or some other level, as instructed) will be considered further. For the quadrat data, figure frequency of species A as

$$\frac{\text{number of quadrats with species A}}{\text{total number of quadrats}}.$$

Multiply by 100 for percent frequency. Express the density as mean number per 100 m^2 (which is the area of one quadrat). The denser and more frequent species should be emphasized.

2. Complete Table 19–1. Use standard plant manuals (such as those listed in the Bibliography) and any other sources available for your region for supplemental information.

RESULTS AND DISCUSSION

1. Why were the edges of the fields excluded from sampling?

2. With these methods, is a species with 25% frequency in the line transect more or less abundant than one with 25% frequency in the quadrat sampling?

3. What is primary succession? Secondary succession? Give an example of each. Which is old field succession?

4. What was the proportion of annuals and perennials at the beginning and at later successional stages?

5. Describe a typical pioneer plant, using at least the criteria in Table 19–1. If coverage was calculated, how did it change in succession?

6. What are *r*- and *K*-selection? Contrast a strongly *K*-selected species with the one you describe in question 5.

7. What proportion of the dominants in the first-year field have wind-dispersed seeds? In the older fields? Do dispersal mechanisms seem to be related to succession?

8. What besides dispersal affects the likelihood of a viable seed being present in an old field at the time of abandonment?

9. With years of cultivation, most vegetative propagules are eliminated from old fields; most of the plants which appear after abandonment grew from seeds. Compare two arbitrary species, X and Y. X seeds arrive each fall, overwinter, and germinate in the spring. Y seeds arrive in the fall and overwinter; some of them germinate in the spring and others delay germination until later, perhaps years later. Each spring previous to abandonment, the field was plowed, killing seedlings.

a. What environmental cues could trigger germination of X seeds? In Y seeds? (Remember, this can affect only some seeds each year.)

b. Which parent plant probably uses more energy per surviving offspring plant?

10. A *reaction,* in ecology, is the modification of a site by an organism already present. For instance, ground squirrels dig tunnels and leave mounds of bare earth in meadows. Give several examples of reactions by pioneer plants which make the field different from when it was cultivated and first abandoned. What reactions are there by the first shrubs or trees as they overtop the herbs? Define *coaction.* Is there always a clear distinction between coaction and reaction?

11. Do coactions and reactions favor the present species, or the following species, or can't such a generalization be made? Discuss from an evolutionary standpoint.

12. Were any of the dominants of the older fields present (though not necessarily dominant) in the younger fields? Traditional views of succession suggest that many species of later stages come in only after reactions of earlier species have changed the site; Egler (1954) emphasized the idea that the species of later stages are present very early although not dominant because they grow slowly. Which view does your data support?

13. What are some natural seres that may resemble the old field sere?

14. What is a climax community? What is the probable climax community for the sites you studied?

15. Does the size of the field have anything to do with the speed of succession?

16. List some animal species that are characteristic of the seral stages you studied. How might some of them influence the species of plants present at various times or the speed with which succession proceeds?

Teaching Notes

It is difficult to sample more than one site adequately in three hours. Early autumn is the best season because many species are in flower or fruit; however, if the class can identify seedlings and basal rosettes, any time from late spring on is suitable. A reference herbarium or color slides that the students can see beforehand will expedite field work. The procedures and most of the questions apply equally well to many priseres, such as those on strip-mined land and sandbars. Increment borers and related supplies including beeswax and phloroglucinol are available from Forestry Suppliers, Inc., 205 West Rankin St., Box 8397, Jackson, Mississippi 39204. Pure beeswax candles will substitute for beeswax in a pinch.

BIBLIOGRAPHY

Bazzaz, F. A. 1970. Secondary dormancy in the seeds of the common ragweed *Ambrosia artemisiifolia*. *Bull. Torrey Bot. Club* 97: 302–305.

———. 1974. Ecophysiology of *Ambrosia artemisiifolia:* a successional dominant. *Ecology* 55: 112–119.

Beatley, J. C. 1956. The winter-green herbaceous flowering plants of Ohio. *Ohio J. Sci.* 56: 349–376.

Clements, F. E. 1907. *Plant Physiology and Ecology*. Holt, New York.

Canfield, R. H. 1941. Application of the line interception method in sampling range vegetation. *J. Forestry* 39: 388–394.

Drury, W. H., and I. C. T. Nisbet. 1973. Succession. *J. Arnold Arboretum* 54: 331–368. (Reprinted in Golley, F. B., ed. 1977. *Ecological Succession*. Benchmark Papers in Ecology vol. 5. Dowden, Hutchinson & Ross, Stroudsburg, Pennsylvania, pp. 287–324.)

Egler, F. E. 1954. Vegetation science concepts I. Initial floristic composition, a factor in old-field vegetation development. *Vegetatio* 14:412–417.

Horton, J. S., T. W. Robinson, and H. R. McDonald. 1964. *Guide for Surveying Phreatophyte Vegetation*. U.S. Dept. Agric., Agric. Handbook no. 266: 1–37.

Jackson, J. R., and R. W. Willemsen. 1976. Allelopathy in the first stages of secondary succession on the piedmont of New Jersey. *Amer. J. Bot.* 63: 1015–1023.

Newell, S. J., and E. J. Tramer. 1978. Reproductive strategies in herbaceous plant communities during succession. *Ecology* 59: 228–234.

Parenti, R. L., and E. L. Rice. 1969. Inhibitional effects of *Digitaria sanguinalis* and possible role in old-field succession. *Bull. Torrey Bot. Club* 96: 70–78.

Raynal, D. J., and F. A. Bazzaz. 1975. Interference of winter annuals with *Ambrosia artemisiifolia* in early successional fields. *Ecology* 56: 35–49.

———. 1975. The contrasting life-cycle strategies of three summer annuals found in abandoned fields in Illinois. *J. Ecol.* 63: 587–596.

Rice, E. L. 1964. Inhibition of nitrogen-fixing and nitrifying bacteria by seed plants (I). *Ecology* 45:824–837.

———. 1974. *Allelopathy*. Academic Press, New York.

Stowe, L. G. 1979. Allelopathy and its influence on the distribution of plants in an Illinois old-field. *J. Ecol.* 67: 1065–1085.

Wieland, N. K., and F. A. Bazzaz. 1975. Physiological ecology of three codominant successional annuals. *Ecology* 56: 681–688.

Old Field Studies in the Northeastern U.S.

Bard, G. E. 1952. Secondary succession on the piedmont of New Jersey. *Ecol. Monogr.* 22: 195–215.

Buell, M. F., H. F. Buell, J. A. Small, and T. G. Siccama. 1971. Invasion of trees in secondary succession on the New Jersey piedmont. *Bull. Torrey Bot. Club* 98: 67–74.

Dayton, B. R. 1975. Early stages of vascular plant succession in a central New York old field. *Amer. Midl. Nat.* 94: 62–71.

Hanks, J. P. 1971. Secondary succession and soils on the inner coastal plain of New Jersey. *Bull. Torrey Bot. Club* 98: 315–321.

Mellinger, M. V., and S. J. McNaughton. 1975. Structure and function of successional vascular plant communities in central New York. *Ecol. Monogr.* 45: 161–182.

Squiers, E. R., and W. A. Wistendahl. 1977. Changes in plant species diversity during secondary succession in an experimental old-field system. *Amer. Midl. Nat.* 98: 11–21.

Old Field Studies in the Southeastern U.S.

Golley, F. B. 1965. Structure and function of an old-field broomsedge community. *Ecol. Monogr.* 35: 113–137.

Keever, C. 1950. Causes of succession in old fields of the Piedmont, North Carolina. *Ecol. Monogr.* 20: 229–250.

McQuilkin, W. E. 1940. The natural establishment of pine in abandoned fields in the Piedmont Plateau region. *Ecology* 21: 135–147.

Odum, E. P. 1960. Organic production and turnover in old field succession. *Ecology* 41: 34–49.

Shontz, J. P., and H. J. Oosting. 1970. Factors affecting interaction and distribution of *Haplopappus divaricatus* and *Conyza canadensis* in North Carolina old fields. *Ecology* 51: 780–793.

Old Field Studies in the Midwestern U.S.

Bazzaz, F. A. 1968. Succession on abandoned fields in the Shawnee Hills, southern Illinois. *Ecology* 49: 924–936.

Evans, F. C., and S. A. Cain. 1957. Preliminary studies on the vegetation of an old field community in southeastern Michigan. *Contr. Lab. Vertebr. Biol. Univ. Mich.* 51: 1–17.

Evans, F. C., and E. Dahl. 1955. The vegetational structure of an abandoned field in southeastern Michigan and its relation to environmental factors. *Ecology* 36: 685–706.

Hopkins, W. E., and R. E. Wilson. 1974. Early old field succession on bottomlands of southeastern Indiana. *Castanea* 39: 57–71.

Quarterman, E. 1957. Early plant succession on abandoned cropland in the central basin of Tennessee. *Ecology* 38: 300–309.

Thomson, J. W., Jr. 1943. Plant succession on abandoned fields in the central Wisconsin sand plain area. *Bull. Torrey Bot. Club* 70: 34–41.

Tramer, E. J. 1975. The regulation of plant species diversity on an early successional old-field. *Ecology* 56: 905–914.

Old Field Studies in the Western U.S.

Booth, W. E. 1941. Revegetation of abandoned fields in Kansas and Oklahoma. *Amer. J. Bot.* 28: 415–422.

Perino, J. V., and P. G. Risser. 1972. Some aspects of structure and function in Oklahoma old-field succession. *Bull. Torrey Bot. Club* 99: 233–239.

Piemeisel, R. L. 1951. Causes affecting change and rate of change in a vegetation of annuals in Idaho. *Ecology* 32: 53–72.

Rice, E. L., W. T. Penfound, and L. M. Rohrbaugh. 1960. Seed dispersal and mineral nutrition in succession in abandoned fields in central Oklahoma. *Ecology* 41: 224–228.

Shantz, H. L. 1917. Plant succession on abandoned roads in eastern Colorado. *J. Ecol.* 5: 19–43.

Tomanek, G. W., F. W. Albertson, and A. Riegel. 1955. Natural revegetation on a field abandoned for thirty-three years in central Kansas. *Ecology* 36: 407–412.

Identification Books

Clark, G. H., and J. Fletcher. 1909. *Farm Weeds of Canada*. 2nd ed., rev. and enlarged by G. H. Clark. Canada Dept. Agriculture, Ottawa. (Reprinted 1923.)

Courtenay, B., and J. H. Zimmerman. 1972. *Wildflowers and Weeds*. Van Nostrand Reinhold, New York.

Cronquist, A., A. H. Holmgren, N. H. Holmgren, J. L. Reveal, and P. K. Holmgren. 1977. *Intermountain Flora: Vascular Plants of the Intermountain West, U.S.A. Vol. 6. The monocotyledons*. Columbia Univ. Press, New York.

Fassett, N. C. 1976. *Spring Flora of Wisconsin*. 4th ed., rev. and enlarged by O. S. Thomson. Univ. Wisconsin Press, Madison.

Fernald, M. L. 1950. *Gray's Manual of Botany*. 8th ed. American Book, New York.

Gilkey, H. M. 1957. *Weeds of the Pacific Northwest*. Oregon State College, Corvallis.

Gleason, H. A. 1963. *The New Britton and Brown Illustrated Flora of the Northeastern United States and Adjacent Canada*. 3 vol. Hafner, New York.

Hitchcock, C. L., and A. Cronquist. 1973. *Flora of the Pacific Northwest: An Illustrated Manual*. Univ. Washington Press, Seattle.

Kearney, T. H., and R. H. Peebles. 1969. *Arizona Flora*. 2nd ed. Univ. California Press, Berkeley.

Knobel, E. 1977. *Field Guide to the Grasses, Sedges and Rushes of the United States*. Rev. by M. E. Faust. Dover, New York.

Munz, P. A., and D. D. Keck. 1963. *A California Flora*. Univ. California Press, Berkeley.

Newcomb, L. 1977. *Newcomb's Wildflower Guide*. Little, Brown, Boston.

Pammel, L. H. 1913. *The Weed Flora of Iowa*. Iowa Geol. Surv. Bull. no. 4, Des Moines.

Parker, K. F. 1972. *An Illustrated Guide to Arizona Weeds*. Univ. Arizona Press, Tucson.

Peterson, R. T., and M. McKenny. 1968. *A Field Guide to Wildflowers of Northeastern and Northcentral North America*. Houghton Mifflin, Boston.

Runnels, H. A., and J. H. Schaffner. 1931. *Manual of Ohio Weeds*. Ohio Agric. Exp. Sta., Wooster, Bull. no. 476.

United States Dept. Agriculture, Agric. Research Serv. 1971. *Common Weeds of the United States*. Dover, New York.

Wilkinson, R. E., and H. E. Jaques. 1972. *How to Know the Weeds*. 2nd ed. Brown, Dubuque.

Exercise 20
FOREST SUCCESSION

Key to Textbooks: Brewer 178–193; Barbour et al. 202–236; Benton and Werner 69–71, 193–223; Colinvaux 71–90; Emlen 246–251; Kendeigh 24–26; Krebs 428–448; McNaughton and Wolf 410–448; Odum 251–267; Pianka 62–64; Richardson 92–146; Ricklefs 751–759; Ricklefs (short) 377–399; Smith 95, 612–631; Smith (short) 176–194; Whittaker 7–8; 171–185.

OBJECTIVES

We interpret the present situation in a forest so that future composition can be predicted. A transition matrix is used for the analysis. Succession is viewed as a process of tree loss and replacement; some of the ways in which these events occur are considered.

INTRODUCTION

As a forest develops from a few seedlings in a field to a mature forest, the kinds of dominant trees change. At first, the site is open so that the earliest trees are in full sun with few neighbors. These grow, and under them the next generation of trees is getting started. These do not have full sun but are shaded by the larger trees. The site has changed in other ways, too—there may be competition for water and nutrients, squirrels nesting in the trees may eat or bury tree seeds, the fallen leaves may have changed the soil. Seedlings of the first trees often do not survive well in the new conditions, but seedlings of other species can. Eventually the replacement stops—or slows considerably—when young trees live successfully under their parent trees. The self-perpetuating result is called the *climax*.

It isn't quite that simple; a forest may be patchy where a tornado came through, where a river flooded and killed trees, where the soil is different, or where some tree that isn't usually there happened to get established. However, there are general patterns related to the effects (direct and indirect) of the species already there and to the requirements of the species invading. One of the most important requirements is for light, rated by foresters as *shade-tolerance*. The first trees to invade are generally intolerant; they won't do well in shade. The later trees can survive in more shade (and generally cast a denser, deeper shadow). The climax trees must be shade-tolerant enough to grow in the shade under their parents.

Knowledge of the requirements of, and effects of, the trees found in the area can provide much information about the likely history and future of the forest. Even if only the present composition is known some statements about the past and future can be made. It is reasonable to say that the present canopy trees were once small and that the present small trees (not every one!) will become, at various times, canopy trees. Thus, a stand table (as in Exercise 16) suggests some of the changes during forest succession. It is also probable that after a canopy tree dies, a smaller tree near it will take its place in the canopy.

Key Words: succession, climax, shade tolerance, transition matrix.

Here we will collect data on these likely replacements to construct a transition matrix. This can be used to make an estimate of the future and past canopy composition. We make a couple of simplifying assumptions, both of which could be overcome by complicating the mathematics. First, succession is viewed as a one-for-one replacement; the total density of canopy trees remains the same although the species composition may shift. Second, we assume that within a time period all the present canopy trees die and all the replacement trees become canopy trees, with no allowance for different longevities.

At least one assumption, though, cannot be overcome mathematically: the species of trees now present are the only species considered possibly present in the past and future. With some knowledge of forests in your area you can decide how much this assumption limits your use of the forest transition matrix.

M. B. Usher (1966) was the first to apply matrices to the study of forest change. Using essentially the approach of Leslie (see Exercise 12), he projected recruitment into larger size classes in a pine plantation. Waggoner and Stephens (1970) first used the approach to study changes in species composition including an attempt to anticipate an eventual steady-state forest.

MATERIALS

compass (1 for each pair of students)

random numbers table (inside front cover)

identification books (see Bibliography to Exercise 16)

PROCEDURE IN THE FIELD

The study site should be a forest of at least a few hectares. Sample well within the forest, avoiding the edges.

1. As a group, identify the common canopy species. Note that some tree species do not reach the canopy in any forest; exclude them from consideration. Divide into pairs to collect data.
2. Locate random sample points in the forest. One way to do this is for each pair to walk a transect line and locate random points from it. The pairs might disperse in random directions from the center of the forest, or the pairs could start at random intervals along, say, the north edge of the forest and walk transect lines due south to the other side. Along your transect line, walk 20 paces, then walk a random direction (say, 1 = south, 2 = west, etc.) and a random number (0 to 9) of paces to your random point. After collecting data at that point, return to where you left your transect line, go another 20 paces, and so on.
3. At each random point, find the nearest canopy tree. Identify it, then identify the tree that seems most likely to replace it when it dies. In general this should be the tallest tree of a potential canopy species that is within 5 m of the canopy tree. Use any forestry expertise available to make your conclusions more accurate; for example, if there are two nearby small trees of about the same size and one is fast growing, choose it over an individual of a slow-growing species.

On the Data Sheet record the likely replacements for each canopy species.

4. Make general notes on fallen and standing dead trees and on the way the canopy gap is filled.
5. Each pair should take about 50 points. Use the pooled class data for computations in the next section. One hundred points will do, but try to take enough that every important tree species is represented by 5 (preferably 10) canopy individuals. In some forests this may not be possible.

Forest Succession Data Sheet

Likely Replacers. Make tally marks in the appropriate box.

Canopy Species

Probable Replacement Species														

PROCEDURE IN THE LABORATORY

For each species of canopy tree you now have a list of probable replacements. Change the raw numbers of replacement trees into proportions. For example, if there were 20 trees of species A in your sample and the probable replacements were

species A	10
species B	2
species C	8

then the corresponding proportions would be species A = 0.50, species B = 0.10, and species C = 0.40. Do this for each species and prepare a table approximately like this:

Probable Replacement Species	Canopy Species		
	A	B	C
A	0.50	0.10	0
B	0.10	0.10	0.75
C	0.40	0.80	0.25

Each column sums to 1.00. This table of probable replacements will be used as a transition matrix.

Next, list in a column the number of each species of canopy tree. Use the same order in the list as in the columns and rows of the transition matrix. This list will be used as a column vector, a matrix of only one column. For this example, say that the present composition of the canopy is 120 of species A, 240 of B, and 600 of C. The next step is to multiply the transition matrix by the column vector; the result is a column vector of the composition at some later time when all of the present trees have died and have been replaced.

To multiply the transition matrix by the column vector, multiply the first column of the matrix by the top number in the column vector, the second column by the second number, and so on. You need to have as many columns in the matrix as numbers in the column vector, so if you have probable replacers A, B, C, and D under a canopy of only species A, B, and C, fill in with zero of species D in the canopy.

$$\underset{\text{Transition Matrix}}{\begin{bmatrix} 0.50 & 0.10 & 0 \\ 0.10 & 0.10 & 0.75 \\ 0.40 & 0.80 & 0.25 \end{bmatrix}} \times \underset{\substack{\text{Column Vector} \\ \text{(a total of 960)}}}{\begin{bmatrix} 120 \\ 240 \\ 600 \end{bmatrix}} =$$

$$\begin{bmatrix} (120 \times 0.50) + (240 \times 0.10) + (600 \times 0) \\ (120 \times 0.10) + (240 \times 0.10) + (600 \times 0.75) \\ (120 \times 0.40) + (240 \times 0.80) + (600 \times 0.25) \end{bmatrix} = \begin{bmatrix} 60 + 24 + 0 \\ 12 + 24 + 450 \\ 48 + 192 + 150 \end{bmatrix} = \begin{bmatrix} 84 \\ 486 \\ 390 \end{bmatrix} \begin{matrix} \text{of species A} \\ \text{of species B} \\ \text{of species C} \end{matrix}$$

(a total of 960)

The resulting column vector is in the same units (numbers per hectare, percent, etc.) as the initial column vector.

Now, do the multiplication using the class data. We suggest that this be done using calculators. Use the new composition as a column vector for a second multiplication. For this, you may wish to use the computer program CHANGE given in Appendix 5 if computer facilities are available. Do this enough times to see what happens if forest succession really proceeds in this way, with set probabilities of one species replacing another.

DISCUSSION

1. Compare canopy composition after one multiplication (one canopy replacement) with that of the present. If you studied the same forest in Exercise 16, how do your predictions using the transition matrix differ from your predictions using the stand table?

2. Categorize the species you recorded by shade tolerance using Table 2–3 in *Principles of Ecology* by Brewer (p. 33) or some other source. Are the species that increased more or less shade tolerant than the ones that decreased? Comment.

3. Are the species now present in the forest the same species you would expect to find earlier in the series? Later? If not, discuss how severely this limits your use of the transition matrix.

4. Starting with the transition matrix and current composition, you did repeated multiplications to get the composition at several times in the future. What happens when this is done?

5. Are there forests in the region like the one you predicted after one and after several multiplications?

6. Is the forest as currently constituted a climax forest? Is the forest as projected after one or several multiplications a climax forest? In terms of the transition matrix model, what is a climax forest?

7. List the ways in which trees are lost from the canopy either singly or in groups.

8. We assumed that the largest nearby tree would replace each canopy tree. List some cases in which this might not hold true. May this sometimes be related to the cause of loss of the canopy trees?

9. A major assumption of this model is that the probability of any one species being replaced by another species is a constant throughout the successional process; for example, the probability of a black oak being replaced by a black cherry is the same in a young forest as in a near-climax forest. Evaluate this assumption.

10. If you think that any features of the predicted forests are unlikely, examine the data and try to identify specifically the sources of the peculiar results. Suggest improvements in the model or the methods.

11. How could you use the replacement data to describe the forest that was (that is, forest composition at an earlier time)?

12. Suppose that you have a map of a forest showing every tree over a foot *DBH* for 1975 and 1980. How could you use that data to predict future successional changes by means of transition matrices?

Teaching Notes

A pair of students can do at least 50 points in three hours.

BIBLIOGRAPHY*

Bormann, F. H., and G. E. Likens. *Pattern and Process in a Forest Ecosystem: Disturbance, Development, and the Steady State Based on the Hubbard Brook Ecosystem Study*. Springer-Verlag, New York.

Bray, J. R. 1956. Gap phase replacement in a maple-basswood forest. *Ecology* 37: 598–600.

Brewer, R., and P. G. Merritt. 1978. Wind throw and tree replacement in a climax beech-maple forest. *Oikos* 30: 149–152.

Horn, H. S. 1975. Forest succession. *Scientific American* 232(5): 90–98.

———. 1975. Markovian properties of forest succession. Pp. 196–211 in M. L. Cody and J. M. Diamond (eds). *Ecology and Evolution of Communities*. Belknap Press, Cambridge.

Pielou, E. C. 1974. *Population and Community Ecology: Principles and Methods*. Gordon and Breach, New York.

*For tree identification manuals see Exercise 16.

Schneider, G. 1966. *A 20-year Investigation in a Sugar Maple-Beech Stand in Southern Michigan*. Agric. Exp. Sta., Michigan State Univ., Research Bull. 15, East Lansing.

Searle, S. R. 1966. *Matrix Algebra for the Biological Sciences (Including Applications in Statistics)*. Wiley, New York.

Stephens, G. R., and P. E. Waggoner. 1970. The forests anticipated from 40 years of natural transitions in mixed hardwoods. *Conn. Agric. Sta. Bull.* 707, New Haven.

Usher, M. B. 1966. A matrix approach to the management of renewable resources with special reference to selection forests. *J. Appl. Ecol.* 3: 355–367.

———. 1979. Markovian approaches to forest succession. *J. Anim. Ecol.* 48: 413–426.

Waggoner, P. E., and G. R. Stephens. 1970. Transition probabilities for a forest. *Nature* 225: 1160–1161.

Williamson, G. B. 1975. Pattern and seral composition in an old-growth beech-maple forest. *Ecology* 56: 727–731.

Exercise 21

LAKE SAMPLING

Key to Textbooks: Brewer 220–228; Benton and Werner 229–261; Colinvaux 247–251, 256–260; Kendeigh 56–75; Krebs 500–502, 524–526, 530–535; McNaughton and Wolf 481–496; Odum 295–314; Pianka 68–71; Richardson 147–168; Ricklefs 99–102, 118–120; Ricklefs (short) 23–32, 56–59, 170–173; Smith 30, 61–63, 202–215; Smith (short) 158–159, 279–294; Whittaker 237–243, 325–334.

OBJECTIVES

Here we examine the plankton, benthic organisms, and certain physical factors of a temperate region lake using standard methods.

INTRODUCTION

Ecologists find it convenient to divide lakes into subunits (littoral, profundal, and so on) which are generally based on depth. Most lake organisms stay in only one of these subunits, and for a good part of the year the lake water likewise does not circulate throughout all the depths.

Although the subunits tend to be based on depth, it is indirect effects of depth which are important. Cold water (down to 4° C) is heavier than warmer water, so deeper water is colder in the summer. In deeper lakes, this leads to *stratification;* the lake is layered. In the summer, the upper, warmer water *(epilimnion)* does not mix with the lower, colder water *(hypolimnion)* because of the sharp density change at the *thermocline*. Changing temperatures in the temperate-zone autumn causes changes in the water density that allow the water to mix *(overturn)*. Under ice cover in winter there is again little mixing; springtime brings another overturn and then stratification occurs again, as the surface is warmed.

The amount of light is also related to depth. Because of reflection, scattering, and absorption, light is dimmer (and bluer) in deep water than at the surface.

Depth, because of the temperature and light changes associated with it, is related to the amount of oxygen dissolved in the water directly from the atmosphere and from photosynthesis. Cold water can hold more oxygen than warm water, but the amount of oxygen used in respiration increases with temperature. Secondly, plants need light for photosynthesis which adds oxygen to the water. As a result, the amount of dissolved oxygen varies with depth and by season. In the summer, there can be plants in the water to the depth of sufficient light. Dissolved oxygen from the air is not available to the hypolimnion because of the stratification. The cold deep water can hold much oxygen, but the

Key Words: thermal stratification, epilimnion, hypolimnion, thermocline, overturn, plankton, benthos (bottom fauna), turbidity, eutrophication, extinction coefficient, compensation depth, oxygen debt, Secchi disk, Sedgwick-Rafter cell, Ekman and Petersen dredges.

organisms there use it without replenishing it. If the lake has many organisms, the decay of the dead ones (which sink to the bottom) and the respiration of the live ones may so deplete the oxygen that the only successful species are ones that can carry on anaerobic respiration or have other respiratory tricks.

In the spring and fall, the overturn mixes the lake so that dissolved oxygen is about even (and near saturation) throughout.

In the winter under ice, oxygen diffusion at the surface is reduced, so oxygen is added mostly by photosynthesis. If the ice and snow cover is so thick that little light penetrates, the amount of dissolved oxygen declines through the winter. In some shallow, productive lakes, there may not be enough dissolved oxygen left over from the fall to carry it through to spring, causing winterkill.

People and their activities tend to add organic matter and nutrients to lakes. This ''enrichment'' often causes abundant growth of plants (both plankton and macrophytes) which die when the growth reduces light too much. Their decay, which mostly occurs in the hypolimnion, uses oxygen and increases the problems of the deep-water organisms.

In this exercise, we will look at two groups of lake organisms: the plankton, which drift about in the water; and the benthic macroinvertebrates, which live in the bottom substrate, sometimes even several meters deep in muck. Their life styles are quite different. The plankton are mostly microscopic and, although some are motile or can change their buoyancy, they are at the mercy of any currents. Plant plankton, or phytoplankton, live where there is sufficient light for photosynthesis to produce more oxygen than respiration uses. The amount of phytoplankton varies quite a bit during the year; ''blooms'' may color the water for a few days, followed by a die-off and then perhaps a bloom of another species. Zooplankton (pronounced ''zoe-oh-plankton'') are mostly grazers on phytoplankton, although there are predatory species. Filtration is the most common method of getting food, but some seize food with their legs or mouthparts.

Benthic organisms live in the muck or sand at the bottom of a lake. They are similar across the bottom except near shore where there may be a zonation resulting from light penetration and plant growth. Food is available to the benthos in the rain of dead organic debris from the upper photosynthetic region of the lake. You will be looking only at benthic macroinvertebrates, but bacteria, fungi, and some smaller invertebrates can be found here.

MATERIALS

ice spud with attached rope about 5 m long, or 1 or more rowboats with life jackets, anchors, and so on

Secchi disk, with attached calibrated rope long enough to reach the lake bottom; or underwater photometer

bucket, with line marking 5 or 10 liters content, with attached rope about 3 m long

plankton net with attached collecting bottle with lid

distilled water in wash bottle (optional)

Ekman and/or Petersen dredge with attached calibrated rope long enough to reach the bottom

lidded bucket to contain dredged material

resistance thermometer with calibrated lead long enough to reach the lake bottom

neutral formalin*

no. 30 sieve (0.6 mm openings)

bucket to use with sieve

sorting trays

---

*Make this by saturating 37% formaldehyde with magnesium carbonate.

forceps
medicine droppers
small petri dishes (about 5 cm diameter)
graduated cylinder
dissecting microscopes
Sedgwick-Rafter counting cells
compound microscope(s) with Whipple eyepiece disk and mechanical stage
stage micrometer(s)
Hensen-Stempel pipette or wide-bore pipette of 1–2 ml capacity
identification books (see Bibliography)

PROCEDURE IN THE FIELD

The exercise as described is done through the winter ice; however, it can be done just as well from rowboats. If done in the winter, be sure the ice is thick enough to support a class standing all in one spot. If done from boats, students should wear life jackets. The chemical tests described in Exercise 22 are not assigned below but may be done here if Exercise 22 is not to be done or for comparative purposes.

Water from various depths for chemical and other tests can be obtained by using underwater samplers such as the Van Dorn sampler or the Kemmerer water bottle (plastic rather than metal, preferably), if they are available. These offer also a way of sampling plankton at various depths that could be substituted for the dipper method described in parts 5 and 6.

1. All equipment should be fastened securely to ropes, and the rope held not only by the person using the equipment but by someone standing back simply holding the end of the rope. Cold hands raising a heavy dredge will slip.

2. With the spud, chop a hole in the ice large enough for the bucket and dredge to pass through—about 0.4 m across. It is better to chop the hole almost through across its whole width rather than to chop a small hole all the way through and then try to enlarge it.

3. Slowly lower the Secchi (pronounced "seck-ee") disk until it can no longer be seen. If working from a boat, do this on the shady side. Record this depth on following page. Lower it about 1 m more, then slowly bring it up until it reappears. Record this depth. Then lower it to the bottom for an estimate of depth at this site. (Alternatively or in addition use an underwater photometer to get light intensity at the surface and at several depths.) If a Hach kit is available you may want to obtain a turbidity reading to compare with the Secchi disk reading.

4. Lower the lead of the resistance thermometer into the water and record the temperature just below the ice, 10 cm below, 1 m below, and at 5 m intervals to the bottom. If this exercise is done during the summer, you may wish to use 0.5 m intervals in the region of the thermocline to locate it precisely.

5. Use the calibrated bucket to bring up lake water from near the surface. Each time, fill the bucket to the calibration mark (pour any excess back into the hole to avoid icing up the area around the hole). Have two people hold the plankton net vertically between them, several meters away from the hole. Pour the water through the plankton net. Note that any organisms in the lake water smaller than the holes of the plankton net mesh will be held inside the net and end up in the vial. Repeat, keeping track of the amount of water used, until the plankton is suitably concentrated. You should be able to see some plankton or color; it is likely to take 30–70 liters of lake water. Record the amount of lake water used: ______________ liters

6. Slosh down the *outside* of the plankton net, so that plankton on the net are rinsed into the collecting vial. Or, use distilled water in a wash bottle to spray inside the net to wash down clinging plankton.

Secchi Disk Readings	Temperature	Depth
Disappears at ______ m	______	just below surface
Reappears at ______ m	______	10 cm
Average: ______ m	______	1 m
	______	______
	______	______
Approximate depth at	______	______
sample site: ______ m	______	______

7. Fasten the Ekman dredge in the open position. Lower it—but not the messenger!—nearly to the bottom, then lower it slowly until it rests on the bottom. Check to be sure the messenger is secure on the rope, then send it down to trip the jaws. Bring up the dredge to just below the surface of the water, then slip a bucket under the jaws to catch the leaks as the dredge is lifted clear. Dump the contents into the bucket. Put a lid on the bucket if it is to be transported to the lab. Rinse the dredge in the lake. If the bottom is sandy or pebbly use a Petersen bottom sampler rather than an Ekman dredge. The Petersen dredge is held open by tension and trips when it lands on the bottom; the force with which it bites into the substrate depends on its weight. You may wish to try both types of dredges (plus others if available) and compare results.

8. The preliminary sorting of the macroinvertebrates is messy and is best done in the field. If done inside, care should be taken to avoid filling drains with dirt. First, add some water to the muck in the bucket and stir it, breaking up large clods. Dipper this onto a no. 30 sieve which (if inside) is held over a catch-bucket or tub. Most of the dirt will pass through the sieve. Discard it and put some water into the catch-bucket. Swish the sieve in this for another washing. Keep the rim of the sieve above the water level, and tilt the sieve to work the debris to one side. Empty the contents of the sieve into a sorting tray. (If sorting cannot proceed within a day, put the organisms and debris into a jar with water and enough neutral formalin to give a 5–10% solution.)

PROCEDURE IN THE LABORATORY

1. Identify and count all of the macroinvertebrates using blunt forceps, medicine droppers, and dissecting scopes. Pool the class data and record in the Benthos Data Sheet.

To figure the density of these macroinvertebrates, first find the area sampled. This is, approximately, the area of the jaw opening of the dredge. (If you sorted muck collected in more than one dredging, multiply.)

Total area sampled by dredging: ______ m^2

2. The plankton concentrate may be stored live at a few degrees above freezing for a few days or preserved by adding neutral formalin to make about a 5% solution. For identification it is best to view at least some live material since the type of movement may be characteristic and because preservation causes some color and shape changes.

3. Measure the volume of concentrate. Figure the concentration factor:

$$\frac{\text{volume of concentrate (ml)}}{\text{volume of water poured through plankton net (l)}} = \underline{\qquad\qquad} \text{ ml/l.}$$

Figure the density (number per m^2) to complete the table.

Benthos Data Sheet

Taxon	Your Count	Class Total Count	Density (Number per m^2) = $\frac{\text{class total count}}{\text{total area sampled}}$

4. Mix the plankton concentrate and divide it into two portions:
   a. Slightly more than 1 ml per student of preserved concentrate at room temperature, to be counted in Sedgwick-Rafter cells.
   b. The rest of the concentrate, chilled to slow movements or preserved. The large plankton in this portion will be counted.

5. Plankton in a measured portion of the Sedgwick-Rafter cell will be counted. The number of counts needed depends on the density of plankton in the concentrate and the statistical validity desired (see Appendix 4). Every student should count at least a part of a strip.

Measure the top to bottom width of the Whipple eyepiece disk with the stage micrometer (with a 10× objective this will be about 1 mm). Record here: ______________ mm.

Fill the Sedgwick-Rafter cell, which holds 1 ml. To fill the cell, first place it on a level surface. Lay a cover glass diagonally across it so that both ends are open. Gently mix the preserved plankton concentrate, then pipette a sample into the cell, slightly overfilling it. Turn the cover glass to enclose the cell without any air bubbles. Place it on the microscope stage.

Arrange the Whipple eyepiece disk so that its edge is parallel to the long sides of the Sedgwick-Rafter cell. Move the cell so that one side is in the field of view. Identify and count all plankton seen between the top and bottom lines of the Whipple grid. Move the Sedgwick-Rafter cell across the stage and continue counting a strip across it. Count as ''in'' plankton on the top line; exclude those on the bottom line. Record the kinds and numbers on the Small Plankton Data Sheet.

Next, figure the density of plankton in the lake water. First calculate the volume of concentrate viewed per strip. This is the width (about 1 mm) × the length (50 mm) × the depth (1 mm) of the cell. If the width is exactly 1 mm, the volume is 50 mm^3 = (50/1000) ml = 0.05 ml. Record the volume you counted here: ______________ ml. Pool the data for a class total volume and number of plankton; record. Then convert to numbers of each kind per liter of lake water, as

$$\frac{\text{number of plankters counted}}{\text{volume of concentrate they were counted in}} \times \text{concentration factor} = \text{density}$$

Small Plankton Data Sheet

Zoo- or Phyto-plankton?	Taxon	Your Count	Class Total Count (in ______________ ml of concentrate)	Density (number per liter of lake water)

6. To count the large plankton (cladocerans, copepods, large colonial algae), first measure the "b" portion of the concentrate and record here: ______________ ml. Pour all of it out into several small petri dishes. Place each dish on a dissecting microscope stage, allow to settle a moment, then use 20–30× magnification to identify and count all the large plankton. (Some kinds may be the same as counted in the Sedgwick-Rafter cells.) Record your count and the class total in the Large Plankton Data Sheet. Figure the density (the number per liter of lake water) as

$$\frac{\text{number of plankters counted}}{\text{volume of the ``b'' portion}} \times \text{concentration factor} = \text{density.}$$

7. Use additional reference material as necessary to classify each species as zooplankton or phytoplankton.

RESULTS AND DISCUSSION

1. It is best to take a Secchi disk reading at sun noon. Why? What other factors can interfere with a Secchi disk reading?
2. What causes turbidity in a lake? What units were used in measuring turbidity? What are some effects of turbidity?

Large Plankton Data Sheet

Zoo- or Phyto-plankton?	Taxon	Your Count	Class Total Count (in ______ ml of "b" concentrate)	Density (number per liter of lake water)

3. Graph water temperature (horizontal axis) against depth (vertical axis). At what depth was water the warmest? The coldest?

4. What physical characteristic of water results in thermal stratification? Describe the seasonal cycle of deeper temperate zone lakes, relating temperature and amount of dissolved oxygen to stratification and overturn. How does thermal stratification result in anaerobic conditions at the bottom? How is oxygen added to and removed from lake water? Define epilimnion, thermocline, and hypolimnion.

5. Describe the seasonal cycle of temperatures in shallow (<2 m) lakes.

6. Compare the number of phytoplankton to the number of small zooplankton to the number of large zooplankton.

7. The zooplankton of a lake is often mostly one species of copepod, one species of cladoceran, and one species of rotifer, with very few other species of each group. Did your sample show this tendency?

8. What is cyclomorphosis and how does it relate to this exercise?

9. What, if any, sizes of plankton were not sampled? What benthic organisms would be missed in the sampling? What other lake biota were not sampled?

10. What sorts of seasonal variation, if any, would you expect in the numbers of benthic insect larvae? In the numbers of *Tubifex* and *Limnodrilus?*

11. *Chaoborus,* the phantom larva, migrates upward at night. What proximate factor do you think causes this? What ultimate factor? How might you test your ideas?

12. *Tubifex, Limnodrilus,* and chironomid larvae have a special physiological feature which enhances their survival in the benthos. What is it? How does it help?

13. Many benthic organisms build up an oxygen debt. What is that? What times of year would they be most likely to have an oxygen debt?

14. What chain of events causes winterkill of fish?

15. Is the lake you sampled oligotrophic, eutrophic, or dystrophic? What do these terms mean?

16. What is eutrophication? Cultural eutrophication?

17. Lake Erie has sometimes been called a "dead lake." What is meant by that? Are so-called dead lakes devoid of life?

18. What among your sample were producers? Consumers? What do the benthic organisms and other consumers eat? For the lake as a whole, what other producers exist that you did not sample? Are there additional sources of organic input (which becomes food for consumers or decomposers)?

19. Do you think that there are more species, or fewer, in a square meter of forest or grassland soil compared to a square meter of lake bottom? Account for the difference in diversity.

20. Among the benthic invertebrates, which were immature stages (larvae, naiads) rather than adults? What are the corresponding adult organisms and where do they live? How do the immature stages get to the lake bottom?

21. (a) Light intensity decreases at an exponential rate with water depth. The rate of decrease is referred to as the *extinction coefficient* and is represented by η (lowercase eta). If we assume that the Secchi disk depth corresponds to the level at which light transmission has been reduced to 10% of surface light intensity (which is approximately true, although actual values may vary from about 1–15%), then we can estimate η very simply:

$$\eta = -2.3/\text{Secchi disk depth (in m)}$$

Using this formula* calculate η for the lake you worked on. $\eta =$ ________________

(b) Knowing η we can calculate the percent transmittance to any depth and, likewise, the depth at which any given reduction of light intensity will occur. The basic relationship is

$$I_d = I_0 \, e^{\eta d}$$

where I is light intensity, with depths d and 0, and e is base of natural logarithms. Specifically, the fraction (F) of light reaching a given depth is

$$F = e^{\eta d}$$

This says to multiply η times depth (in meters) and then raise e to that power (the exponent will always be negative since η is always negative). The resulting figure is the fraction of light transmitted to that depth; multiply it by 100 to get percent transmittance. For example, if $\eta = -0.4$, percent transmittance to 8 m $= 100\, e^{(-0.4 \times 8)} = 4.08$.

Using this formula calculate percent transmittance to the bottom of the lake you studied.

% transmittance = ________________

(c) To calculate the depth at which light is reduced to a specified level (for example to 50% or 10% of surface light) use the rearranged formula

$$d = \frac{\ln F}{\eta}.$$

F is transmittance expressed as a decimal fraction (e.g., 10% transmittance is 0.1). For example, if $\eta = -0.4$ and you want to know the depth at which light intensity is half that at the surface,

$$d = (\ln 0.5) \div (-0.4) = (-0.693) \div (-0.4) = 1.73 \text{ m}.$$

Define *compensation depth*. It occurs approximately at the level where light intensity is reduced to 1% of the surface light. Calculate this depth for the lake you studied.

Depth of 1% light transmission = ________________ m

---

*If you have actual photometer readings for the surface and any given depth you can and should calculate η as

$$\eta = \ln (I_d/I_0)/d$$

where the terms are as defined in part (b) of this question.

Teaching Notes

The field work can be done in a three-hour period but identification and counting will take another hour or so. The exercise can readily be reduced or expanded to match the interests of the class and the equipment available. Most of the equipment is carried by Wildlife Supply Co., 301 Cass St., Saginaw, Michigan 48602. Directions for calibrating a microscope are in Wetzel and Likens (1979) and microscopy manuals. A reference collection for the bottom fauna will speed things up.

BIBLIOGRAPHY

American Public Health Association, American Water Works Association, and Water Pollution Control Federation. 1975. *Standard Methods for the Examination of Water and Wastewater*. 14th ed. Am. Publ. Health Assoc., Washington, D.C.

Cole, G. A. 1979. *Textbook of Limnology*. 2nd ed. Mosby, St. Louis.

Hutchinson, G. E. 1967. *A Treatise on Limnology. Vol. II. Introduction to Lake Biology and the Limnoplankton*. Wiley, New York.

Lund, O. T. 1979. *Handbook of Common Methods in Limnology*. 2nd ed. Mosby, St. Louis.

Pennak, R. W. 1957. Species composition of limnetic zooplankton communities. *Limn. and Ocean*. 2: 222–232.

Welch, P. S. 1952. *Limnology*. 2nd ed. McGraw-Hill, New York.

Wetzel, R. G. 1975. *Limnology*. Saunders, Philadelphia.

———, and G. E. Likens. 1979. *Limnological Analyses*. Saunders, Philadelphia.

Identification Books

Eddy, S., and A. C. Hodson. 1961. *Taxonomic Keys to the Common Animals of the North Central States, Exclusive of the Parasitic Worms, Insects, and Birds*. Burgess, Minneapolis.

Edmonson, W. T., ed. 1959. *Fresh-water Biology*. 2nd ed. Wiley, New York.

Lehmkuhl, D. M. 1979. *How to Know the Aquatic Insects*. Brown, Dubuque.

Merritt, R. W., and K. W. Cummins. 1978. *Aquatic Insects of North America*. Kendall-Hunt, Dubuque.

Needham, P. R., and J. G. Needham. 1962. *A Guide to the Study of Fresh-water Biology*. 5th ed. Holden-Day, San Francisco.

Pennak, R. W. 1978. *Freshwater Invertebrates of the United States*. Ronald Press, New York.

Prescott, G. W. 1962. *Algae of the Western Great Lakes Area with an Illustrated Key to the Genera of Desmids and Freshwater Diatoms*. Rev. ed. Brown, Dubuque.

———. 1970. *How to Know the Freshwater Algae*. Brown, Dubuque.

Exercise 22

COMPARISON OF POLLUTED AND UNPOLLUTED AREAS OF A STREAM

Key to Textbooks: Brewer 164–169, 228–230, 259–261; Barbour et al. 139–142; Benton and Werner 262–274, 459–466, 472–479; Colinvaux 265–267; Kendeigh 36–55; McNaughton and Wolf 470–481, 632–634; Odum 148–154, 318–320, 443; Pianka 286–290; Ricklefs 768–772; Smith 61–63, 215–225, 588–590, 707–709; Whittaker 340–343.

OBJECTIVES

This exercise has five objectives. (1) We learn what a clean stream is like and how its two main microhabitats differ. (2) Then we see how the biological and chemical traits of the stream change with pollution. (3) In this connection the concept of biological indicators is introduced. (4) Methods of studying stream ecosystems differ from terrestrial and even from lake methods, so the exercise is also an opportunity to learn some of these methods. (5) Finally, we see how to measure diversity and give some thought as to what produces higher or lower diversity in communities.

INTRODUCTION

Many of our cities are located on rivers, and the waterways have been used for waste disposal. Originally most of the waste was organic and the stream organisms used it much as they used leaves and other organic inputs. Organisms below the outfall removed the material and farther downstream the stream was about as clean as before. As human populations grew, there came to be more and larger cities, closer together, and industry added new kinds of pollutants. A clean stretch of river became a rarity. Recently, waste treatment programs have helped, but even now classes in most regions will have little trouble finding a polluted stream for field trip purposes.

Many different wastes pollute rivers. Here we are mostly concerned with organic wastes such as sewage or wood fibers which require oxygen as they are decomposed. Because oxygen is abundant in a natural stream, many stream organisms are very sensitive to a lack of it. As decomposer animals respire, they use dissolved oxygen; if there is an excess of these organisms (due to an excess of material to be decomposed) dissolved oxygen will be decreased. The amount of oxygen in the water is

Key Words: water pollution, dissolved oxygen, organic pollution, biological indicators, diversity, diversity indices, Cairns' sequential comparison index, Simpson's index (*C*), Shannon-Wiener index (*H'*), current, biochemical oxygen demand (BOD), pools and riffles.

influenced by several other factors (such as temperature, turbulence, and photosynthesis) and, in addition, is only one aspect of stream pollution.

Organisms may be used as *biological indicators* of water pollution in several ways. Test organisms, often small fish, may be caged and kept in the water and their health monitored. This doesn't show whether that animal could actually live in the stream, unprotected from predators and foraging on its own, or if it could reproduce there. Generally this sort of test is done with organisms particularly sensitive to some material suspected of being present.

Another approach is the presence or absence of various indicator species. Slow or sessile organisms, which cannot swim away from intermittent pollution, will reflect conditions which may not otherwise be evident when the researcher checks the site. Macroinvertebrates such as worms, mollusks, and various arthropods are well-suited for such study as they are relatively easy to work with and show various tolerances to pollution. Gill-breathing stonefly and caddisfly larvae, for instance, can live only where there is abundant oxygen in the water. Some other invertebrates can tolerate low-oxygen water, because they breathe atmospheric oxygen via breathing tubes or have some other special adaptation. They may be in well-oxygenated water, too, but competition there is likely to keep their numbers low.

We could also consider the *diversity* of macroinvertebrates in the community. There is low diversity where only a few kinds of organisms are found (though there may be thousands of them), as we might find in a stream so polluted that only very tolerant forms survive. Diversity is higher where many kinds of organisms live, and where no species far outnumbers the others. This is usually true of clean streams. Diversity is more than the number of species (species richness); it also involves the evenness or equitability of the numbers (or biomass) of each kind.

We include here three of many ways of calculating a number value for the diversity of a collection: Cairns and co-workers' (1968) "runs" or sequential comparison index, which requires very little mathematics, and the more widely used Simpson's (1949) and Shannon-Wiener indices.

Notice that to calculate diversity, you don't need to identify the organisms. You merely have to separate them into different kinds, which may be different species or different stages of the same species (which are ecologically different if not taxonomically different). You do, however, have to collect every organism in a sampled area, even if you have 50 stonefly larvae to every caddisfly larva.

MATERIALS

metric tape
meter stick
float or cork
stopwatch or watch with second hand
fluorescent dye
thermometer
Surber stream-bottom sampler
small stiff brush, such as a nail brush
buckets
garden trowel
flat-bottomed dip nets
no. 30 sieve (0.6 mm openings)
wide-mouth quart jars with lids
20 small jars, such as baby food jars, with lids
grease pencils or labels

trays for sorting

sorting tray marked with lines 2 cm apart

medicine droppers

blunt forceps

hipboots

Hach portable engineer's water test laboratory (or a similar portable kit)

Hach BOD kit

Millipore coli-count or Hach coliform tube assembly

dissecting microscopes

topographic map (such as USGS quadrangle map*) which shows both sites and the area between them

map measure (optional)

6-inch rulers

manuals for identification of aquatic macroinvertebrates (suggestions in bibliography)

calculators with ln ($\log_e$) key or tables of natural logarithms

neutral formalin† if the organisms will be kept beyond a day

PROCEDURE

Ideally, this exercise is done at two sections of the same stream, the upstream site relatively clean and the downstream site polluted. It is best if the sites are similar in width and depth (mostly less than knee deep). Each should have riffles and pools. If there is a deep channel, a bridge is handy.

There are chemical tests, measurements, and biological collections to do at each of the two sites. To avoid stirring up sample areas, take the water samples for the chemical tests upstream with no walking about in the water, do the biological collecting in the middle, and take the stream measurements downstream. Divide the tasks and switch roles at the second study site. At each site:

1. Make general notes about the site. Is the stream straight or meandering? Is it shaded? Is there much vegetation on the banks? Many underwater plants? Is the substrate rocky or muddy? Are there varied microhabitats?

2. Take three measurements of stream temperature a few centimeters below the surface, stream width, and stream depth; average and record on the next page.

3. Time a float or cork as it travels a measured distance (about 30 m) in the main channel. Repeat for three trials and take the average.

4. Note that current varies in different microhabitats within the stream. If time permits use some fluorescent dye in a medicine dropper to examine these differences. For example, put a few drops at various places in a pool and watch the rate at which it moves downstream. Put a few drops at various points in an eddy and watch the movements. Hold the medicine dropper underwater and inject some dye into a bed of plants and watch the rate at which the current removes it. (Be careful that you do not pollute the stream. In some states, you may need permission before adding the dye.)

5. Following the directions in the Hach portable engineer's water laboratory, do the following tests and record the results (add others if appropriate to your polluted stream):

---

*Available at local business supply stores or directly from the USGS. For areas west of the Mississippi the address is Branch of Distribution, Central Region, USGS, Box 25286, Denver Federal Center, Denver, Colorado 80225. For areas east of the Mississippi and including Minnesota, write Branch of Distribution, Eastern Region, USGS, 1200 South Eads St., Arlington, Virginia 22202.

† To make concentrated neutral formalin, saturate 37% formalin with magnesium carbonate.

	Cleaner Site				Polluted Site			
	Trial 1	Trial 2	Trial 3	Aver-age	Trial 1	Trial 2	Trial 3	Aver-age
Temperature, °C								
Width, m								
Depth, m								
Seconds to travel								
________ m	Velocity, m/sec				Velocity, m/sec			

	Cleaner Site	Polluted Site
Dissolved oxygen (DO)		
Turbidity		
pH		
Copper		
Phosphate		
Iron		
Chromium		
Sulfate		
Total hardness		

6. Figure the percent oxygen saturation. To do this, divide your value for dissolved oxygen by the value read from the graph (Fig. 22–1) for the same temperature and multiply by 100.

Cleaner Site	Polluted Site

7. Use the Hach BOD (biochemical oxygen demand) kit according to its directions. These results will be available in five days.

Cleaner Site	Polluted Site

8. Do a coliform count as directed by the kit.

Cleaner Site	Polluted Site

9. Select a representative stony site in a riffle where the water is no deeper than the Surber sampler frame, or at least shallower than arm's length. Wade to the site from downstream and place the sampler with the mouth of the net facing upstream. Lower it to the bottom and hold it down firmly. Pick up rocks and so forth from within the frame. Place the rocks in a bucket and pass them to shore. Next, use the trowel to stir up the substrate and float organisms into the net.

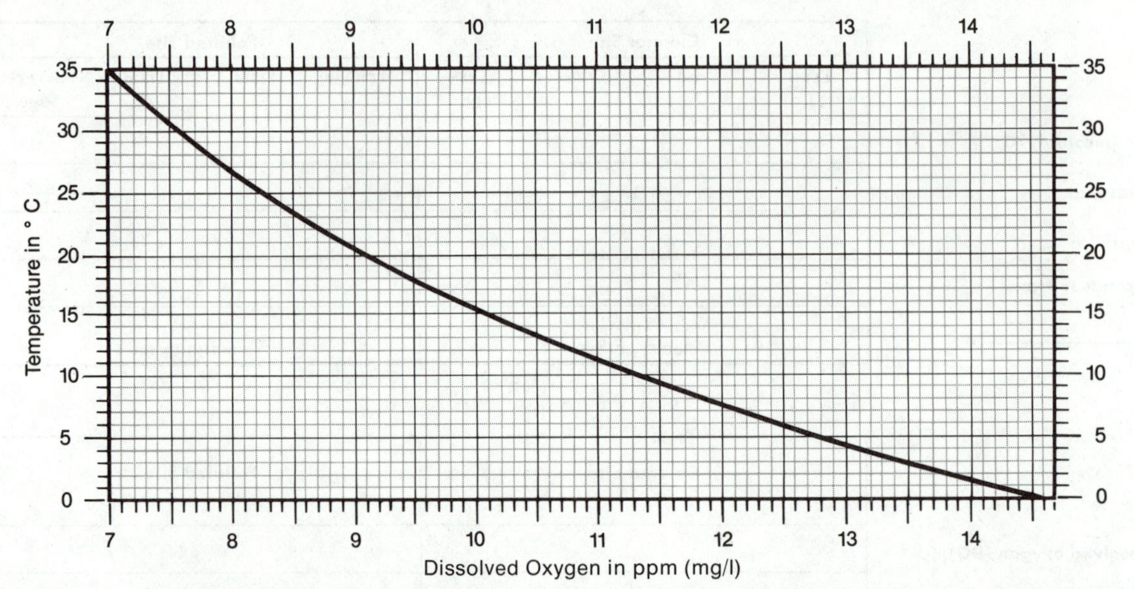

Figure 22–1. The amount of dissolved oxygen contained in water saturated with oxygen at a given temperature. The values plotted are for an atmospheric pressure of 760 mm Hg. At high elevations a correction for the lower air pressure should be calculated. If the amount of oxygen in saturated water is S at 760 mm Hg and the amount at a different pressure (P) is S', the formula is

$$S' = S\left(\frac{P - p}{760 - p}\right)$$

where p is the vapor pressure of water (in mm) at the appropriate temperature. You will have to look this up in a table in some source such as the *Handbook of Chemistry*.

On shore, several people should work over the collected rocks with forceps and nail brush to dislodge macroinvertebrates. Place them into a wide-mouth jar. Replace the rocks and sample another spot if instructed. Transfer the organisms captured in the Surber sampler net to the wide-mouth jar, turning the net inside out to pick off any clinging animals. Label the jar with the site, number of samples, and ''Surber Sampler.''

10. The organisms collected should be sorted from the debris while they are still alive. If they are transported to the lab before sorting, at least pick out any rocks, clams (keep these), and so forth which would mash them. To sort, pour the contents of the jar into sorting trays with less than 0.5 cm of water. Use blunt forceps and eyedroppers to separate the macroinvertebrates from the debris. If organisms cannot be identified and counted within a day, preserve them by adding one part concentrated neutral formalin to nine parts organisms and water.

11. Sample a square foot (about 930 cm^2)* area of mud in a pool with a dip net; if the net is 1 foot wide, scoop an area 1 foot (30 cm) long (adjust accordingly if your net is not 1 foot wide). Scoop 5 cm deep and move the net into the current. Put the collected mud into a bucket and rinse the net by pouring water over it into the bucket. Repeat the sampling if so instructed. Do at least some preliminary sieving of animals from mud at the site; final sorting can be done as in number 10. Place a portion of the mud onto a no. 30 sieve and swish it around in a bucket of water. Transfer the organisms and remaining debris into a collection bottle and repeat until all the collected mud has been sieved. Label the jar with site, number of samples, and ''Dip Net.''

12. Take ten mud samples from an undisturbed pool at each site in the following manner: Scoop a small jar into the mud about 3 cm deep and make a path 5 cm long. Fill the jar with water. Label. Take these samples back to the laboratory and let them settle, uncovered, for several hours to a few days.

---

*A square foot is used because this is the usual size of the Surber sampler.

Then, look at the organisms (particularly tubificid worms) at the water-mud interface. Describe their actions. With good light and magnification, count and identify the organisms in place. Total the number of each kind for each site. Figure the total area sampled, and calculate the number of each kind per square foot.

13. Figure the sequential comparison index, a measure of diversity, for each of the two sets of riffle invertebrates and the two sets of pool invertebrates collected with the dip net. To do this, put each well-mixed collection into the sorting tray marked with lines with a small amount of water (not so much that they move around). If the pan is crowded, divide and do this in separate batches. Look at each specimen, then the one next to it, in a systematic way—left to right across the top row, right to left in the next row, and so on. (If more than one pan is used, figure it as one long continuous pan or as completely separate samples. There is a small difference.) Compare adjacent animals; are they alike? If they look alike—same size, shape, color—they are in the same "run." If they are different, that marks the end of a "run" and the beginning of another. Runs can have only one member or can include all the individuals in the pan if they are all the same. The sequential comparison index *(SCI)* is the number of runs ÷ the number of individuals in the collection. A collection where each individual is different from its neighbors has an *SCI* of 1; if all are alike, the *SCI* is 1 ÷ the number of individuals.

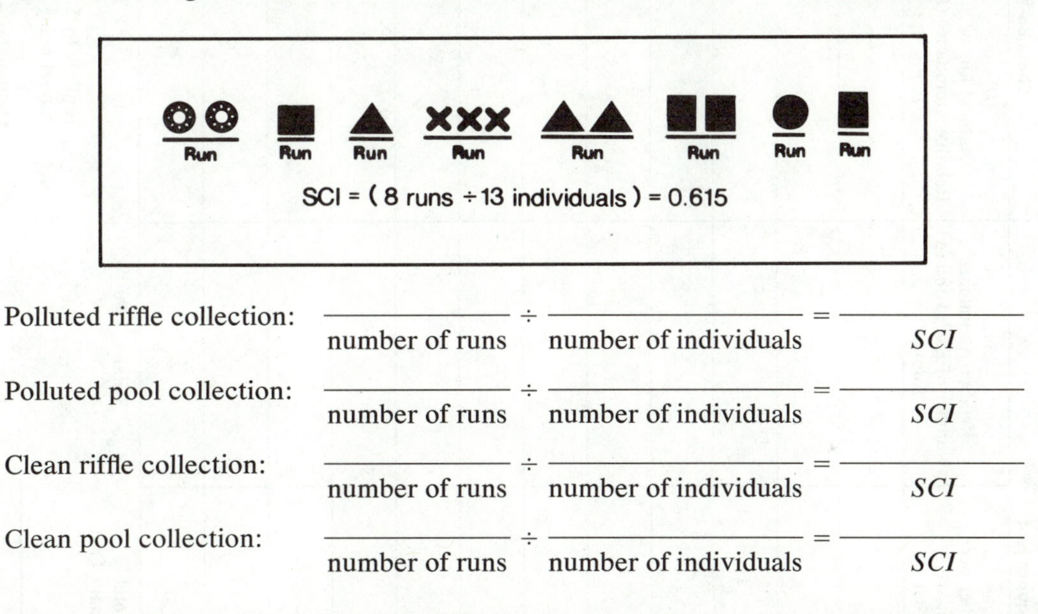

Polluted riffle collection:	______	÷	______	=	______
	number of runs		number of individuals		*SCI*
Polluted pool collection:	______	÷	______	=	______
	number of runs		number of individuals		*SCI*
Clean riffle collection:	______	÷	______	=	______
	number of runs		number of individuals		*SCI*
Clean pool collection:	______	÷	______	=	______
	number of runs		number of individuals		*SCI*

14. In the laboratory with the aid of dissecting microscopes, sort the collected invertebrates into like kinds. Keep the four samples separate: clean and polluted Surber and dip net. Assume that organisms that look different are different and keep them separate. (Even if they are just different stages of a single species, they may have different ecological roles, which is the important thing here.) Identify them to the taxonomic level specified by the instructor; give each kind a name, such as Ephemeroptera A, Ephemeroptera B, etc. Count them and list on four separate Work Sheets as number per square foot. If desired, group members may copy the Work Sheets at this point and do the calculations independently.

15. Calculate the Shannon-Wiener diversity index for each of the four lists of organisms. This diversity index is based on information theory; it shows how successful one would be at guessing what the next bit of information (in this case, the next kind) will be, knowing the first bit. Where all the bits of information are the same, that is, where diversity is low, it is easy to predict the next bit. Where the bits of information are all different, it is hard to predict the next bit and diversity is high.

The basic formula is

$$H' = -\Sigma p_i \log p_i$$

where H' is the Shannon-Wiener diversity and p_i is the proportion of individuals of a kind, or the

Diversity Work Sheet

Site/Date: ______________________

	Area Sampled: ________ square feet			Shannon-Wiener Index (H')		Simpson Index (C)
Kind	Number Found in Sample	(A) Number per Square Foot	(B) Relative Abundance Number of Each Kind (A) ÷ Total (N) = p_i	(D) Natural Log of Relative Abundance (B) = $ln\ p_i$	$B \times D$ = $p_i\ ln\ p_i$	$B \times B$ = p_i^2
Total Number of Kinds ________	Total Number of Individuals ________	Total per Square Foot N= ________	Total = 1.00		Total ($\Sigma\ p_i\ ln\ p_i$) ________	Total (Σp_i^2) ________

Multiply by 1.4427 to convert to log_2 ________

Multiply by −1 to make positive H' = ________

Subtract from 1 C = ________

Permission is granted to photocopy this page.

number of a single kind ÷ the total number collected. Any log can be used. It was developed using $\log_2$; we'll use ln ($\log_e$), then convert the result to $\log_2$.

Follow the steps on the Work Sheet to calculate the Shannon-Wiener diversity for each of the four collections.

H' polluted riffle ________ H' clean riffle ________

H' polluted pool ________ H' clean pool ________

H' using $\log_2$ is usually between 0 and 5, but may be larger. If your answers are above 5 check your calculations.

16. Calculate the Simpson diversity index, following the Work Sheet directions. It is a measure of the probability of interspecific encounter—in a diverse collection, the probability that two of the same kind will meet is lower. The usual formula for the index is

$$C = 1 - \Sigma p_i^2$$

where C is Simpson diversity and p_i is as defined above. The Simpson index can have values from 0 (when the whole sample consists of a single species) to 1; if your answers are outside that range, check your calculations.

Simpson diversity:

polluted riffle ________ clean riffle ________

polluted pool ________ clean pool ________

17. Use the topographic map to determine the evaluation of each site and the distance between them.

Elevation of cleaner site ________ of polluted site ________

Subtract to find the drop in elevation ________ feet = ________ m

Straight-line distance between the sites ________ miles = ________ km

Stream distance between the sites (do this with a map measure if available or roughly by moving the edge of a ruler along the curves) ________ miles = ________ km

What is the land use between the sites?

18. Calculate the discharge, the amount of water flowing through each site:

$$\text{Discharge} = \frac{\text{average}}{\text{depth}} \times \frac{\text{average}}{\text{width}} \times \frac{\text{average}}{\text{velocity}} \times \text{a correction factor: use 0.9 for a muddy substrate and 0.8 for a rocky substrate}$$

Discharge, cleaner site ________ m /sec polluted site ________ m /sec

RESULTS AND DISCUSSION

1. Which site had the greater discharge? What is likely to account for the difference?
2. What is the effect on the velocity of a stream of straightening the stream bed ("channelization")? Which can carry more sediment, a fast or a slow stream? How does channelization affect habitat diversity?

3. What are some effects of high amounts of suspended solids? Of dissolved solids? Of turbidity?

4. What does BOD measure?

5. What is *alkalinity*? What is its relation to *hardness*? What is meant (in terms of mg/l $CaCO_3$) by soft water?

6. Did you find more than the acceptable amounts of anything?* What?

7. Discuss the effects of and probable causes of differing amounts of the tested chemicals between the sites.

8. Why are chemical tests for pollution sometimes not as reliable as biological assessments?

9. How does percent oxygen saturation change with temperature? If there was a difference in temperature, which site was warmer? What do you suppose made the difference? How does increased temperature affect the amount of dissolved oxygen in the water (each 10° C rise in temperature roughly doubles respiration rate)? How is BOD affected by increased temperature?

10. What factors not mentioned above could make a stream warmer? What could be done to reduce the temperature of an urban stream?

11. How is dissolved oxygen added to stream water? How is the amount of dissolved oxygen affected by turbulence? By suspended solids?

12. What is an indicator organism? Why do benthic macroinvertebrates make good indicators of water quality? Include methods of respiring in your discussion. What is ecological amplitude, and how does it relate to this exercise?

13. Compare, between sites, the fauna collected with the Surber sampler, the dip net, and the small jar.

14. Were any indicator organisms collected? What were they? What do they indicate? Discuss the two sites, and compare the pools to the riffles, basing your interpretation on the indicator organisms.

15. Which of the study sites had a more diverse benthic fauna? If more than one diversity index was calculated, did they agree? At a single study site, did the pools or riffles have a more diverse fauna? Rank the clean pool, polluted pool, clean riffle, and polluted riffle as to number of species, number of individuals per square foot, and each diversity index figured. Comment on these rankings. Was the biggest difference between polluted and unpolluted sites in richness or equitability?

16. Why are diversity indices often used in assessing stream pollution?

17. What features of a habitat tend to make diversity high?

18. Discuss the relative cleanliness of the sites, basing your discussion on the diversity of macroinvertebrates collected.

19. In general, what is a pollutant?

20. What is a point source of pollutants? What is a non-point source? Which pollutants in the study stream are probably from point sources? From non-point sources? List some likely sources.

21. After the invertebrates collected in the Surber sampler were separated from the debris, they probably clung to each other. They weren't eating each other or mating. Why were they clinging? Explain how this could be adaptive.

22. What are some adaptations to current?

23. What is *physiographic succession* in streams? Were any of the differences between the two study sites probably dependent on this factor rather than pollution? Devise ways of testing your answer.

---

*Quality criteria (EPA 1976) for fresh water aquatic life are DO—at least 5.0 mg/l; pH—6.5–9.0 considered harmless to fish; copper—varies with species; no more than 0.1 mg/l for brook trout in soft water; iron—no more than 1.0 mg/l; chromium—no more than 100 mg/l. For the other factors measured, quality criteria are complex or not applicable.

Teaching Notes

If travel time is short and with division of labor, the physical and chemical work and collecting can be done in a three-hour lab with the identifications done later; or use two lab periods, each including a trip and the lab work for one site. A reference collection speeds up identifications. For information of Hach equipment and addresses of regional offices write to Hach Chemical Co., P.O. Box 389, Loveland, Colorado 80537, or telephone 1-800-525-5940. It will save time if students have a chance to practice with the Hach kit before going into the field. Fluorescent dye is available from Forestry Suppliers, Inc., 295 West Rankin St., P.O. Box 8397, Jackson, Mississippi 39204. Surber samplers are available from Wildlife Supply Co., 301 Cass St., Saginaw, Michigan 48602. The other equipment is widely available; we suggest using the sturdiest dipnets you can find.

BIBLIOGRAPHY*

American Public Health Association, American Water Works Association, and Water Pollution Control Federation. 1975. *Standard Methods for the Examination of Water and Wastewater*. 14th ed. Am. Publ. Health Assoc., Washington, D.C.

Cairns, J., Jr., D. W. Albaugh, F. Busey, and M. D. Chanay. 1968. The sequential comparison index—a simplified method for non-biologists to estimate relative differences in biological diversity in stream pollution studies. *J. Water Poll. Contr. Fed.* 40: 1607–1613.

Environmental Protection Agency, U.S. 1976. *Quality Criteria for Water*. EPA, Washington, D.C.

Golden, J., R. P. Ouelette, S. Saari, and P. N. Cheremisinoff. 1979. *Environmental Impact Data Book*. Ann Arbor Science, Ann Arbor, Michigan.

Hurlburt, S. H. 1971. The non-concept of species diversity: a critique and and alternative parameters. *Ecology* 52: 577–586.

Hynes, H. B. N. 1970. *The Ecology of Running Waters*. Univ. Toronto Press, Toronto.

———. 1974. *The Biology of Polluted Waters*. Univ. Toronto Press, Toronto.

Kaesler, R. C., J. Cairns, Jr., and J. S. Crossman. 1974. Redundancy in data from stream surveys. *Water Res.* 8: 637–642.

McIntosh, R. P. 1967. An index of species diversity and the relation of certain concepts to diversity. *Ecology* 48: 392–404.

Office of Water Data Coordination, Geological Survey. 1977– . *National Handbook of Recommended Methods for Water-Data Acquisition*. U. S. Dept. Interior, Reston, Virginia.

Simpson, E. H. 1949. Measurement of diversity. *Nature* 163: 688.

Warren, C. E. 1971. *Biology and Water Pollution Control*. Saunders, Philadelphia.

Wetzel, R. G., and G. E. Likens, 1979. *Limnological Analyses*. Saunders, Philadelphia.

*Suitable identification books are listed in Exercise 21.

Exercise 23

THE TROPHIC ECOLOGY OF HUMANS

Key to Textbooks: Brewer 12–17, 122–138, 263–269; Benton and Werner 101–120; Colinvaux 145–180, 229–245; Emlen 341–349; Kendeigh 164–166; Krebs 488–493, 517–577; McNaughton and Wolf 97–148; Odum 37–85; Richardson 260–271; Ricklefs 626–680; Ricklefs (short) 124–155; Smith 25, 117–151; Smith (short) 47–71; Whittaker 192–231.

OBJECTIVES

With the overall aim of studying energy flow in ecological systems, we estimate energy intake in humans by keeping track of food consumption. We then try to calculate energy flow at lower trophic levels, in this case in the agricultural ecosystems from which most of our food comes. A third section tries to broaden the context still further to consider the energy subsidies involved in current agricultural practice. Some features of energy systems analysis are included.

INTRODUCTION

All organisms function by using energy. To the biologist the most important energy source for organisms is the food they burn in the complicated cellular processes of respiration. This energy moves the organism's muscles, builds its tissues, sends its nerve impulses, and does most of the rest of the work of the organism. Although a carnivore may obtain its food by eating other animals, the eventual source of virtually all of the food of animals is the green plant. Green plants make their food by photosynthesis, a process in which energy from sunlight is captured and stored in the chemical bonds of carbohydrates, proteins, and fats.

Human food chains can be looked at in exactly the same way as food chains of other animals. Sun⟶corn⟶beef⟶you and sun⟶grass⟶mouse⟶hawk are each reducible to sun⟶producer⟶primary consumer⟶secondary consumer. However. the food chains to which humans belong, except in hunter-gatherer or very primitive agricultural societies, receive energy subsidies—energy from outside the organisms' metabolism—to a greater degree than most other food chains.

Most energy subsidies in human food chains come from energy stored in fossil fuels. Whether used directly, converted to electricity, or used in the manufacture of machinery and chemicals, this energy subsidizes human consumption in many ways. For example, it increases yields per acre,

Key Words: energy flow, agricultural ecosystems, energy systems analysis, energy subsidies, gross energy intake, vegetarianism, gross production, net production, trophic levels, fossil fuel, food chains and webs, photosynthesis, respiration.

makes possible cultivation of more acres by fewer people, allows the transportation of food produced in one place to persons living in another and keeps food edible long past the time it would have rotted without refrigeration.

In this exercise we will try to obtain some quantitative estimates of the amount of energy involved in your own metabolism, in the food chains leading to you, and in the subsidies connected with your food.

MATERIALS

Calorie counter, 1 per student if possible

calculators

PROCEDURE

Calculating Your Gross Energy Intake

1. During one 24-hour period keep careful track of the foods you eat and their caloric content. Try to choose a typical day. If possible each student should be supplied with a Calorie counter—a list of foods and their energy value—so that each meal can be analyzed on the spot. Both a standard list that includes bread, potatoes, and vegetables, and a brand-name list for Twinkies, Sara Lee, etc., are useful. If not enough Calorie counters are available for each student to check them out, it will suffice for students to record amounts of each food and later convert them to Calories by consulting the appropriate lists in the laboratory. Labels on food packages are another source of caloric information.

Record foods in the categories given on the Work Sheet. Note that the food Calorie as given in the books and on labels is the kilocalorie. Assigning the various components of some mixtures will be a problem that you will have to solve by reading labels and using common sense. For example, the Calories in most kinds of commercial salad dressings can be divided about half and half between vegetable oil and sugar. If some food that you eat is not listed, include it with a related one.

2. The sum of the caloric values of all the food you consumed is your gross energy intake for the 24-hour period. (For most persons it will be in the range from about 1,500 to 3,000 kcal.) A part of this food will be indigestible; the energy in the indigestible fractions of the food will pass through the body and out in the feces. The indigestible portion of common foods varies from 60% for wheat bran to zero for pure sugar. Current human diets, high in meat, fat, and highly processed foods, contain only a small percentage, possibly 5%, of indigestible material. For vegetarians, 20% might be a reasonable figure.

The rest of the energy, the energy in food that is digested and assimilated, goes to supply your metabolic activities for a day. This amount of energy is released within your body and dissipated as heat to your surroundings. If, on a given day, you eat food containing slightly fewer Calories than you use in your energy-requiring activities you will lose weight as you draw on energy in your fat stores. A pound of fat contains about 3,600 kcal, so a total deficit of this amount over a period of time will cause you to lose a pound of body weight. Contrariwise, if you eat food containing more Calories than you use, you will gain weight, at the same rate.

3. For our later calculations, your energy intake on a yearly basis is needed. Obtain this by multiplying totals for each food on the Work Sheet by 365, entering the products on the Analysis Sheet, and then summing the results.

Enter here this sum expressed in millions (10^6) of kcal:

Your gross energy intake ______________ $\times 10^6$ kcal/year

Work Sheet for Trophic Ecology of Humans*

Food	Breakfast	Lunch	Dinner	Between Meals	Total
Bread					
Wheat cereal					
Citrus fruits					
Orange juice					
Coffee					
Tea					
Peanut butter					
Rice or rice cereal					
Potatoes					
Carrots, other vegetables					
Apples, other fruits					
Vegetable oil					
Margarine					
Wine					
Beer					
Beet sugar					
Cane sugar					
Soft drinks					
Corn cereal					
Sweet corn					
Milk					
Cheese					
Eggs					
Chicken					
Pork					
Beef					
Tuna					
Perch					
Shrimp					
Total	Your 24-hour gross energy intake ________ kcal				

*Record kilocalories (= Calories) if possible. If not, record servings or weights, and later convert to kilocalories.

Energy Flow in the Agricultural Ecosystems that Support You

1. Figure 23–1 is a simplified energy flow diagram of which you are a part. You have just calculated one value for the diagram, your gross energy intake. Enter it in the appropriate blank at the upper right on the diagram. We will now follow the flow of energy prior to your place in the system. If you have looked ahead and the series of computations looks forbidding, do not despair. Taken a step at a time it is easy.

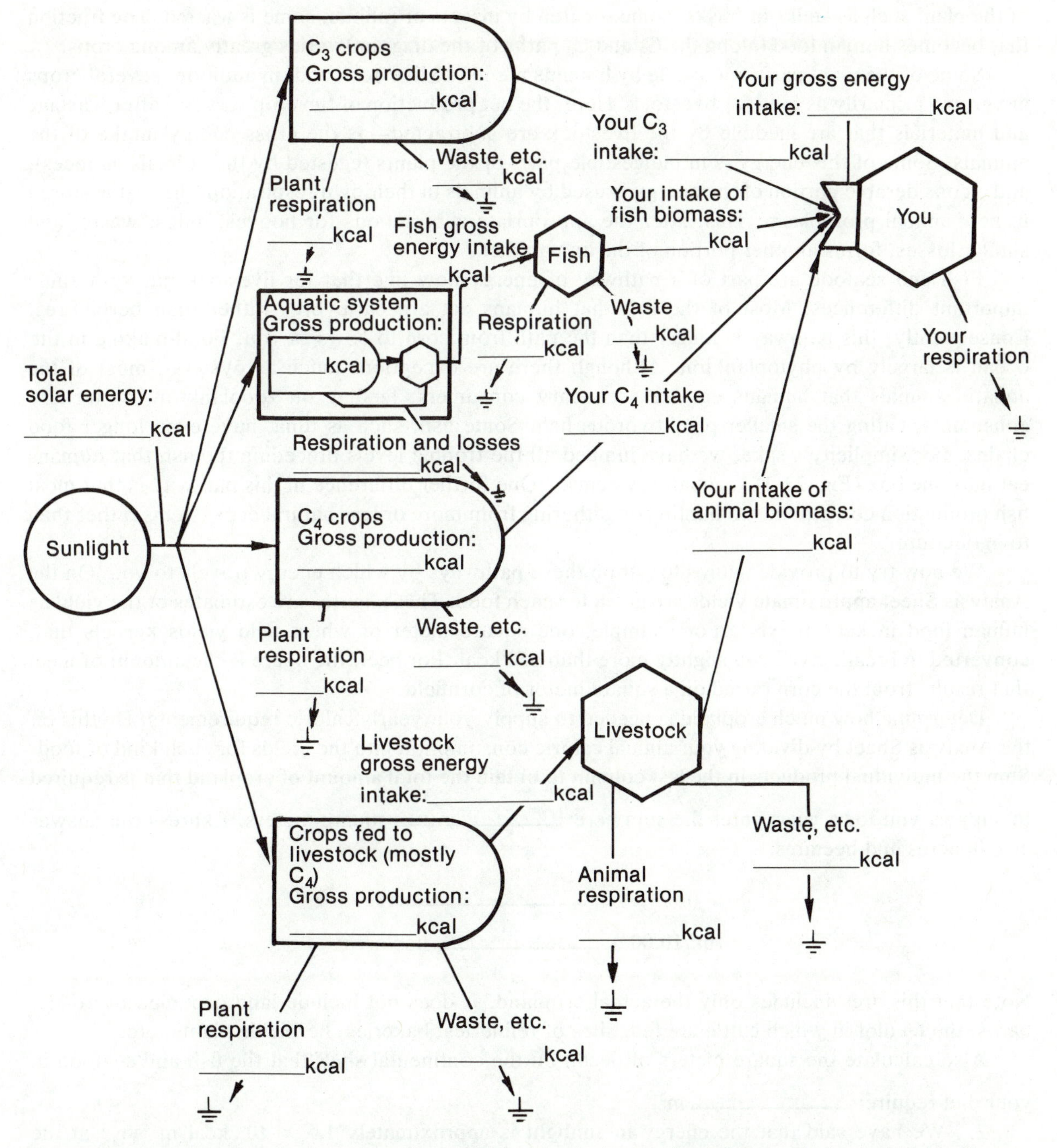

Figure 23–1. Energy flow in ecosystems from which you obtain your food. The symbols used in this diagram were developed by H. T. Odum (1971). Energy sources are circles, production systems (such as plants) are bullets, and consumers (such as animals) are hexagons. The heat sink (electrical ground) symbol shows energy lost from a system. The notched projection on the hexagon representing you indicates an interaction of energy sources. The rectangle is a general purpose symbol for systems you do not want to analyze in detail. In this case it includes all the aquatic trophic levels below the fish eaten by humans.

Solar energy strikes the earth in areas that provide your food at the rate of about 1.5 million kilocalories per square meter per year (henceforth, we will give such rates as 1.5×10^6 kcal $m^{-2}yr^{-1}$). The amount of solar energy that plants are able to store in photosynthesis is small. On a yearly basis, including both growing and non-growing seasons, rates as high as 1% are rare. We are speaking here of *gross production,* the total energy storage by the plant. A considerable fraction of this energy is used by the plant itself in its own respiration. The energy left after the plant's respiration is *net production*. This is the energy stored in new biomass. The percentage that this represents of the energy in the sunlight hitting the field varies; probably 0.2–0.3% is a fair estimate for temperate zone crops.

Some of the plant's net production cannot be used by humans as foods. Some is in inedible parts of the plant such as stalks or husks, some is eaten by insects or rabbits, some is wasted. The fraction that becomes human food (along the C_3 and C_4 paths of the diagram) varies greatly among crops.

Some of the plant parts not usable by humans are fed to livestock and, in addition, several crops are grown primarily as food for livestock. Here, the net production of the crop serves—after wastage and materials that are inedible by the livestock are subtracted—as the gross energy intake of the animals. Some of this energy is in indigestible parts of the plants (egested by the animals as feces), and a considerable portion of the energy is used by animals in their own respiration; the rest is stored in new animal protoplasm. This, after the appropriate subtractions for hooves, hides, waste, and similar losses, forms another portion of the human diet.

Fish and seafood are part of a pathway of energy flow like that for livestock but with some important differences. Most of the fish that humans eat are carnivores rather than herbivores. Consequently, this pathway is longer than the path from corn to cows to you. Food-making in the ocean is largely by phytoplankton. Although there are exceptions (such as oysters), most of the aquatic animals that humans eat are secondary consumers, feeding on zooplankton, or tertiary consumers, eating the smaller planktivorous fish. Some fish, such as tuna, have even longer food chains. For simplicity's sake, we have lumped all the trophic levels preceding the fish that humans eat into one box (Fig. 23–1), "aquatic systems." One further difference in this pathway is that most fish production corresponds to hunting or gathering from more or less natural ecosystems rather than to agriculture.

We now try to provide figures to put on these pathways by which energy travels to you. On the Analysis Sheet approximate yields are given for each food. These figures are estimates of the yield as human food in kcal $m^{-2}yr^{-1}$. For example, one square meter of wheat field yields kernels that, converted to bread, gives you slightly more than 650 kcal. For beef, the figure is the amount of meat that results from the corn raised on a square meter of cornfield.

Determine how much cropland is needed to supply your yearly caloric requirements. Do this on the Analysis Sheet by dividing your annual caloric consumption into the yields for each kind of food. Sum the individual products in the last column to obtain the total amount of cropland that is required to support you for a year; enter the sum here: ____________ square meters. Express the answer also in acres and hectares:

(m^2/4,047): ____________ acres

(m^2/10,000): ____________ hectares

Note that this area includes only the actual cropland. It does not include land occupied by roads, barns, the feedlot in which cattle are fed, sites of refineries, bakeries, herbicide plants, etc.

Also calculate the square meters of ocean on the continental shelf that the fish and seafood in your diet require: ____________ m^2.

2. We have said that the energy in sunlight is approximately 1.5×10^6 kcal $m^{-2}yr^{-1}$ at the earth's surface (the actual figure varies somewhat; if you have a more accurate one for your region, use it). Multiply the square meters of land required to support you by 1.5 million to obtain the solar energy input to that land. Enter the value here and at the left of the diagram:

Total solar energy: ____________________ kcal yr^{-1}

Analysis Sheet for Trophic Ecology of Humans

Food	Your Annual Consumption (kcal yr^{-1})	Yield (kcal $m^{-2}yr^{-1}$)	Square Meters of Land Required to Support You (Consumption ÷ Yield)
		C_3 Plants	
Bread	________	650	________
Wheat cereal	________	810	________
Oranges, grapefruit	________	1000	________
Frozen orange juice	________	410	________
Coffee	________	4	________
Tea	________	40	________
Peanut butter	________	920	________
Rice or rice cereal	________	1250	________
Potatoes	________	1600	________
Carrots	________	810	________
Other vegetables	________	200	________
Apples	________	1500	________
Pears, peaches	________	900	________
Vegetable oil	________	300	________
Margarine	________	300	________
Wine	________	600	________
Beer	________	300	________
Beet sugar	________	1990	________
		C_4 Plants	
Cane sugar	________	3500	________
Soft drinks	________	3500	________
Corn cereal	________	1600	________
Sweet corn	________	250	________
		Animal Products	
Milk	________	420	________
Cheese	________	40	________
Eggs	________	200	________
Chicken	________	190	________
Pork	________	190	________
Beef (feedlot)	________	130	________
Frozen fish fillets	________	2*	________

*Per square meter of continental shelf

3. We will now work from each end to fill in some other details of energy flow. First, sum the fraction of your gross energy intake (Analysis Sheet) that comes from C_3 plants, from C_4 plants, from animal biomass, and from fish. Enter these separately at the appropriate places on the diagram.

4. Now add up the square meters required to support you in the same four categories:

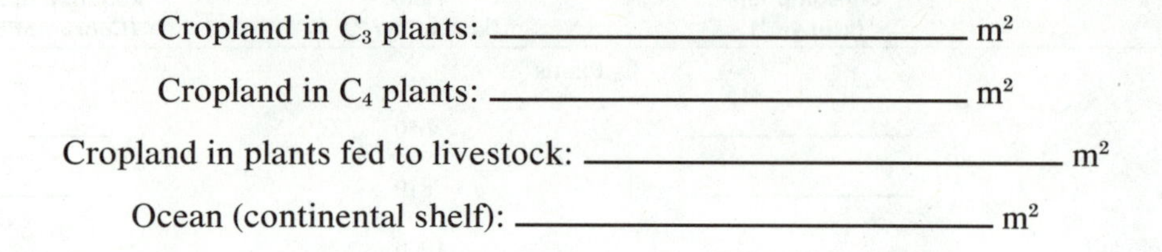

Cropland in C_3 plants: ______________________________ m^2

Cropland in C_4 plants: ______________________________ m^2

Cropland in plants fed to livestock: ______________________________ m^2

Ocean (continental shelf): ______________________________ m^2

We will assume that gross production on the farmland that supplies your food is 12,000 kcal $m^{-2}yr^{-1}$. This is an oversimplification since actual gross production probably varies by growing season, by crop, by year, etc. Multiply the square meters in each cropland category by 12,000 and enter the values in the appropriate place, inside the three bullets on the diagram. We will assume that gross production in the aquatic systems of the continental shelf is 2,000 kcal $m^{-2}yr^{-1}$. Multiply the square meters in this category by 2,000 and enter the value in the small bullet in "aquatic systems." Add the four gross production values together and enter below as the total yearly gross production that forms the base of the trophic pyramid that supports you.

Gross production required to support you: ______________________________ kcal yr^{-1}

5. We have separated the C_3 and the C_4 plants without explanation up to this point. The explanation is that C_3 plants lose much more energy in their own respiration. We will estimate the respiration of the C_3 plants as being 67% of their gross production. Accordingly, multiply the values for C_3 gross production by 0.67 and enter this on the diagram as C_3 plant respiration.

We will estimate C_4 plant respiration as 50% of their gross production. Multiply C_4 gross production by 0.5 and enter the result as C_4 plant respiration.

Assume that the meat and other animal products you consume come from animals fed entirely on C_4 plants and multiply the gross production along this pathway also by 0.5 to obtain plant respiration.

6. We can now fill in the intermediate values in the diagram. From gross production for the C_3 crops subtract both respiration and your C_3 intake. Enter the remainder in the blank for waste. This includes stalks and other indigestible material as well as some material lost in harvesting, wasted in processing, etc.

Do the same for the C_4 crops.

7. Along the fish pathway, make calculations as follows: Assume that half the fish catch is waste (spoilage, heads, scales, etc.). Accordingly, enter an amount equal to your intake in the blank for waste. Fish are efficient at converting their energy intake into new protoplasm; for fish respiration, enter also an amount equal to your intake. Add your intake, waste, and fish respiration together to obtain the value for fish gross energy intake. Subtract this value from gross production in the aquatic system and enter the remainder as "respiration and losses." This is a large, catch-all category that includes respiration by both plants and animals in the trophic levels below the fish and energy going to food chains other than the one to which the fish you consumed belong.

8. For the livestock pathway, assume that your intake of animal biomass contains 10% of the Calories that were in the feed provided to the animals. (This is another oversimplification in that the efficiency of conversion of animal feed to meat varies; beef, for example, has an efficiency below 10% and chickens and pork, above.) Multiply your intake of animal biomass by 10; enter this figure as the gross energy intake of the livestock.

Now subtract this value and also the value for plant respiration on this pathway from the value for gross production for crops fed to livestock. The difference is the value to insert for waste for those plants. It also includes wasted and indigestible feed for the cattle.

9. Finally multiply the gross energy intake of the livestock by 78% and enter that as animal respiration. Multiply gross energy intake of the livestock by 12% and enter that as waste. Here waste is bone, hide, trimmings, etc.

This completes the energy flow diagram for you. It resembles the energy flow diagrams in your ecology text for other ecosystems. You can see the large amount of photosynthetic production and the still larger amount of energy in sunlight required to yield the calories in the food that you eat. Unlike most natural ecosystems, however, this purely biological flow of energy is only a small part of the energy that is involved in the trophic system. In the next section we try to estimate the energy subsidies that are involved.

Fossil Fuel Subsidies to the Food System

1. Figure 23–2 is a simplified view of energy flow in the food system that supports you. To the diagram that you just completed have been added the energy subsidies on which human food production depends. Although some of these are specifically indicated (for example, pesticides and fertilizers) most are simply indicated as coming from fossil fuel. The fossil fuel inputs can be put in several categories:

Production of seeds, equipment, and supplies used on the farm;

Direct energy use on the farm for applying fertilizer and pesticides, irrigating, planting, cultivating, harvest, maintaining buildings and equipment;

Transportation of materials from, to, and on the farm;

Processing, packaging, and storage of food products in factories, packing houses, etc;

Transportation, storage, and sale of food item in stores.

These are all fairly straightforward, although some items are less obvious than others. For example, the energetic costs of growing the seeds to be planted is an obvious cost; less obvious and harder to estimate is the energy that goes into the development of new agricultural and horticultural varieties. The energetic cost of maintaining veterinarians and county agricultural agents needs to be considered. The energy in the form of human labor that goes for hoeing, driving tractors, and planning next year's crops is, in effect, a subsidy to the individual crops.

It is far too complicated and time consuming for you to try to estimate the subsidies for every food in your diet—or, in most cases, even for one of them. Instead we will use figures developed by others who have made such calculations for various parts of the food system. We will break down the food system into production, processing, sales, and consumption.

2. Copy onto the Calculation Sheet your annual consumption (from the Analysis Sheet) of each food. List manufactured products under the basic crop from which they come (bread under wheat, for example).

3. Estimates of the kcal of fossil fuel required to produce 1 kcal of food on the farm are given in the column of figures under agricultural production. This is the energy required to grow the crop (labor, fuel, fertilizer, pesticides, processing and depreciation of machinery and buildings). For meat it is the energy for raising the animals to the point where they go to the slaughterhouse. For fish it is the energy to catch them (fuel, ice, depreciation of boats and nets, etc.). To obtain the annual subsidy in these forms to your food supply, simply multiply your annual consumption by these figures.

For home gardens, use 0.8 kcal fossil fuel subsidy for each kcal of food from a garden which you work by hand and on which you put modest amounts of fertilizer. Most of the fossil fuel subsidy in this case is indirect, through your labor. If you fertilize heavily, use pesticides, have a garden tractor, irrigate, etc., use a 2.0 kcal subsidy for each food kcal.

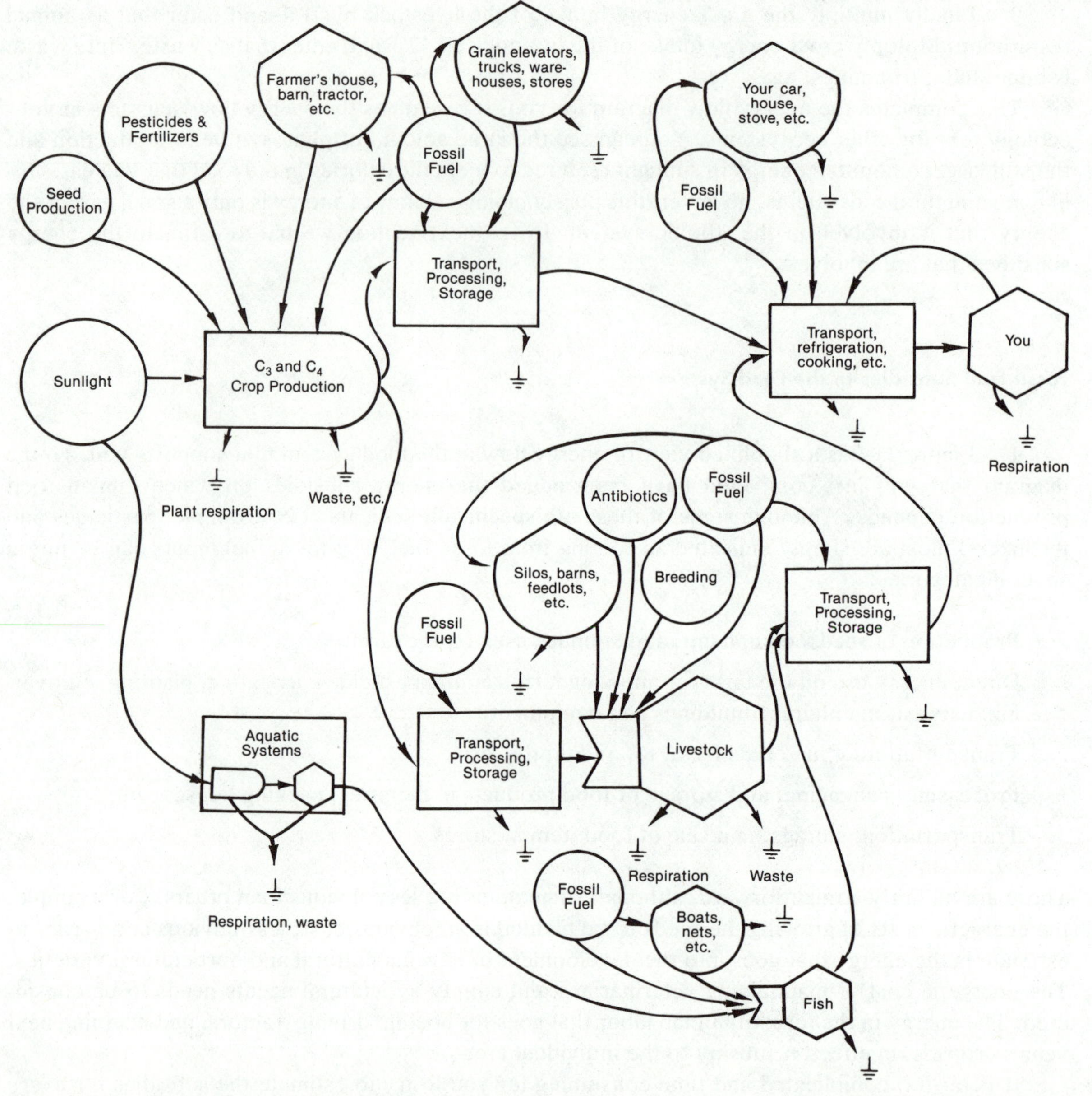

Figure 23–2. Energy flow for the whole food system from which you obtain your food. The symbols are the same as in Figure 23–1. The notched projection on the livestock hexagon indicates the interaction of energy sources, specifically the energy in food, that needed to develop and produce antibiotics, to transport heifers to bulls or vice versa, to transport food and provide shelter, water, etc. The water tanks are storage symbols. The grounds on the rectangles include such sources of loss as waste and spoilage; on the water tanks they refer to depreciation.

4. To calculate processing energy requirements, multiply your annual consumption for each food by one of the following three factors:

0, if little or no processing is involved
(for vegetables, fruits, etc. that are eaten as is).

10, if fairly simply processing is involved
(for example, baking bread, slaughtering and packing meat,
making cheese, canning or freezing vegetables).

40, for highly processed foods (convenience foods, for example).

Calculation of Fossil Fuel Subsidy to Your Food Supply

Crop	Your Annual Consumption (kcal yr^{-1})	Your Annual Fossil Fuel Subsidy (kcal yr^{-1})		
		Agricultural Production	Processing	Sales
		(kcal per kcal*)		
C_3 Plants				
Wheat		(0.3)		
Oranges, grapefruit		(3.0)		
Peanuts		(0.7)		
Rice		(0.7)		
Potatoes		(0.7)		
Vegetables		(2.0)		
Apples		(0.8)		
Pears, peaches		(1.0)		
Soybeans (vegetable oil, margarine)		(0.4)		
Barley (beer)		(0.4)		
Grapes (wine)		(1.0)		
Sugar beets		(0.8)		
C_4 Plants				
Sugar cane (soft drinks)		(0.2)		
Corn		(0.2)		
Animal Products				
Milk		(2.8)		
Eggs		(6.0)		
Chicken		(10.0)		
Pork		(10.0)		
Beef (feedlot)		(15.0)		
Tuna		(8.0)		
Perch		(1.3)		
Shrimp		(60.0)		
Total				

Total Fossil Fuel Subsidy as of Food Purchase: ____________

*The figures in parentheses are the fossil fuel subsidy (in kcal) to agricultural production for every kcal of various kinds of food in your diet. Multiply these figures by your annual consumption of each food to determine total subsidies. Corresponding values for subsidies (kcal per kcal) to processing and sales are given in the text of the exercise.

For some crops you may have entries in two or three categories. For example, if you eat potatoes that you peel and cook yourself, frozen potatoes for french-frying, and instant mashed potatoes, you would keep these separate and multiply your consumption by 0, 10, and 40 respectively.

5. To calculate the energy used in wholesale and retail sales (transportation, lighting, refrigeration, etc.), multiply your annual consumption by 10 kcal. Omit this subsidy for food you raise yourself.

6. Sum the three columns of subsidies (production, processing, sales) and then add the three together to obtain your total fossil fuel subsidy up to point where you buy the food.

7. We will now try to estimate, as a final input to your food supply, the energy involved in your shopping, storage, and preparation of food.† (If you eat mostly in dormitory cafeterias, read over this section, do part d, and then go to question 4 in the Discussion.)

a. If you drive to do your shopping, enter in the blanks below the roundtrip distance to the store you patronize, the average number of trips per week, and the average number of miles per gallon of gasoline your car gets (use 15 for this if you do not know).

(________ trips per week × ________ miles roundtrip distance × 52 weeks) ÷

(________ miles per gallon of gasoline) × 34,800 kcal

The last figure (34,800) is the energy content in kcal of a gallon of gasoline. Carry out the arithmetic. The answer is the amount of energy you use in going to the store to buy food and bringing it home.

Transportation of food to your home: ____________ kcal yr^{-1}

b. To estimate the energy you use in refrigerating food, look up the type of refrigerator and the type of freezer, if any, in the table below. The values are approximate kilowatt hours of electricity used per year. (A few models of refrigerators are especially energy efficient. If you have actual figures for your own refrigerator use them.)

Type of Refrigerator	Kilowatt Hours	Type of Freezer	Kilowatt Hours
Single door, manual defrost	600	Less than 12 cubic feet	
Two door, automatic defrost		upright	900
14 cubic feet	1,500	chest	720
17 cubic feet	1,740	12–22 cubic feet	
20 cubic feet	1,800	upright	1,440 (auto defrost, 1,720)
		chest	1,260
		More than 22 cubic feet	
		upright	1,800 (auto defrost, 2,340)
		chest	1,560

Add the kilowatt hours for the refrigerator and the freezer and multiply the sum by 13,543, the number of the kilocalories in a kilowatt hour.

Refrigeration of food in your home: ____________ kcal yr^{-1}

c. To estimate the energy you use in cooking food, fill in the first equation on the next page if you have an electric range and oven or the second equation if they are gas.* Determine the average number of minutes that you use the oven per week and the total number of minutes in an average week that each burner is on.

†In parts a, b, and c divide energy totals by the number of people involved (e.g., if you do the shopping for three people, divide your transportation energy use by 3).

*The figures for watts or Therms used in the calculations are typical, but if you have more exact values for your own range, use them.

Electric

[(____________________ minutes per week of oven use ÷ 60 minutes) × 52 weeks × 3 kw] +

[(____________________ minutes per week of burner use ÷ 60 minutes) × 52 weeks × 2 kw].

Gas

[(____________________ number of minutes per week of oven use × 52 weeks) × 0.25 Therm] +

[(____________________ number of minutes per week of burner use × 52 weeks) × 0.1 Therm].

Carry out the arithmetic. The answers you obtain are in kwh yr^{-1} for the electric range and Therm yr^{-1} for the gas range. Convert them to kcal by multiplying either by 13,543 (electric) or 25,202 (gas).

Cooking food in your home: ____________________ kcal yr^{-1}

d. Sum the values for transportation, refrigeration, and cooking to obtain a total for the in-home portion of the food system. (For dormitory dwellers use the average for the rest of the class unless otherwise directed.):

____________________ kcal yr^{-1}

Add this to the fossil fuel subsidy for the rest of the food system calculated under point 6 of this section. The two together estimate (with a few omissions) the part of the total subsidy to the U.S. food system that can be assigned to you.

Total fossil fuel subsidy to you: ____________________ kcal yr^{-1}

What is the ratio of fossil fuel Calories to your intake in the form of food?

Total fossil fuel subsidy ÷ Gross energy intake = ____________________

Most of the subsidy is originally in the form of petroleum. Divide the total subsidy by 34,800 to express it in terms of gasoline.

Energy subsidy to you: ____________________ gallons of gasoline

DISCUSSION

1. Define *food chain, food web,* and *trophic level.*
2. To what trophic level do humans belong? What are some problems with the trophic level concept?
3. What is photorespiration?
4. (Primarily for students who eat in cafeterias.) Indicate the information you would need and how you could obtain it for determining energy usage for transportation, storage, and preparation of food at the cafeteria where you eat. (You need not actually obtain the information and make the calculations unless your instructor so directs.)
5. Choose one crop that is grown in your region and try to list the energy subsidies it receives. (If feasible, quantify these and express them as kilocalories, determine the average yield of the crop in your region, express the yield in kcal and calculate the local ratio of food Calorie output:fossil fuel Calorie input.)
6. List some energy-requiring activities involved in the home food system that were not considered in the exercise. How could you go about quantifying them?

7. Worldwide, the caloric intake of humans is about 90% from plants and 10% from animals (meat, milk, eggs). How did your own diet or the class average compare?

8. Give energetic and other arguments for and against vegetarianism.

9. What are some ways of reducing the fossil fuel subsidy in the U.S. food system? Be concrete. Include both culture and technology. Where are the biggest savings possible?

10. What is the general relationship between fossil fuel energy input into the agricultural system (up to the point of crop production and not including the food system beyond that) and yield?

11. Popular writers on the U.S. agricultural system sometimes suggest that, since the food Calories of many U.S. foods are fewer than the petroleum Calories that went into its production, we are simply throwing away the sunlight. To what extent is this true? Would we be better off to quit subsidizing agriculture and convert the petroleum to synthetic food? Why or why not?

12. Specifically, what happens to all the fossil fuel energy used to subsidize food production? Is it in the food?

13. Each member of the class, including the instructor, should list (anonymously) the number of pounds he or she should lose to be at an ideal weight. Each pound represents 3,600 kcal of food intake. Use a class average for ratio of fossil fuel to food Calories and calculate how many kcal in fossil fuel would be saved during the period of dieting in which this weight was lost. Also express the result as gallons of gasoline. Project the savings to the U.S. population of 230 million.

14. What does the arrow going back from you to the rectangle—"transport, refrigeration, cooking, etc."—represent? If the last hexagon represented the whole U.S. population rather than just you where should other arrows go?

15. As shown in the energy subsidy Work Sheet, the subsidy for the production of some foods is much greater than for others. Specifically, what makes the subsidy high for some foods and low for others?

EXPLANATORY NOTE

The yield figures in the Analysis Sheet are based on calculations from "Agricultural Statistics" of the U.S. Dept. of Agriculture and various other sources. The figures on fossil fuel subsidies to agricultural production are largely from Heichel (1974, 1976), Leach (1976), and Rawitscher and Mayer (1977). Calculations of energy subsidy for processing and sales assume a total subsidy of 112 kcal per food Calorie (Hannon and Lohman 1978) distributed as follows: production 18%, processing 30%, sales 8%, preparation and consumption 44% (U.S. Office of Industrial Programs 1976). Although we believe our figures for yields and subsidies are basically correct, we would welcome better values and additions from others interested in the subject.

Teaching Notes

A few minutes can be used in one period to assign the exercise; most of it can then be done out of class. One or two weeks later most or all of a period can be devoted to completing calculations and discussing findings. The volume of computations in this exercise looks a bit daunting but students seem to have little trouble, and the exercise is valuable in getting across the fundamentals of energy flow as well as teaching about human energy systems.

BIBLIOGRAPHY

Cox, G. W., and M. D. Atkins. 1979. *Agricultural Ecology.* Freeman, San Francisco.

Hannon, B. M., and T. G. Lohman. 1978. The energy cost of overweight in the United States. *AJPH* 68: 765–767.

Heichel, G. H. 1974. Comparative efficiency of energy use in crop production. *Conn. Agric. Expt. Sta. Bull.* 739: 1–26.

______. 1976. Agricultural production and energy resources. *Am. Sci.* 64: 64–72.

Komarik, S. L., et al. 1974–1976. *Food Products Formulary, Vol. 1–3.* Avi Publ., Westport, Connecticut.

Lappé, F. M. 1976. *Diet for a Small Planet.* Rev. ed. Ballantine Books, New York.

Leach, G. 1976. *Energy and Food Production.* IPC Science and Technology Press, Guildford, England.

Lockeretz, W., ed. 1977. *Agriculture and Energy.* Academic Press, New York.

Odum, H. T., and E. C. Odum. 1981. *Energy Basis for Man and Nature.* 2nd ed. McGraw-Hill, New York.

Pimentel, D., ed. 1980. *Handbook of Energy Utilization in Agriculture.* CRC Press, Boca Raton, Florida.

Pimentel, D., et al. 1973. Food production and the energy crisis. *Science* 182: 443–449.

Rawitscher, M., and J. Mayer. 1977. Nutritional outputs and energy inputs in sea food. *Science* 198: 261–264.

Slesser, M. 1973. Energy subsidy as a criterion in food policy planning. *J. Sci. Fd. Agric.* 24: 1193–1207.

Stanhill, G. 1974. Energy and agriculture: a national case study. *Agro-Ecosyst.* 1: 205–217.

Steinhart, C., and J. Steinhart. 1974. *Energy: Sources, Use and Role in Human Affairs.* Duxbury Press, North Scituate, Massachusetts.

U.S. Office of Industrial Programs. 1976. *Energy Use in the Food System.* U.S. Federal Energy Administration, Washington, D.C.

Exercise 24

NOISE

Key to Textbooks: Odum 448–450.

OBJECTIVES

We will examine noise as a variety of pollution; examine methods and instruments for measuring loudness; and determine the noise levels under which we live.

INTRODUCTION

Noise is defined as unwanted sound. This is, of course, subjective. There is an old joke about a visitor to fox-hunting country whose host proposed to take him to watch a hunt. As they drew near the scene, the host said, "Listen to that music." The visitor listened for a minute and shook his head. "I'm sorry," he said; "I can't hear a thing for those damned dogs."

Several factors, including pitch, rhythm or the lack of it, and the taste and state of mind of the listener, interact to determine whether a sound is unwanted. Probably the one most influential factor, however, is *loudness* (Fig. 24–1).

Loudness is easily measured using a decibel meter. The decibel scale is not especially straightforward and is worth an explanation. The loudness of a particular sound in decibels is equal to ten times the logarithm of the ratio between the intensity of that sound and the faintest audible sound. The faintest audible sound is taken to be 10^{-12} watts/m^2 (there is that much energy in the sound waves striking a square meter of wall or eardrum). Putting the statement above in equation form, then,

$$\text{loudness of sound S in decibels} = 10 \times \log_{10} \frac{\text{intensity of sound S in watts/m}^2}{10^{-12}}.$$

A sound having 10^{-12} watts/m^2 of power will measure zero decibels (because the logarithm of 1 is zero). A sound having 100 times as much power as the faintest audible sound will measure

$$10 \times \log 100 = 10\ (2) = 20 \text{ decibels.}$$

A log scale is used because of the great range of energy involved from very faint to very loud sounds. A rocket engine may reach 10^6 watts/m^2. In ordinary numbers, then, the power of the faintest audible sound is 0.000000000001 watts/m^2 and the power of the rocket engine is 1,000,000 watts/m^2. Expressed in decibels these are 0 and 180. The logarithmic nature of the scale needs to be kept constantly in mind in interpreting loudness figures. A doubling of decibels means not a doubling in power but an enormously greater change than that. A doubling of power occurs with every increase of 3 decibels and this is true for any point on the decibel scale. Finally, what we subjectively perceive

Key Words: noise, noise pollution, loudness, decibel, sound, stress.

as a doubling of loudness does not correspond either to a doubling of decibels or of power. Generally, a sound that we regard as twice as loud as another sound is about ten decibels higher (corresponding approximately to a tenfold increase in energy).

The meter you will use will probably have three networks, A, B, and C. These differ in their response to different frequencies or pitches. Network C has a response curve that is nearly flat over a wide range of frequencies. Network B is less responsive to lower pitches and network A still less responsive. Although you might suppose that a uniform response would be desirable, low-pitched sounds tend to be less annoying and, at least as far as hearing loss goes, less potentially hazardous to humans, so that the A network is generally used. The network used is indicated in giving sound levels as, for example, 50 dBA, indicating a sound level of 50 decibels using network A.

Research on the effects of noise is just well underway. It is known that sounds of around 80 dB can cause some temporary hearing loss lasting hours or even days. Sounds this loud are unpleasant to most humans, and the hearing loss may be a protective response on the part of the body to make the exposure tolerable. Permanent hearing loss follows prolonged exposure to noise. In general, the risk of deafness increases with both the loudness and the duration of the sound; after 20 years at a job with noise levels of 90 dBA, about 20% of the workers have impaired hearing; at 105 dBA, 20% have impaired hearing after only five years.

Physiological and behavioral effects of noise are even less well known, but a great variety of effects have already been found, including increased blood pressure, adrenalin and ACTH secretion, decreased work efficiency, and increased rates of the stress-related ailments such as ulcers. The California Supreme Court recently held that noise is a legal nuisance and permitted neighbors of the Los Angeles International Airport to recover damages for mental and emotional distress without having to prove impaired hearing or physical damage.

Interference with communication is another effect of noise that may be important. With background noise of 50 dB, normal speech cannot be relied on to transmit information at distances greater than 8 feet; at 60 dB, the distance drops to 2 feet. But in hospitals, where accurate transmission of information may be a matter of life or death, sound levels over 60 dB may regularly be found.

Effects of noise on the laboratory rat have been studied but little is known about wild animals or about plants. This is a worthwhile field for research.

MATERIALS

sound level (decibel) meter with range from 30–120 decibels (It is best to have at least one per section.)

PROCEDURE

1. Familiarize yourself with the sound-measuring instrument. In so doing, obtain an approximate sound level for a classroom or laboratory.
2. Work in groups of two or more as directed. During the next few days each group should check out a meter and measure sound levels in two situations. One will be assigned from the list below and one will be some other situation that the group chooses (and the instructor approves).

Possibilities, depending on ingenuity and access, include various areas in hospitals; different kinds of factories; the noise levels around aircraft, chain saws, etc.

Situations to be included:

A busy period next to the largest nearby highway

Inside the living area of a dormitory, house, or apartment at the quietest time of day

Inside the living area of a dormitory, house, or apartment at a time of more activity

The most remote natural area in the vicinity

Noise Data Sheet

Record below the median dBA level for the various situations investigated.

Situation	Median dBA	Situation	Median dBA

From the data above, calculate the percentage of an average 24-hour day that you spend in the various sound ranges given below.

Sound Range	Time in Each Sound Range	
	Hours	Percent
Below 30 dBA (very quiet)		
30–50 dBA (quiet)		
51–75 dBA (moderately loud)		
76–100 dBA (very loud)		
Above 100 dBA (uncomfortably loud to painful)		
Totals	24	100

A supermarket

A restaurant

A residential neighborhood

The busiest intersection in the city

3. At each site record location, date, and time of day. Take a series of nine readings (or as many as feasible) using all three networks (A, B, C). Note that you must give the meter a few seconds to respond for each change of network. If any activity unusual for the situation is occurring, make a note of it and indicate how it might bias the reading.

4. Determine the median, low, and high values for the locations you visited. Also record the median dBA level for the situations studied by the other groups (on Data Sheet).

DISCUSSION

1. How were dBB and dBC levels generally, compared with dBA levels? What does this mean?

2. What were the highest and lowest readings (single readings, not medians) recorded?

3. Calculate the approximate percent of the 24-hour day you spend in various sound ranges (on Data Sheet).

4. Do any of the recorded levels or the general sound regime under which you live seem too high for comfort, health, or maximum efficiency? The Occupational Safety and Health Administration (OSHA) currently sets permissible noise exposure levels as given in the table. Did your noise regime exceed the OSHA standards? Do you think the OSHA standards are strict enough to prevent unfavorable physiological and psychological effects?

Sound Level, dBA (Slow Response)	Hours Permissible Exposure per Day
90	8
92	6
95	4
97	3
100	2
102	1½
105	1
110	½
115	¼ or less

5. What do you see as reasonable sound level limits for various public and semi-public situations such as the workplace, libraries, highways, residential neighborhoods, etc.? What do you see as the best ways to achieve these limits? Include both social (legislation, litigation, education) and technological (sound-proofing, equipment design, etc.) aspects.

6. Suggest some ways in which individual animals and animal populations in natural habitats might be adversely affected by high sound levels. Design a study that would investigate one possibility.

7. The noise levels produced on takeoff by a Concorde SST and a subsonic Boeing 707 were measured at Fairbanks International Airport. At a small building 500 feet past and 250 feet to one side of the runway, the following data were obtained:

	Outside Building	Inside Building
Concorde SST	121.5 dBA	92.5 dBA
Boeing 707	109.5 dBA	80 dBA

How many times as loud is the SST perceived by humans compared with the 707?

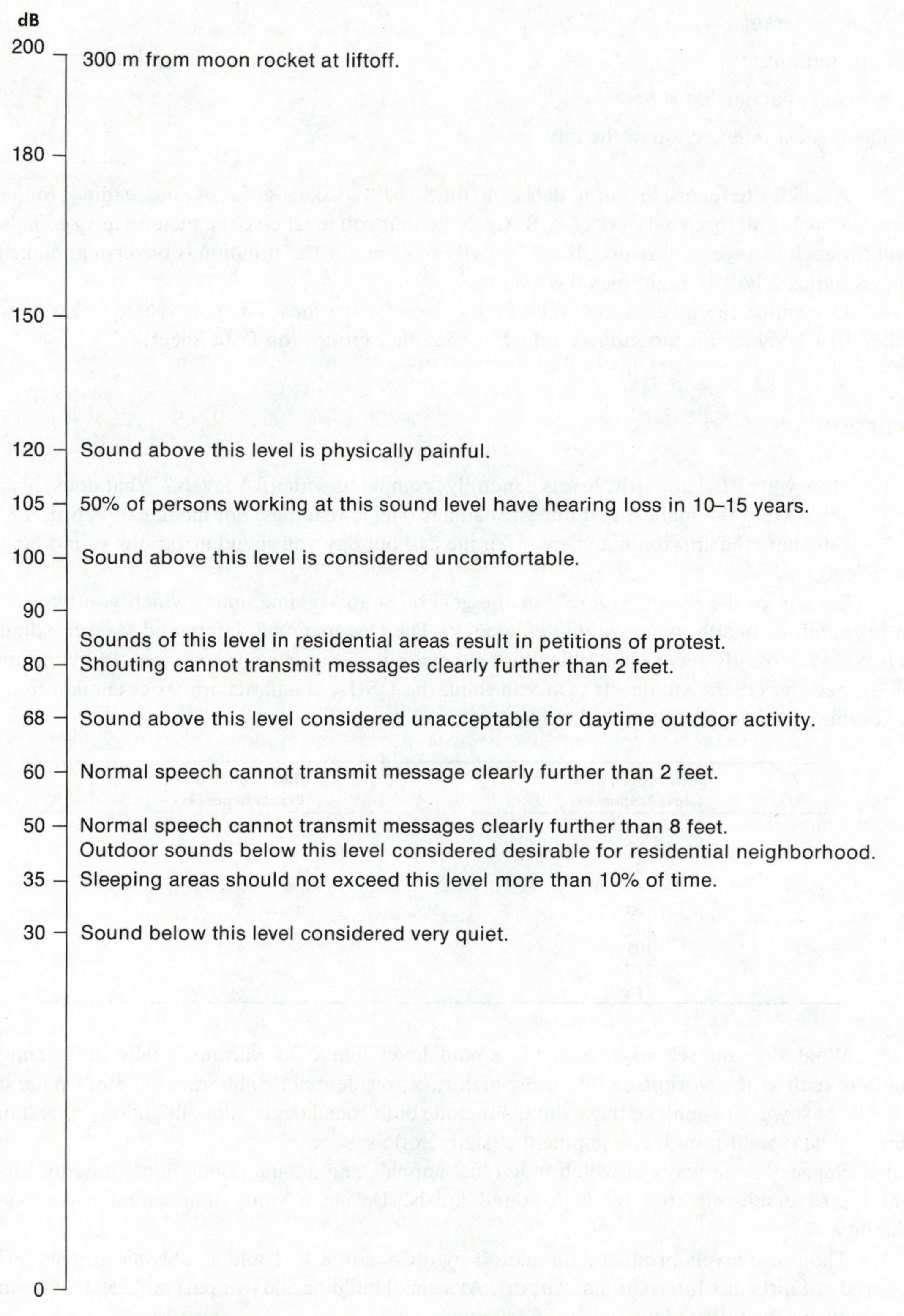

Figure 24–1. Effects and characteristics of various sound levels.

Teaching Notes

Operation of the decibel meter can be demonstrated in a few minutes in the laboratory; then pairs of students can sign up for various times to use the meters out of class. Putting together class results, final calculations, and discussion will take an hour or two the next week.

BIBLIOGRAPHY

Britton, P. 1980. How science makes war against noise. *Popular Science* 215(2): 43–52.

Cook, D. I., and D. F. Van Haverbeke. 1974. Tree-covered land-forms for noise control. *U.S. Forest Service Res. Bull.* 263: 1–47.

Cheremisinoff, P. M., and P. P. Cheremisinoff. 1977. *Industrial Noise Control Handbook.* Ann Arbor Science Pub., Ann Arbor, Michigan.

Keveler, L. 1975. *Noise: The New Menace.* John Day, New York.

Smith, R. J. 1980. Government weakens airport noise standards. *Science* 207: 1189–1190.

Taylor, R. 1970. *Noise.* Penguin Books, Baltimore.

Walton, S. 1980. Noise pollution: environmental battle of the 1980s. *BioScience* 30: 205–207.

Wesler, J. E. 1975. Noise and induced vibration levels from Concorde and subsonic aircraft. *Sound and Vibration,* Oct., 18–22.

Appendix 1

SCIENTIFIC THOUGHT IN ECOLOGY

"How came he, then?" I reiterated. . . .

"You will not apply my precept," he said, shaking his head. "How often have I said to you that when you have eliminated the impossible, whatever remains, however improbable, must be the truth? We know that he did not come through the door, the window, or the chimney. We also know that he could not have been concealed in the room, as there is no concealment possible. Whence, then, did he come?"

"He came through the hole in the roof!" I cried.

"Of course he did. He must have done so. . . ."

SIR ARTHUR CONAN DOYLE, *The Sign of The Four,* 1890

The term "science" refers to both a product and a process. Science is systematic, verified knowledge but it is also the process that generates such knowledge. The dual use of the word should not bother anyone; there are many other examples of the same thing—"art," for example.

Formalized as the *"scientific method,"* the activities involved in scientific thought have tended to be mentioned the first week of freshman science classes and otherwise consigned to the shadowy world of the philosopher of science. In ecology this situation has changed in the past few years. Ecologists have increasingly realized that conscious use of sound methods of scientific inference is the fastest way to advance the field, just as blueprints and measuring are a more efficient way to build a barn than the cut and try method.

As it is generally described in textbooks, the scientific method consists of a series of steps. *Observations* of the natural world disclose a pattern. *Hypotheses* to explain the pattern are devised and *tested.* Hypotheses that are disproved are discarded. Those that survive can be subjected to more testing and, in time, may form a part of the body of *theory* of the subject.

The problem with this formulation is that it makes science seem dull and uncreative to most people. In fact, science is as creative as music and as exciting as exploration, and any description that fails to convey these qualities is defective.

The creativity comes at several points. Discerning order in the natural world—in the colors of flowers that bloom at different seasons, in the sizes of litters of animals, in the numbers of species on different-sized islands—requires both curiosity and ingenuity. Thinking of hypotheses that might explain the patterns requires knowledge of what others have already learned about related matters but it also requires creativity of a high order.

Ingenuity again enters in deriving testable predictions from the hypotheses. Relatively few new hypotheses are testable directly. Instead the scientist has to reason, "If this idea is true, and I do such-and-such, then thus-and-such will happen." The hypothesis that the world is round can today be tested in reasonably direct fashion as the astronauts did, by looking. Through much of human history, however, anyone wishing to test this hypothesis (or the competing hypotheses of flat, cuboidal, helical, etc.) would have had to make a series of fairly ingenious predictions and then go out and make the observations to see if they were fulfilled.

Not only the predictions but the methods or equipment for their testing may be the occasion for great creativity. One of the elegant experiments of biology was an indirect test of a hypothesis concerning what colors of light are effective in photosynthesis. With modern equipment we could answer this question more or less directly in several fairly laborious ways. T. W. Engelmann, working in the late 1800s, was able to answer it quickly and simply with no equipment other than a microscope and a prismlike device.

He reasoned that algae subjected to light of different colors would give off different amounts of oxygen, depending on the rate of photosynthesis. He needed, then, a way of providing uniform algal cultures with different kinds of light and of measuring the oxygen production. What Engelmann did—it might be worth pausing here and thinking about what sort of jury-rigged setup you or I would have tried—was this: he placed a single strand of filamentous algae on a slide, added to the water some bacteria that he knew were motile only in the presence of oxygen, and then projected along the length of the algal filament a miniature spectrum. Where the bacteria were moving were the places where oxygen was produced and, therefore, the parts of the spectrum effective in photosynthesis.

Empirical Testing

An explanation for some phenomenon may occur as a flash of intuition to a scientist just as to a non-scientist. What happens after the flash is usually very different. The next thought of the scientist is likely to be, ''How can I test that?'' The next thought of the poet, the prophet, or the politician may well be, ''How can I convince everybody else of that?''

Scientific ideas are, in other words, tested *empirically*. The scientist goes into the field or laboratory and looks, feels, smells, measures or, in some other way, checks his or her idea against reality. There are other ways of settling questions—theological (''it is written''), democratic (''the ayes have it''), occult (''the tea leaves say . . .''), commonsensical (''it's only logical'')—but these have nothing to do with science.

The idea that the truth is arrived at by empirical testing is central to science. It is also what often makes conversation difficult with educated non-scientists. Although they may agree in principle, in practice they often revert to trying to settle issues verbally. Which of two conflicting ideas is correct is decided by the skill with which the ideas are advocated. Hot debates with bombast and polemic occur in science, and most scientists enjoy them (a few think they are unseemly). But everyone knows that in the end the rhetoric means nothing. The erroneous hypothesis, no matter how appealing and no matter how persuasively put, will lose out.

It is important that observations and experiments be *repeatable* by scientists other than the one who originally reported them. Scientists are not immune to mistakes of observation and inference: they can misread instruments and miscalculate numbers; they can convince themselves that a set of data that disagrees with a pet hypothesis is the result of mistakes in technique and another set that agrees came when the technique was straightened out. Startling results, contradictory of current theory, are unlikely to be accepted until other scientists get the same results.

Skepticism and Open-mindedness

Skepticism is an important part of science. Scientists must doubt any hypothesis until it has been subjected to testing and, in fact, are skeptical of the testing until it has been replicated. It is scientific to doubt ideas that conflict with established theory—such ideas as clairvoyance, water dowsing, and psychic spoon bending, for example. But open-mindedness is an equally important trait for a scientist. Any proposition that has not been tested may be right, and this may be true even if the proposition does conflict with current theory.

The belief that the continents are fixed, for example, pervaded scientific thought through most of this century. Most scientists were skeptical of, and few were open-minded toward, other ideas such as Wegener's continental drift hypothesis. The idea of stable continents and its consequences had the status of a *paradigm,* a generally accepted model. Facts that seemed in conflict with this hypothesis—the Permian glossopteris flora of India, Africa, Australia, and Antarctica; the distribution of the ratite birds (such as ostriches, rheas, and cassowaries); and even geological data—tended to be explained away in various ways. Improbable land bridges were postulated, for example, over which tropical organisms made long marches through temperate or arctic climates.

In the 1960s, studies on sea floor spreading suggested a way in which continental drift might occur, and the accumulation of additional data, such as paleomagnetic observations, that were difficult to explain given continental fixity culminated in the rapid discrediting of the former dogma. By the end of the 1960s, few scientists still held to the idea of continental fixity. In its place was a new paradigm, that of plate tectonics.

Scientists must resist half-baked ideas but they must be receptive to observations that conflict with current theory. This is a difficult combination. Most scientists find skepticism an easier prescription to follow than open-mindedness, at least in their own specialty.

Falsifying Hypotheses

Although we may sometimes be hard put to come up with even a single credible explanation for some phenomenon, more often we can devise several that might account for it. If we cannot, our friends working on the same subject usually can. Our task then is to determine which hypothesis is correct. The approach of the scientist is to attempt to eliminate the incorrect hypotheses. Science advances by disproving hypotheses—*falsifying* them, in the terminology of Karl Popper. This is for the simple reason that no hypothesis (except trivial ones, such as the proposition that everyone in the room has two ears) can ever be proved. We are reasonably confident that the earth revolves around the sun in accord with a particular set of rules that have been worked out. Every year that this happens we are a little more confident; however, it would take only one day in which the earth did something else to falsify that hypothesis thoroughly. That is the point: one contradictory observation is sufficient to falsify a hypothesis but nothing short of examining every possible case could prove it.

The hypothesis for which no disproof is forthcoming is taken as the correct explanation but not in the sense that the matter is settled and done with. Some new observation may contradict the hypothesis or theory (the latter term refers to a hypothesis that has survived some testing). Complete and utter rejection of the old theory may be the result, but what usually happens is that the new observation merely makes necessary some refinement. The old theory may come to be seen as a special case of a more general theory. Liebig's Law of the Minimum, for example, is a special case of Blackman's Law of Limiting Factors.

Hypotheses (or predictions from hypotheses) that cannot be disproved are of no interest to science. The first question to ask about any proposed explanation is how it could be disproved. If there is no observation you could make (leaving aside purely technical problems) that would disprove the idea, it needs to be re-formulated—or forgotten. Statements may be unfalsifiable by nature, as in "Our destiny is ruled by the stars," "two species cannot coexist indefinitely in the same ecological niche," and "ghosts are visible only to people who believe in the occult."

The first statement is not falsifiable because, if our destiny is ruled by the stars, the stars must rule our hypothesis-testing destiny no less than other aspects. We cannot set up observations in which star rule is absent. The second hypothesis is unfalsifiable because "indefinitely" would allow two species to coexist any length of time and because "ecological niches" have *n* dimensions, so that after checking 80 with overlap, we are still not sure that looking at the 81st will not eliminate it. The third hypothesis is immune to disproof because open-minded persons attempting to compare believers and skeptics would be unable to assess the validity of any claimed visions.

A potentially testable hypothesis may be unfalsifiable in practice because its proponents expand it to take into account any refutation. Karl Popper gives Marxism as such a case. Marx's predictions, from an avowedly scientific model, have proved false, but Marxists have reacted by expanding the model into a religion rather than rejecting it. There are plenty of expansible hypotheses within ecology. Clements's monoclimax hypothesis is one of the best (see Brewer 1979, pp. 190–193). Exceptions to the idea that all land surfaces become occupied by the same climax community were accommodated by the recognition of relatively permanent subclimaxes, disclimaxes, preclimaxes, and so forth.

Strong Inference

Trying to state all the likely explanations for a phenomenon and generating tests that will distinguish between them has been called "strong inference." The alternative, coming up with a single credible hypothesis and testing its predictions, is weaker because the same prediction may stem from two or more hypotheses. For example, a team of ethologists offered an explanation for eye-lines, the dark or light lines that run back toward the eye from the beak of certain birds. They suggested that the lines are aiming or sighting devices so that a bird looking at a possible food item is in the same position as a hunter peering down a gun barrel. They reasoned that accurate sighting would be more important for insect eaters than for seed eaters (insects might move) and made the prediction that a higher percentage of insect-eating species should have eye lines than should seed eaters. They counted the number of birds in each category and found that the prediction was fulfilled.

J. G. Engemann (not the same person who worked on algal action spectra) supports a different explanation for eye lines. He thinks they serve to camouflage the eye and also mask the movement of the head of the bird toward an insect, thus making the insect less likely to try to escape. The idea is that the perceived motion will be to the side, outward along the stripe, rather than forward. Clearly, this hypothesis yields exactly the same prediction: more insect eaters than seed eaters should have eye lines. What is needed is a prediction that will be true for one and not the other.

Not all possible explanations for a phenomenon are mutually exclusive. Two or more may be correct; in fact, this is generally the case if we lump together both evolutionary (ultimate) and physiological or behavioral (proximate) explanations (see Brewer 1979, pp. 18–19). Evolutionarily, snowshoe hares may turn white because of camouflage; this being true does not negate the hypothesis that photoperiod acting on the endocrine system via the eye is the proximate cause of their molting to white fur. Similarly, a hypothesis that humans avoid close interbreeding because of the importance of genetic diversity in their offspring can be true along with any of the following modes of incest avoidance: persons have an inherited ability to recognize and avoid mating with close relatives; avoidance is produced by some innate sexual aversion to persons one has grown up with; or avoidance is produced by learning cultural precepts against incest.

Experimental Method

One way of testing hypotheses is by *experiment,* defined as controlled observation. We suspect that a certain chemical is a carcinogen; in the simplest experiment we divide a bunch of rats into two groups which we treat exactly the same except that we administer the chemical to the *experimental* group and do not administer it to the *control* group. If the experimental group develops no more tumors than the control, the hypothesis is refuted. If there are more cancers, then the likelihood is that they are related to the chemical since it was the only difference between the groups.

Comparative Method

Classic experimentation is a powerful tool but it is not the only approach. Experiments are always open to the objection that the results apply only to the special set of conditions in the experimental protocol. With different temperatures, different diets, or different concentrations of the chemical, the results might be different; similarly for different species—the results may tell us nothing about carcinogenicity in humans.

Of course, it is rarely possible to set up experiments involving carcinogens in humans, and there are many other cases in which it is unethical, illegal, too expensive, too time consuming, or otherwise impossible to carry out a classical experiment. We might wish to test the hypothesis that plant diversity at first rises after a forest fire and then slowly drops. It might be possible to locate several similar forests, sample them, set aside half as controls, burn over the others, and then re-sample each at 10- or 20-year intervals to follow trends in plant numbers. It would probably be better, however, to study several forests on similar sites that historical or other information showed had been burnt over 1, 5, 20, 50, and 100 years in the past. Such applications of the comparative method are often called *natural experiments*.

To return to the carcinogen example, we might proceed by predicting that there would be no difference in cancer rates between factory workers manufacturing the chemical and other persons living in the region (a *null hypothesis* set up for the purpose of trying to reject it; see Appendix 4). From the general population we would try to select a sample that matched the factory workers in such traits as age, sex, smoking history, and diet. If we found a higher cancer incidence in the factory workers, we would reject the hypothesis of no difference; this would be evidence on the side of carcinogenicity.

Rarely would a single comparative study settle the matter. An individual study is often open to the possibility that some additional factory you did not, or could not, randomize between the groups is producing the results. It might be argued, for example, that the cancer was caused by something else associated with working in the factory instead of the chemical. We would then have to search around and find similar factories making something else and see if their workers had higher cancer rates also.

Descriptive Studies

Not every scientific study need have in it every element we have discussed. It may be so laborious or time consuming to test predictions of even a single hypothesis that a requirement for "strong inference" would slow rather than help scientific progress. Some claim that the difficulty of even listing every conceivable explanation for some phenomena makes strong inference in a strict sense impractical.

There is nothing wrong, and much right, with studying the life history of a bird or the plant composition of a relict prairie simply to find out what they are like. Such studies are generally called "descriptive," a term that is derogatory in some quarters but not here. Descriptive studies are science; their conclusions are systematic, empirically based, and verifiable. The information in descriptive studies is often invaluable for later testing of hypotheses. For example, a hypothesis relating life history strategies to habitat in birds might take years and hundreds of thousands of dollars to test if the person who came up with the hypothesis had to travel from place to place studying each species. If good studies of many species are already available, he or she may be able to pull together the information for the test by a few days or weeks in the library.

Descriptive studies often turn up patterns that suggest hypotheses; this is perhaps their greatest scientific value to the person doing them. He or she finds that the bird being studied occurs only in forests with conifers and hypothesizes that needle-leaved foliage is required in the bird's habitat selection. The hypothesis may be true, partly true, or false but whichever it is cannot be shown by the data of the original study. It was the existence of a correlation between bird and conifer that suggested the hypothesis; it is circular now to use the correlation as evidence for the hypothesis. Other studies, such as going to another region to see if the correlation still held, would be necessary to test the hypothesis.

Levels of Organization

There was a time when the idea was fairly prevalent that science's aim was to explain the universe in terms of physics; some people, mostly chemists, still believe this. But one of the beauties of the natural world is its hierarchical structure. Ecologists may study organisms, populations, or communities. Each of these levels has its own properties that can be studied and described with little reference to other levels. For analysis of the working of each level we may need to look at processes at the next lower level but rarely do we need to practice reductionism further than that.

To understand change in population size, for example, we may have to include studies of birth and death processes of individual organisms. Few ecologists, however, find it useful to explain predation or competition in terms of elementary particles or even organic molecules. This sometimes comes as a surprise to (in Peter Medawar's phrase) many rather simple-minded scientists who think that population biologists or ecologists *ought* to be interested in topics important at a lower level of organization, such as stereoisomerism or Avogadro's number. A more productive kind of reductionism in ecology has been the recent attempts to analyze population and community features, such as social interactions and succession, in Darwinian terms.

The aim of science is to understand the workings of the universe. In this grand attempt, ecosystems are no less important than atoms, except numerically.

QUESTIONS

1. A bright but skeptical child of your acquaintance is unwilling to accept the widespread popular belief that the earth is spherical but suspects instead that it is flat but round, rather like a large pizza. Produce some predictions that could be tested with a minimum of equipment (no satellites or space ships) that would serve to distinguish between these two hypotheses.
2. Based on observations made or data gathered in one or more of the exercises, come up with an ecological ''problem,'' a question that requires a causal explanation. Suggest at least two biologically credible hypotheses to account for the phenomenon and suggest one or more observations that could potentially disprove one or both.
3. Look through some recent ecological journals. Find (a) a primarily descriptive study; (b) a study that seems to seek confirmation rather than falsification of a hypothesis; (c) a study that uses strong inference; (d) a study that you suspect was begun without any clear hypotheses and that uses data from the study to support or negate hypotheses *a posteriori*. Extra points for the most blatant unfalsifiable hypothesis.
4. Choose some controversial idea, such as the claim that talking lovingly to plants makes them grow better. Select two or three students to take the pro side and the con side. Give them time to think about the subject but do not permit them to do any actual substantive preparation, such as reading any of the literature. Then have them debate the issue and have the rest of the class vote on which side is right. What classes that you have taken does this remind you of?
5. Propose some non-experimental tests of the proposition that dietary salt, beyond that occurring naturally in foods, causes high blood pressure in humans. In class, have one person suggest a specific comparative study and assume that the results are positive. The rest of the class should point out some alternative hypothesis that could have accounted for the seemingly positive results. Suggest a second test that will distinguish between these two explanations. Examine this test in the same way, and continue until no one can think of any loophole for the whole set of tests taken together.
6. Certain groups advocate the teaching of ''scientific creationism'' as an alternative hypothesis to evolution by natural selection for explaining the existence of the many different kinds of plants and animals. State each position as a hypothesis or model from which you can derive several testable (falsifiable) predictions. Assemble evidence from such fields as paleontology, comparative anatomy, physiology, and behavior to make the tests. Are some of the predictions falsified? How might creationists or evolutionists explain away the apparent falsifications? Are the explanations themselves testable, or do some leave the realm of science?

BIBLIOGRAPHY

Ayala, F. J., and T. Dobzhansky, eds. 1974. *Studies in the Philosophy of Biology: Reduction and Related Problems*. University of California Press, Berkeley.

Brewer, R. 1979. *Principles of Ecology*. Saunders, Philadelphia.

Callaghan, C. A. 1980. Evolution and creationist arguments. *Am. Biol. Teacher* 42: 422–425, 427.

Engelmann, T. W. 1882. On the production of oxygen by plant cells in a microspectrum. *Botanische Zeitung* 40: 419–426. (Engl. trans. in M. L. Gabriel and S. Fogel (eds.). 1955. *Great Experiments in Biology*. Prentice-Hall, Englewood Cliffs, New Jersey.)

Engemann, J. G., and R. W. Hegner. 1981. *Invertebrate Zoology*. 3rd ed. Macmillan, New York.

Kiefer, H. E., and M. K. Munitz, eds. 1970. *Mind, Science, and History*. State Univ. New York Press, Albany.

Kuhn, T. S. 1970. *The Structure of Scientific Revolutions*. 2nd ed., enlarged. In *Foundations of the Unity of Science,* volume 2. Univ. Chicago Press, Chicago.

Lakatos, I., and A. Musgrave, eds. 1970. *Criticism and the Growth of Knowledge*. Cambridge Univ. Press, London.

Marks, D., and R. Kamman. 1980. *The Psychology of the Psychic*. Prometheus Books, Buffalo, New York.

Nisbett, R., and L. Ross. 1980. *Human Inference: Strategies and Shortcomings of Social Judgement*. Prentice-Hall, Englewood Cliffs, New Jersey.

Northrop, F. S. C. 1947. *The Logic of the Sciences and the Humanities*. Macmillan, New York.

Platt, J. R. 1964. Strong inference. *Science* 146: 347–353.

Popper, K. R. 1962. *Conjectures and Refutations*. Basic Books, New York.

______. 1963. *The Open Society and Its Enemies*. 4th ed. Princeton Univ. Press, Princeton, New Jersey.

______. 1968. *The Logic of Scientific Discovery*. 2nd ed. Hutchinson, London.

Snow, C. P. 1963. *The Two Cultures: And a Second Look*. A Mentor Book published by the New American Library, New York.

Strong, D. R., Jr. 1980. Null hypotheses in ecology. *Synthese* 43: 271–285.

Tricker, R. A. R. 1965. *The Assessment of Scientific Speculation*. American Elsevier, New York.

Willson, M. F. 1981. Ecology and science. *Bull. Ecol. Soc. America* 62: 4–12.

Appendix 2

SURVEYING ECOLOGICAL LITERATURE

Suppose you need some sound information on an ecological topic for a research project, for a letter to your legislator, or just because you want to know what the stuff the local factory is putting in the air will do—how do you get it? You get it by searching the literature; how wide and deep your search goes depends on how much information you need. (You do not get it, at least past the sixth grade, by sending an earnest letter to all the ecologists whose names you can find, asking them to send you everything they know about pollution.) The following suggestions for a literature search assume that you have access to a reasonably good library.

Good textbooks are a place to start. They may summarize in a general way what is known about your topic and may give you a citation or two to some other books or articles. You might also check the card catalogue of the library (by looking under "Forestry," for example, if your topic had something to do with growing trees) to see if there are some books dealing specifically with the topic. It may be that what you find in the textbooks or in a specialized book or two will be enough. You may find, however, that the information is too skimpy or too old and decide that you need to go further. Books and articles that you have found will generally have bibliographies of some sort. You can go to the books and articles you find cited and read them; then go to the books and articles that you find cited in the bibliographies of those books and articles; and so on. You might envisage the process as leading you backward through a stack of references increasing to infinity but, in fact, the bibliographies will soon begin to contain more and more articles that you have already seen.

This process is good for filling in the details of a subject but does not get you up-to-date. To do this you need to use some specialized periodicals that contain indexes to recent literature. The most familiar of these is probably the *Readers' Guide to Periodical Literature* (1900– . H. W. Wilson Co.). It is basically a subject (and author) index to the articles published in popular and semi-popular magazines like *Atlantic, Audubon, Holiday,* and *National Geographic*. You do not get into the research literature with *Readers' Guide*.

Probably the index of greatest all-around value to biologists is *Biological Abstracts* (1927– . Biosciences Information Service of Biological Abstracts). Among the other indexes useful in the field of ecology and the environment are

Applied Ecology Abstracts (1975– . IRL)

Ecological Abstracts (1974– . Geo Abstracts Ltd.)

Environment Abstracts (1971– . Environment Information Center, Inc.)

Forestry Abstracts (1939– . Commonwealth Agricultural Bureaux)

Pollution Abstracts (1970– . Pollution Abstracts Inc.)

Sport Fishery Abstracts (1955– . U.S. Fish and Wildlife Service)

Wildlife Review (1935– . U.S. Fish and Wildlife Service)

Zoological Record (1864– . Zoological Society of London)

Each of these has its own idiosyncrasies but most have a main body consisting of abstracts of articles and books followed by a large index. The abstracts are arranged in categories, which vary among the different index journals. For example, *Biological Abstracts* has *ecology* with subcategories, *pest control* with subcategories; it also has many other categories covering the whole field of biology. *Ecological Abstracts* has categories for *terrestrial ecology, theories, methods and techniques* plus others and subcategories under each. *Environment Abstracts* has 21 categories such as *air pollution, energy,* and *population planning and control.* The others use other divisions based on the specific fields they cover.

The indexes in all the sources are similar but each has its own special features that you will have to learn by experience. Basically, each abstract is listed in the index under the many different subjects to which it pertains. A paper on pesticides and reproduction in birds, for example, would be listed at least under pesticides, reproduction, and birds and probably under several other subjects (DDT and egg-shell thinning, for example). So you look in the index for your topic, find the abstract number, flip forward to the abstract and read it. You will probably be able to tell from the abstract whether you want to read the article itself or not. If you do, the abstract includes the author; date; title of the article; name, volume, and number of the periodical; and the pages.

Finally, there is an ingenious index that works like the method we already mentioned of one bibliography leading to several articles each of which have bibliographies, etc., except that it goes forward instead of backward. This is *Science Citation Index* (1961– . Institute for Scientific Information). If you find some really good article on your topic published, say, in 1970, you can look up that article (under the author) for the various periods of 1970, 1971, and so on to the latest issue and find out what articles have cited that really good article. Some of these may be good too and all are new and will have at least some connection with your topic. (Through its ''permuterm subject index'' *Science Citation Index* can also be used in the same way as the other index journals.)

Ecology has two excellent reviews, which publish articles providing up-to-date summaries of published research on specific topics. If the specific, generally limited, subject covered in one of these articles corresponds with your interest, it should provide a virtually complete bibliography up to the time the article was prepared. The two review serials are *Advances in Ecological Research* and *Annual Review of Ecology and Systematics.*

These bibliographic aids will lead you to the research literature, that is, to the journals publishing papers that report original scientific findings. There are a few thousand serials in the field of biology and a good share of these publish at least an occasional article of ecological interest. The following list of journals reporting original basic research in ecology is arbitrary but includes most major sources:

American Naturalist
Animal Behaviour
Ecology
Ecological Monographs
Forest Science
Human Ecology
Japanese Journal of Ecology
Journal of Animal Ecology
Journal of Ecology
Journal of Range Management
Journal of Wildlife Management
Limnology and Oceanography
Nature
Oecologia
Oecologia Plantarum
Oikos
Science
Vegetatio

Some of the following also publish basic research but lean more toward practical or applied studies:

Biological Conservation
Environmental Conservation
Environmental Management
Environmental Pollution
Environmental Science and Technology
Journal of Applied Ecology
Journal of Environmental Health
Pesticides Monitoring Journal

Depending on how broad one's view of ecology is, many more "applied" journals could be listed in such fields as waste water treatment, law, solar energy, etc.

The following may occasionally have reports of original research but are aimed primarily at communicating biological, ecological, or environmental information to the educated layman.

Ambio
American Scientist
Audubon
BioScience
The Ecologist
Environment
EPA Journal
Mazingira
Natural History
Scientific American

BIBLIOGRAPHY

Bottle, R. T., and H. V. Wyatt, eds. 1971. *The Use of Biological Literature*. 2nd ed. Archon, Hamden, Connecticut (especially Chapter 12 by S. L. Sutton).

Smith, R. C., and R. H. Painter. 1966. *Literature of the Zoological Sciences*. 7th ed. Burgess, Minneapolis.

Appendix 3

PREPARING SCIENTIFIC MANUSCRIPTS

It is good practice in both thinking and writing for students to prepare one or more laboratory reports in the style and format that would be used if the paper were to be submitted to a scientific journal. Publication no less than the research that precedes it is an essential part of science; by such publication the results of research are communicated. This appendix gives some comments that will help the novice writer prepare an acceptable manuscript. The paper can be one reporting an individual research project or it can be based on one of the exercises. Exercises that are particularly appropriate for this purpose are 4 (Time and Energy Budgets), 8 (Territoriality), 9 (Predation), 11 (Population Growth), 14 (Biomass in a Grassland), 16 (Forest Composition), 17 (Island Biogeography), 19 (Succession on Abandoned Fields), 20 (Forest Succession), 21 (Lake Sampling), and 22 (Comparison of Polluted and Unpolluted Areas of a Stream).

Each manuscript should be written with a particular journal in mind and details of format (capitalization, whether headings are centered, etc.) should conform to its editorial policy. In the absence of other instructions the student may follow the format for *Ecology,* the bimonthly journal of the Ecological Society of America. Most journals, including *Ecology,* publish suggestions for authors at regular intervals. You should study these suggestions and also examine recent issues of the journal for details of format. The *CBE Style Manual,* distributed by the American Institute of Biological Sciences, should also be available to consult.

The manuscript should be typewritten on paper 8½ × 11 inches. Do not use erasable paper (the typing and also the editor's penciled marks for the typesetter will smear). Leave margins of at least an inch on every side. The entire manuscript, including tables and the literature cited section, should be double-spaced.

Scientific writing should be as concise as possible without slighting precision, clarity, and correct grammar. Use plain English wherever possible.

Scientific names, and generally nothing else, should be underlined.

Errors accumulate with successive re-writings of a manuscript. For this reason, the final copy should be carefully checked against the original data. Literature citations should also be verified.

Typical sections of a scientific paper are Title, Abstract, Introduction, Procedure, Results, Discussion, and Literature Cited.

Title. The title should briefly denote the contents of the paper. A title should be as specific as possible.

Abstract. The abstract should briefly describe the study and state the important findings and conclusions. Sometimes a terminal Summary is used instead of an Abstract.

Introduction. In this section should be given the justification for conducting the study and writing the paper. Related studies may be mentioned to place the work in context. Acknowledgments may be included here or in a separate section just before the Summary. Include only those persons or organizations who really helped in conducting the study or writing the paper.

Procedure. Give here the materials and methods used and, if not included in the Introduction, the time and place of the study. Methods already in the literature need not be re-described; just include a citation to an appropriate paper or book. Innovations should be described thoroughly enough to allow their repetition by other workers.

Results. Herein should be stated the salient findings of the study. The text may be documented with tables and figures summarizing the relevant data.

Discussion. This section includes discussion of the findings of the study such as syntheses, interpretations, and enumeration of additional problems.

Literature Cited. Included here should be all of the articles and books mentioned in the paper and no others. Do not cite papers that you have not seen, unless you indicate the source from which you obtained the citation. The following are typical forms for literature citations in *Ecology*:

Brown, R. T., and J. T. Curtis. 1952. The upland conifer-hardwood forests of northern Wisconsin. Ecological Monographs 22: 217–234.

Nice, M. M. 1941. The role of territory in bird life. American Midland Naturalist 26: 441–487.

———. 1943. Studies in the life history of the song sparrow II. Transactions of the Linnaean Society of New York 4: 1–247.

Toumey, J. W., and C. F. Korstian. 1947. Foundations of silviculture upon an ecological basis. Second edition, revised. John Wiley and Sons, New York.

Tables and Figures

All tables and figures should be cited in the text. Tables should be numbered consecutively in Arabic numerals throughout a paper. They should be typed, double-spaced, and placed one to a page at the end of the manuscript. It is better to cite tables (and figures) by making a statement about the information presented in them, such as:

Germination averaged 87% for the first 40 days and 56% for the next 60 days (Table 2).

rather than by indicating their existence

Table 2 presents germination results for . . .

A typical form for the title and headings of a table is the following:

Table 3. Age-specific litter sizes for prereduction (1965–67) and postreduction (1969–70) years.

Mother's age (years)	Mean litter size	
	Before reduction	After reduction

Figures should be numbered consecutively throughout a paper. Legends for figures should be typed consecutively on a separate page, as

Fig. 1. Drainage pattern of the Raritan River.
Fig. 2. Results of the mechanical analysis of . . . .

The number of the figure should be penciled lightly on the back of the figure. Figures should be placed together at the end of the manuscript before the tables.

Points That Are Frequently Troublesome

1. Remember that your aim is to prepare a manuscript from which a typesetter can produce printed pages like those in the journal you are copying; you are not trying to reproduce the look of the journal yourself.
2. Include sample sizes at an appropriate point in the paper, generally either in Methods or in tables in Results.
3. Lists of organisms should have a logical order—taxonomic, alphabetical, and by abundance are three possible sequences.
4. The title should be descriptive of the study, not of the laboratory exercise.

5. For each journal there is one right format for the literature cited section, the tables, etc. You should find out what that is and follow it exactly.

6. Most tables and many figures have the functions of (a) documenting the findings stated in the Results and (b) presenting some aspects of the Results in slightly greater detail. They should not simply repeat material stated in prose nor should they be raw data.

7. Proofread and correct the manuscript to eliminate typing errors and misspellings. It helps to have someone else proofread, too.

BIBLIOGRAPHY

Council of Biology Editors Style Manual Committee. 1978. *Council of Biology Editors Style Manual*. 4th ed. Council of Biology Editors and American Institute of Biological Science, Arlington, Virginia.

Strunk, William, Jr., and E. B. White. 1979. *The Elements of Style*. 3rd ed. Macmillan, New York.

University of Chicago Press. 1969. *A Manual of Style*. 12th ed. Univ. Chicago Press, Chicago.

Appendix 4
STATISTICAL ANALYSIS

INTRODUCTION

A knowledge of experimental design and statistical analysis is essential for an ecologist or any scientist. This appendix cannot substitute for a statistics course; its aim is to give some understanding of the statistical procedures called for in the exercises. Although the appendix is mainly a reference, it is designed so that it can be used as an exercise if this is desired.

Generally scientific studies deal with *samples*. Samples are parts, or subsets, of *populations*. We may wish to compare the beak length of some species of bird on the island of Bermuda with its beak length on the mainland. The beak lengths of all the birds of this species on Bermuda and on the mainland form the populations.* Ordinarily it will be too time consuming or too expensive (or impossible) to catch and measure every individual. Instead, we sample; we catch a fraction of the population, preferably randomly, and reason from this sample to the whole population. (Some principles of sampling are dealt with in Exercise 5.) Values calculated from samples are called *statistics;* they estimate corresponding traits of populations, which are called *parameters*. For example, the mean beak length of our sample of Bermudan birds is a statistic; the actual mean beak length of all the individuals is a parameter.

In using samples we need to interpret them properly, which is the job of statistical analysis. We also need to gather them in such a way that they can be sensibly interpreted; assuring this is the job of experimental design. The major aim in designing our data gathering is to eliminate variables other than the one we are interested in. It would not do, for example, to have mostly immature female birds from Bermuda and adult male birds from the mainland (unless we knew beforehand that there is no difference in beak length with sex or age). Experimental design, in short, is the way we protect ourselves against the criticism that something else instead of the factor we are studying produces the effects we find. It is primarily a matter of biological knowledge of our subject, so as to anticipate likely problems, and of logical thought, so as to allow for them.

Measures of Central Tendency and Spread

Suppose we wish to describe some sample quantitatively. Just as a way of getting some figures to work with, members of the class might run in place for a minute and then take their pulse for ten seconds.† Multiply the ten-second rate by six to express pulse rate as beats per minute. Each estimate of pulse rate is an observation (designated by x). A set of observations might look like this: 76, 81, 87, 88, 91, 93, 93, 95, 98, 101.

How can we treat this data? We can, to begin with, tabulate it, and learn quite a bit by looking at the table. Thus:

Class	Number in Each Class
75.0–79.9	1
80.0–84.9	1
85.0–89.9	2
90.0–94.9	3
95.0–99.9	2
100.0–104.9	1

*Note that *population* in the statistical sense refers to a set of data—beak lengths in this case. The birds themselves form a biological population—organisms belonging to the same species and occurring in the same area at the same time.

†If this appendix is used as a class exercise it would be desirable to record smokers and non-smokers separately and use both sets of data for computations up through the t test.

We could then treat the data graphically and see most of the same things slightly more clearly. If we use bars to show the number of observations in each class, such a graph is called a *histogram*. Thus:

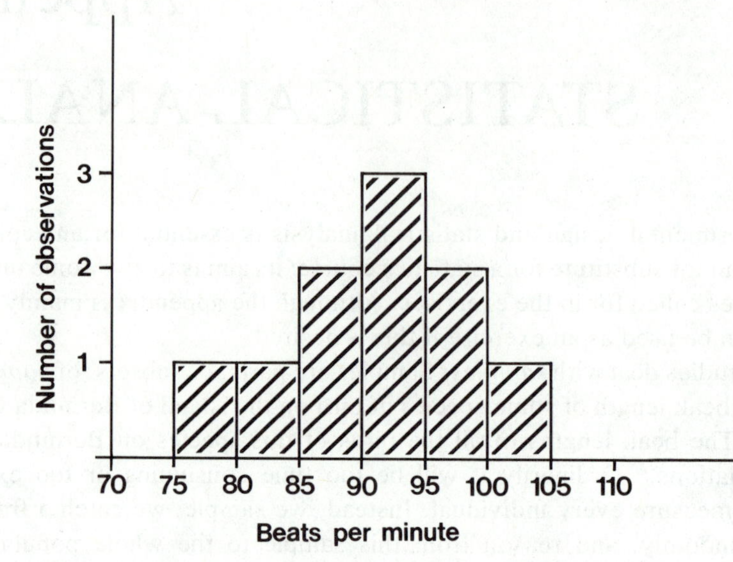

If, rather than using bars, we drew a smooth curve through the data we would have an approximation of the bell-shaped, normal curve that most biological variables tend to show.

Graphs can be helpful but it is often more useful to be able to express the features that we can see in the graph as numbers. One of the features that we will want to specify is the middle of the curve—what is an average observation in this sample? The most widely used average (or *measure of central tendency,* to use the statisticians' phrase) is the *arithmetic mean*. It is most peoples' idea of an average, computed by adding up all of the observations and dividing by how many of them there are. In mathematical symbols this is

$$\bar{x} = \frac{\Sigma x}{N}$$

which is read "x bar is equal to the sum of the x's divided by N" (Σ, which stands for "the sum of," is a capital sigma). The mean ($\bar{x}$) of the sample of heart rates is 90.3.

Another measure of central tendency, often a more useful statistic than the mean because it is less affected by oddities, is the *median*. The median is the middle observation, the observation above and below which half the observations lie. You find it by putting the observations in order from low to high and then counting up or down until you come to the middle one. If there were 11 observations the median would be observation 6. Since there are 10, an even number, in our sample, there is no one middle observation. Observations 5 and 6 bracket where the median would be, so we add these together and divide by 2. In this case the median is (91+93)/2, or 92.

Another measure of central tendency is the *mode,* defined as the most common observation. Here the mode is 93.

The mean, median, and mode give us information about the midpoint of the data. Another interesting feature is the *spread* of data around the midpoint. The simplest measure of spread, or *variability,* is the *range*. The range is the largest value minus the smallest or, more often, it is given by simply stating the two values rather than carrying out the subtraction. The range for the set of data we are using is 76 − 105 or 29. The range is not a particularly valuable statistic because it depends strongly on sample size. If we had 100 observations instead of 10, the range would probably extend quite a bit further in each direction.

The usual measure of spread is the *variance* (S^2) or its square root, the *standard deviation* (S). If our sample corresponds reasonably closely to a normal curve the standard deviation tells us this: From 1 standard deviation below the mean to 1 standard deviation above the mean will be included 68% of all the observations. This is illustrated in the graph below. Likewise, the mean ± 2 standard deviations includes 95% of the observations and the mean ± 3 standard deviations includes 99.7% of the observations.

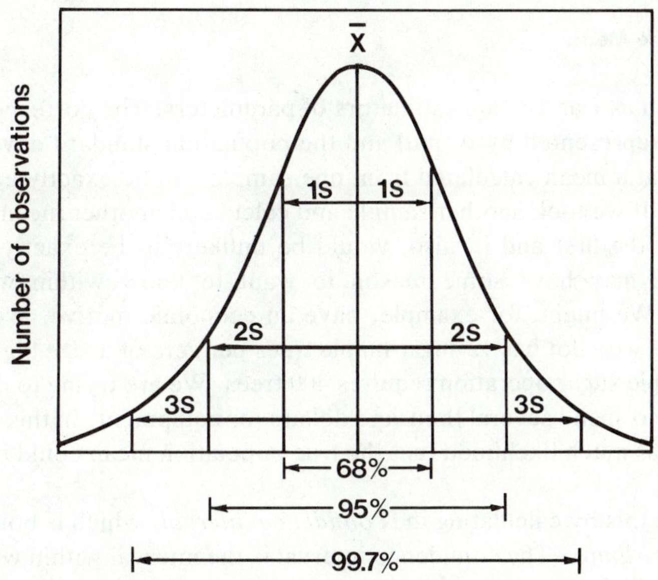

The basic calculation of the standard deviation is as follows for a very simple example in which the sample consists of three observations: 1, 2, and 3.

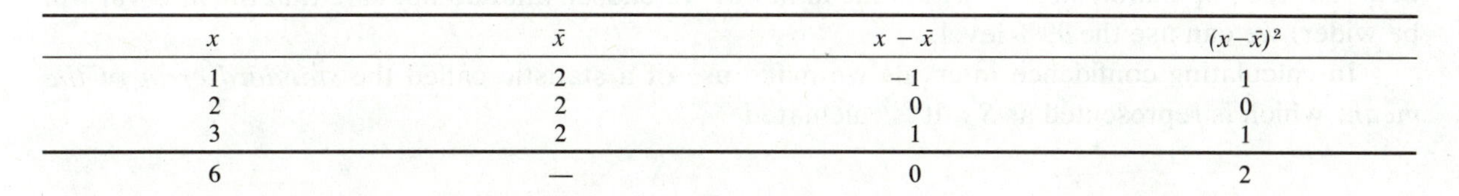

x	$\bar{x}$	$x-\bar{x}$	$(x-\bar{x})^2$
1	2	−1	1
2	2	0	0
3	2	1	1
6	—	0	2

The variance (S^2) is calculated as

$$S^2 = \frac{\Sigma(x-\bar{x})^2}{N-1}.$$

In this case, $\Sigma(x-\bar{x})^2$ is 2; $(N-1)$ is 2; and S^2, accordingly, is 2/2 or 1.

As can be seen, the variance is approximately the average squared deviation from the mean. ($\Sigma(x-\bar{x})^2/(N-1)$ has the form we used for calculating a mean, except that $N-1$ rather than N is used. The reason for using one less than the sample size can be pursued in any statistics text.) Note that the variance is calculated by squaring the deviations from the mean. To obtain a statistic that is in the same units as the mean, we must take the square root of the variance and obtain the standard deviation, thus:

$$S = \sqrt{S^2}.$$

In using calculators it is generally simpler to use the following formula for computing the variance instead of the one shown earlier

$$S^2 = \frac{\Sigma x^2 - (\Sigma x)^2/N}{N-1}.$$

This is mathematically equivalent and slightly easier to use. Many calculators now have automatic computation of mean, variance, and standard deviation that require you to do little more than enter the individual observations.

Confidence Limits for the Mean

Our statistics, such as $\bar{x}$ and S, are estimators of parameters. The corresponding parameters are the population mean (represented by μ, mu) and the population standard deviation (σ, sigma). It is unlikely, of course, that a mean calculated from one sample will be exactly equal to the population mean, i.e., that $\bar{x} = \mu$. If we took another sample and calculated another mean, it would probably be slightly different from the first and it, also, would be unlikely to be exactly the same as the true population mean. We may have some reason to want to know within what interval the true population mean lies. We might, for example, have an economic motive. We have determined by sampling that a 40-acre woodlot has 12 sugar maple trees per acre of a size big enough to tap. A successful small-scale maple sugar operation requires 400 trees. We are trying to decide for ourselves or someone else whether to spend several thousand dollars for equipment. In this case we might want to know whether there was much likelihood that the true population mean could be lower than ten trees per acre.

We can determine this by calculating the *confidence interval,* which is bounded by high and low values called *confidence limits*. The confidence interval is the interval within which we are confident, with a specified probability, that the population value (parameter) lies. Usually the 95% confidence interval is chosen but if we want a narrower interval (with, of course, a greater chance that the population mean will fall outside the limits) we could use the 90% level or if we want to be very confident that the population mean is within the limits we've chosen (and do not care that the interval will be wider) we can use the 99% level.

In calculating confidence intervals we make use of a statistic called the *standard error of the mean,* which is represented as $S_{\bar{x}}$. It is calculated

$$S_{\bar{x}} = \frac{S}{\sqrt{N}}.$$

It may be useful to some students to realize that $S_{\bar{x}}$ is an estimate of the standard deviation of a sample of means. That is, we could draw a bunch of samples from one population; we could calculate a mean for each and use those means as individual observations in drawing a normal curve. $S_{\bar{x}}$ is our estimate of the standard deviation that this curve would have.

The true population mean (μ) should lie within $\bar{x} \pm 2\, S_{\bar{x}}$ 95% of the time. Suppose we have a sample such that $N = 81, \bar{x} = 20, S = 4.50$. $S_{\bar{x}}$ is $4.50/\sqrt{81}$ or 0.50. The lower confidence limit then will be $20-(2 \times 0.50)$ or 19 and the upper confidence limit will be $20 + (2 \times 0.50)$ or 21. The 95% confidence interval is 19 to 21; we may also state it as 20 ± 1.00.

To show the process more generally (for other probabilities), the confidence interval for the mean is

$$\bar{x} \pm t \times S_{\bar{x}}.$$

The t value corresponds to the 2 in the example but will vary according to sample size and the degree of confidence desired. The t value is looked up in a t table (such as Appendix Table 4–1). For example, if your sample size is 11 and you want to calculate the 90% confidence interval you enter the table at 10 on the side ($n = N - 1$) and at 0.10 (that is, $1.00-0.90$) at the top. The table value is 1.812.

In the sample above we can say that we are 95% confident that the population mean lies between 19 and 21. This means if we do this same sort of thing 100 times we will include the population mean in the computed interval 95 times (on the average). Five times out of a hundred it will be outside the interval.

APPENDIX TABLE 4–1. Critical Values of the *t* Distribution

n	Probability That *t* Will Be Exceeded (α)			
	0.10	**0.05**	**0.02**	**0.01**
1	6.314	12.706	31.821	63.657
2	2.920	4.303	6.965	9.925
3	2.353	3.182	4.541	5.841
4	2.132	2.776	3.747	4.604
5	2.015	2.571	3.365	4.032
6	1.943	2.447	3.143	3.707
7	1.895	2.365	2.998	3.499
8	1.860	2.306	2.896	3.355
9	1.833	2.262	2.821	3.250
10	1.812	2.228	2.764	3.169
11	1.796	2.201	2.718	3.106
12	1.782	2.179	2.681	3.055
13	1.771	2.160	2.650	3.012
14	1.761	2.145	2.624	2.977
15	1.753	2.131	2.602	2.947
16	1.746	2.120	2.583	2.921
17	1.740	2.110	2.567	2.898
18	1.734	2.101	2.552	2.878
19	1.729	2.093	2.539	2.861
20	1.725	2.086	2.528	2.845
21	1.721	2.080	2.518	2.831
22	1.717	2.074	2.508	2.819
23	1.714	2.069	2.500	2.807
24	1.711	2.064	2.492	2.797
25	1.708	2.060	2.485	2.787
26	1.706	2.056	2.479	2.779
27	1.703	2.052	2.473	2.771
28	1.701	2.048	2.467	2.763
29	1.699	2.045	2.462	2.756
30	1.697	2.042	2.457	2.750
40	1.684	2.021	2.423	2.704
60	1.671	2.000	2.390	2.660
120	1.658	1.980	2.358	2.617
∞	1.645	1.960	2.326	2.576

Condensed from *Biometrika Tables for Statisticians,* 1966, Cambridge University Press. Used by permission of the *Biometrika* Trustees.

Comparing Two Means (Student's *t* Test)

Suppose we wish to see whether heart rate after exercise is different in smokers compared with non-smokers. Or, suppose we sample the biomass of a grassland during a wet year and a dry year and want to know whether the drop we found is enough that we can conclude that it is real, that it is statistically significant. These questions involve deciding whether the two means are the same or different. We could get a general idea by calculating confidence intervals. If the mean for the dry year lay outside the 95% confidence interval for the wet year and vice versa, this would be a pretty good indication that they were not just two samples representing the same population. The actual test we use to test the significance of a difference between two means, the *t* test, formalizes this general approach.

Our statistical hypothesis in this case is that $\bar{x}_1 = \bar{x}_2$. We calculate t as

$$t = \frac{(\bar{x}_1 - \bar{x}_2)}{S_d} .$$

S_d is the standard error of the difference (between the two means). It is approximately the average of the standard errors that we would calculate from the two separate samples.

If we are starting from scratch, it is simplest to calculate S_d in two steps. First we calculate a value called the *pooled variance*. It is

$$S_p^2 = \frac{\left[\Sigma(x_1^2) - \frac{(\Sigma x_1)^2}{N_1}\right] + \left[\Sigma(x_2^2) - \frac{(\Sigma x_2)^2}{N_2}\right]}{(N_1-1)+(N_2-1)}.$$

From this we calculate S_d

$$S_d = \sqrt{S_p^2\left(\frac{1}{N_1} + \frac{1}{N_2}\right)}.$$

We then plug this into our original formula to calculate t.

We will go through a t test and illustrate the steps that will be used in testing any statistical hypothesis. We have samples such that $N_1 = 10$, $N_2 = 15$ and $\bar{x}_1 = 3.5$ and $\bar{x}_2 = 4.5$. Assume that the pooled variance, S_p^2, has already been calculated and is 20.

1. We first state the hypothesis that we wish to test. In this case, it is $\bar{x}_1 = \bar{x}_2$. (Statistical hypotheses are always of the *null* form: that no difference exists between the two values being considered.)

2. We then select the level of significance (often indicated by α, alpha). This is a probability level. Specifically, it is the probability of rejecting the hypothesis and thus concluding that the means are different when they are actually equivalent (the null hypothesis true). Ordinarily a significance level of 5% or 1% is selected. We will use 5% for this example.

3. We then compute the statistics.

4. We accept or reject the hypothesis.

In this case,

$$S_d = \sqrt{20\ (1/10+1/15)}.$$

Carrying out the arithmetic, this is $\sqrt{3.33}$ or 1.826. We calculate t as (4.5−3.5)/1.826, or 0.548.

To accept or reject the hypothesis we consult the t table (Appendix Table 4–1) and compare our calculated value of t with the table value. We enter the t table with n equal to $[(N_1-1) + (N_2-1)]$ or in this case 23. In this situation, this value is referred to as degrees of freedom. From the other direction we enter the table at an alpha of 0.05. The table value of t, then, with 23 degrees of freedom and an alpha of 0.05 is 2.069. We compare computed value with table value and accept the hypothesis if the computed value is smaller. We reject the hypothesis if the computed value is larger. In this case, our computed value of 0.548 is smaller than the table value of 2.069. We cannot reject the hypothesis; this large a difference between means could occur more often than five times in a hundred if the samples were actually drawn from the same population.

Chi Square (χ^2) Test of Goodness of Fit

Biologists fairly often need to decide whether some distribution of results conforms to some theoretical or expected distribution. Genetics students learn to do χ^2 tests to decide whether the corn or beans or fruit flies they study show a 3:1 ratio. One way an ecologist might use such a test is to determine whether a series of individual isopods (for example) when put in an experimental chamber with a dark end and a light end choose one over the other. If they have no preference between dark and light, logically they should go to one end about as often as the other; half the observations should be in the dark end and half in the light. Our hypothesis will be that the observed distribution is the same as this expected distribution. χ^2 is given by

$$\chi^2 = \Sigma\ \frac{(O-E)^2}{E}$$

where O is the observed number (not percentage) and E the expected number.

It is simplest to compute χ^2 if we set up a table such as shown in the example below. Suppose we had, in 75 trials, 50 isopods at the dark end and 25 at the light end. The expected result (assuming no preference) would be 37.5:37.5. Our hypothesis is that the observed and expected ratios are the same. Before computing χ^2 we should choose an alpha which, as with the *t* test, will ordinarily be 5% or 1%. Let us again use 5%.

Categories	Observed (O)	Expected (*E*)	(O−*E*)	$(O-E)^2$	$(O-E)^2/E$
Dark	50	37.5	12.5	156.25	4.17
Light	25	37.5	−12.5	156.25	4.17
Totals	**75**	**75**	—	—	**8.34**

χ^2 thus is 8.34. To accept or reject the hypothesis we consult the χ^2 table (Appendix Table 4–2). We use it basically in the same way as the *t* table; here, however, degrees of freedom (down the side of the table) are equal to the number of categories minus 1. In this case there are two categories (dark and light), so there is one degree of freedom. Entering the table with one degree of freedom and an alpha of 0.05, we find a value of 3.84. Our calculated value is greater than this, so we reject the hypothesis. We would get this big a difference between our distribution and a 1:1 ratio fewer than five times out of a hundred if only chance were involved. We think that this is unlikely enough to say that our results show a significant difference (at the 5% level). We are, in effect, affirming that isopods select a darker rather than a brighter site under the conditions we provided.

Calculating χ^2 from a 2 × 2 contingency table is described in Exercise 6.

APPENDIX TABLE 4–2. Critical Values of the χ^2 Distribution

Degrees of Freedom	Probability That Chi-square Value Will Be Exceeded (α)			
	0.100	0.050	0.010	0.005
1	2.71	3.84	6.63	7.88
2	4.61	5.99	9.21	10.60
3	6.25	7.81	11.34	12.84
4	7.78	9.49	13.28	14.86
5	9.24	11.07	15.09	16.75
6	10.64	12.50	16.81	18.55
7	12.02	14.07	18.48	20.23
8	13.36	15.51	20.09	21.96
9	14.68	16.92	21.67	23.59
10	15.99	18.31	23.21	25.19
11	17.28	19.63	24.72	26.76
12	18.55	21.03	26.22	28.30
13	19.81	22.36	27.69	29.82
14	21.06	23.68	29.14	31.32
15	22.31	25.00	30.58	32.80
16	23.54	26.30	32.00	34.27
17	24.77	27.59	33.41	35.72
18	25.99	28.87	34.81	37.16
19	27.20	30.14	36.19	38.58
20	28.41	31.41	37.57	40.00
30	40.26	43.77	50.89	53.67
40	51.80	55.76	63.69	66.77
50	63.17	67.50	76.15	79.49
60	74.40	79.08	83.38	91.95
70	85.53	90.53	100.4	104.22
80	98.58	101.9	112.3	116.32
90	107.6	113.1	124.1	123.3
100	118.5	124.3	135.3	140.2

Condensed from *Biometrika Tables for Statisticians,* 1966, Cambridge University Press. Used by permission of the *Biometrika* Trustees.

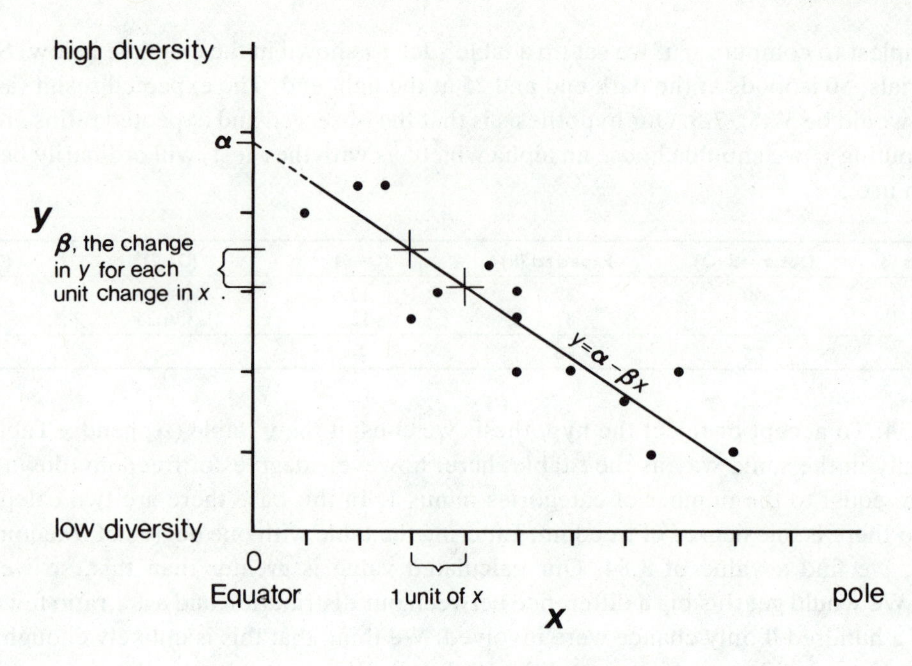

Linear Regression

We sometimes wish to describe the relationship between one variable and another. For example, we may wish to relate diversity to latitude and see if it really decreases toward the poles. We may plot the data on a graph and get a bunch of dots (as shown above) corresponding with some measure of diversity for various latitudes. We might then have a good look at the graph and draw a line through the points that fits the data pretty well. We would then be able to predict what diversity ought to be at some latitude where we had no points and we would also be able to specify the rate at which diversity changed with latitude (this would be the slope of the line). If the data can be described by a straight line, we can use a statistical technique that gives us an objective rather than subjective (eyeball) way to draw such a line.

The equation for a straight line (any straight line) is

$$y = \alpha \pm \beta x$$

where y is the dependent variable (diversity in the example), x the independent variable (latitude), α the origin of the line (the value that y would have if x were zero), and β the slope of the line (that is, the change in y for each unit change in x). What we need then are values for α and β.

The method statisticians use for calculating the values to draw this regression line is called the *least squares method*. It produces the line that best fits the data in the sense that the (squared) deviations of the individual data points from the line are minimized. The two values, α and β, are calculated as follows:

$$\beta = \frac{\Sigma x_i y_i - \dfrac{(\Sigma x_i)(\Sigma y_i)}{N}}{\Sigma x_i^2 - \dfrac{(\Sigma x_i)^2}{N}}$$

$$\alpha = \bar{y} - \beta\bar{x}$$

To obtain α, then, we first calculate β and also the averages (mean values) of y and x and use these three values to solve for α. Solving for β and α is simplified by constructing a table thus (for some simple, made-up data):

x_i	y_i	x_i^2	x_iy_i
1	2	1	2
2	4	4	8
3	6	9	18
4	8	16	32
$\Sigma x_i = 10$	$\Sigma y_i = 20$	$\Sigma x_i^2 = 30$	$\Sigma x_i y_i = 60$
$\bar{x} = 2.5$	$\bar{y} = 5$		
$(\Sigma x_i)^2 = 100$			

$$\beta = \frac{60 - 50}{30 - 25} = \frac{10}{5} = 2$$

$$\alpha = 5 - (2 \times 2.5) = 0$$

In this example, β, the slope of the line, is 2; in other words, y changes 2 units for every unit change in x. And α is zero; that is, when $x = 0$, y, also, is zero.

To predict some value of y we use the formula for a straight line. For example, we might wish to predict a diversity value for Hawaii using mainland data and then assemble the Hawaiian data and see if it conforms to the prediction (which, as an island, it probably would not).

$$\hat{y} = \alpha + \beta x$$

$\hat{y}$ will be our estimate of y for a specified x. In the example given, we could estimate y for $x=20$, thus:

$$\hat{y} = 0 + (2 \times 20) = 40.$$

Other Statistical Procedures

We have in the foregoing considered some of the important elementary statistical methods useful to ecologists. There are, however, many other useful procedures, some simple and some so complex that only the availability of high-speed computers makes their general use feasible. We mention below a few additional methods, without going into the details of calculation.

Analysis of Variance (ANOVA). We sometimes want to see whether any of several different factors influence a variable. For example, in agriculture (where much of the early development of statistics occurred) someone may wish to see whether wheat yield varies with fertilizer, strain of plant, and herbicide. By an appropriate experimental design, it can be determined in one analysis of variance test whether any of these (alone or in interaction) caused significant differences. The advantage of this procedure is in reducing the number of tests we need to perform. This is, in a sense, labor-saving but there is a more fundamental reason for preferring to do one analysis of variance rather than 20 t tests. Recall that the probability that we specify in performing a test (such as a t test or a chi-square test) is the probability of rejecting a true hypothesis. Using an alpha of 5% we say, in effect, that if we perform 100 tests leading to rejecting a hypothesis, on five of those occasions we will expect to make the mistake of rejecting a hypothesis that is actually true. If we can reduce the number of tests we must perform we can reduce the number of times we will make such mistakes.

Non-parametric Tests. Many of the foregoing procedures depend on assumptions about the parameters of the population from which we are drawing samples; specifically, most assume a normal distribution. Certain statistical procedures make no such assumptions. The chi-square test is one such procedure. As general rules, non-parametric tests can be used in situations where we cannot assume that the population is normally distributed and they use easier arithmetic than parametric statistics. Their disadvantage is that they tend to be less powerful; that is, a larger sample size is needed to reject a false hypothesis. Some widely used non-parametric tests, in addition to the chi-square test, are the sign test, the rank sum test, the Mann-Whitney U test, and the runs test.

PROBLEMS

1. A randomly chosen sample of annual ragweed plants from a bare area in the first year of succession had the following heights (cm): 40, 46, 58, 50, 85, 46, 49, 44, 53, 51, 38.

a. Prepare a table grouping the data into height classes and from this prepare a histogram of the data.

b. Give the mean, mode, and median.

c. Give the variance and standard deviation.

d. Give the 90% confidence limits for the mean.

2. A randomly chosen sample of annual ragweed plants from the same area in the second year of succession had the following heights (cm): 40, 28, 52, 33, 39, 27, 31, 32, 21, 31. You wish to determine whether mean height has changed from the first to the second year. Perform the appropriate test.

3. In one area white-tailed deer does produced 42 single fawns and 6 sets of twins. In another area, the numbers were 83 singles and 20 sets of twins. Was there any difference between the areas in the tendency to twin?

4. The intrinsic rate of natural increase (expressed as r per week) of an insect varied with relative humidity as follows: 30% *RH*—0.30; 40%—0.31; 50%—0.42; 60%—0.51; 70%—0.76.

a. Calculate α and β.

b. Calculate r for a humidity of 90% assuming the same trend held.

c. What humidity marks the lower limit of permanent occurrence for the insect (that is, at what relative humidity does $r = 0$)?

d. Draw the straight-line graph relating r to humidity.

BIBLIOGRAPHY

Fisher, R. A., and F. Yates. 1957. *Statistical Tables for Biological, Agricultural, and Medical Research*. Hafner, New York.

Harnett, D. L. 1975. *Introduction to Statistical Methods*. Addison-Wesley, Reading, Massachusetts.

Hope, K. 1968. *Methods of Multivariate Analysis*. Univ. London Press, London.

Kempthorne, O., and L. Folks. 1971. *Probability, Statistics, and Data Analysis*. Iowa State Univ. Press, Ames.

Moroney, M. J. 1956. *Facts from Figures*. Penguin Books, Baltimore.

Odel, R. E., D. B. Owen, Z. W. Birnbaum, and L. Fisher. 1977. *Pocket Book of Statistical Tables*. Marcel Dekker, New York.

Siegel, S. 1956. *Nonparametric Statistics for the Behavioral Sciences*. McGraw-Hill, New York.

Simpson, G. G., A. Roe, and R. C. Lewontin. 1960. *Quantitative Zoology*. Rev. ed. Harcourt, Brace, New York.

Zar, J. H. 1974. *Biostatistical Methods*. Prentice-Hall, Englewood Cliffs, New Jersey.

Appendix 5

COMPUTER USAGE AND SIMULATION

Computers are an important enough tool in ecology that some exposure to their use is desirable. Circumstances for different classes and schools differ and the field is changing rapidly so that this section's aim is simply to suggest some general guidelines.

The level at which computer usage is approached in an ecology course can vary anywhere from a demonstration of some of the potentialities of the machine to asking the class to learn a computer language during the first week or two and then using the computer in every exercise thereafter. We suspect that instructors will make increasing use of computers in the future and that the trend will be strongly toward the powerful new microcomputers, such as the TRS-80 from Radio Shack, Heathkit's Z89/H89, and the Apple II and III with their remarkable graphics capabilities. Instructors will increasingly find that students—but not all of them—already have experience with computers, often beginning in high school or earlier.

As a beginning computer language, our preference is for BASIC. The language is simple enough that a student can write useful programs after only an hour or two of instruction. Pascal has its advocates as an introductory language for those who intend to delve more deeply into computer science and, of course, FORTRAN remains useful for long, complex programs and those where large amounts of data are to be processed.

There are three general categories of computer use that ecology students should know. First, they should be aware of the great potential for speeding up data handling through the use of programs specifically written for data of the sort routinely gathered in ecological studies. The program LIFE, included in this Appendix, is an example of such a program. Having students work up one of the life tables for Exercise 10 using calculators and then use LIFE for the other three will give a good idea of the number-crunching capability of computers.

Secondly, students should be aware of the existence of standard library programs. These are of particular interest for statistical analysis; programs are available for virtually any statistical procedure from calculating means to very complex multivariate procedures. Many other library programs also exist for such purposes as sorting, mapping, editing and printing manuscripts and also, in some cases, for analyzing specifically ecological data as in the first category and for modeling ecological situations, discussed below. Library programs are generally available in great variety at college or university computer centers; lists and directions for use are usually available. The number of library programs for microcomputers is smaller but increasing.

The third category is ecological modeling. A *model* is a simplified representation of a real system. A *system* is any unit with interacting components. A system may be a part of a bigger system and may contain smaller systems. A population, for example, is part of an ecosystem and is composed of organisms (which are composed of organ systems). In most modeling, an ecological situation—for example, a population, two competing populations, or an ecosystem—is represented mathematically and the outcome is investigated with time or when parameters of the system are changed.

Models may be simple or complex. If they are simple enough, we may not need a computer to work with them but, even then, the speed of the computer may make its use desirable. The logistic is a simple model of population growth, consisting of a single equation. If we wanted to look at the effect on the shape of the growth curve of varying r_m while keeping K constant, we could generate the necessary values in a dull hour or so at a desk calculator. The program GROW, given here, will do the same thing or evaluate any combination of r_m and K about as quickly as the parameters can be typed in.

Where the computer shines, though, is in modeling complex situations where we can specify something of the behavior of the components but find it impossible to develop analytical solutions to our questions about the system. The kind of modeling involved in such cases is generally referred to as *systems analysis* or *simulation*.

We may wish, for example, to be able to specify the effect on productivity of a grassland resulting from several years of cool summers. Or we may wish to know the effect on soil humus levels in cornfields of removing the cornstalks for alcohol production. The systems analyst would specify the pertinent variables within the system, such as plant biomass, herbivore biomass, litter, and soil organic matter, and pertinent variables outside the system such as solar energy, temperature, and rainfall. He or she would specify quantitatively the effects of each variable on the others. The end result would be a series of arithmetic operations that, when translated into a computer language and carried out, would give biomass production or amount of humus at the end of each time period having the specified conditions.

Systems analysis, and modeling in general, can have several aims. Ecologists often profit just by constructing models. We are forced to clarify our thinking on how the components of the system act and interact; often, it becomes clear that more research has to be done before a reasonably realistic model can be produced. The basic aim of modeling, though, is to produce predictions; these may be used in several ways. If the predicted outcome of the model, when tested in nature, is wrong, this indicates that the model is incorrect or incomplete. We then try to improve our understanding of the behavior of the system's components and test the predictions of an improved version of the model.

Models that, within the limits for which they are intended, are accurate representations of reality can be used to make predictions that tell us what would happen if conditions were actually changed in certain ways. This second predictive use of models, in which we explore the consequences of altering variables, is referred to as *simulation analysis*. Such operations are, in a sense, experiments done on the computer rather than in the real world. They can tell us, accurately insofar as our models are correct and complete, what will happen if a short grass prairie encounters ten years of cold summers, even though it may be quite outside our abilities to produce ten cold summers on a study plot. The predictions can be used in the management of the system. A computer simulation might tell us, for example, that allowing the hunting of does would improve deer habitat and not lower the shootable surplus. A management recommendation to allow doe hunting might then be in order. In the real world, of course, we may find then that there are adverse affects the model did not anticipate. This would improve our knowledge of the deer herd, while generating some ire on the part of the hunters. On the other hand, we might find that even though our prediction was perfectly correct, the hunters simply would not shoot does, preferring to follow their own intuition that killing mommies is bound to lead to lower populations.

The three programs given in this appendix were designed to be used with exercises in this book. They are written in a minimum BASIC that is usable with only minor modifications on most systems, including microcomputers. They will give the student some idea of the use of computers in data handling. GROW can also function as a model of population growth and CHANGE as a model of plant succession.

More complicated simulations could also be included. One place where these could be used would be in a lab on interspecific competition. This manual contains no exercise that deals directly with this topic. The instructor might consider using COEXIST from CONDUIT or Two Populations from COMPress. Sources of these and other ecologically oriented programs are given in the next section.

Sources of Ecologically Oriented Programs

COMPress, Inc., P.O. Box 102, Wentworth, New Hampshire 03282.
Has programs for the APPLE II on the Lotka-Volterra competition and predation models, population growth, and energy.

CONDUIT, P.O. Box 388, Iowa City, Iowa 52244.
Has programs in FORTRAN or BASIC for larger systems plus APPLE II, PET, and TRS–80 on animal and plant competition, spacing of plants, the Leslie matrix, life table, and mark-and-recapture methods.

Creative Computing, P.O. Box 789–M, Morristown, New Jersey 07960.
Has programs for TRS–80 on population growth, pest control, managing a buffalo herd, the mark-and-recapture method, water pollution, controlling rat populations, and the world models used in *Limits To Growth* (Meadows et al. 1972).

Peters, James A. 1971. Biostatistical programs in the BASIC language for time-sharing computers. Coordinated with the book *Quantitative Zoology*. Smithsonian Contributions to Zoology no. 69: 1–46. Has programs on most elementary statistical procedures including *t* tests, regression, Poisson probabilities, partial correlation, ANOVA, χ^2, and allometry.

Three Sample Programs

LIFE

This is designed to be used with Exercise 10, Life Tables. Sample output is given below, followed by the program. Note that the sample output is for an organism with a shorter life span than humans (those with a sharp eye for vital statistics will recognize the Dall sheep).

SAMPLE OUTPUT

```
AT EACH QUESTION MARK TYPE THE LOWER AGE BOUND,
THE UPPER AGE BOUND, AND THE MORTALITY, SEPARATED
BY COMMAS. FOR THE UPPER AGE BOUND IN THE LAST 'AND
OVER' GROUP, TYPE A NEGATIVE NUMBER. THE PROGRAM WILL
TAKE THAT AS THE LAST POINT, AND REQUEST THE MIDPOINT
TO BE USED IN CALCULATIONS FOR THE LAST GROUP
AFTER YOU HAVE CORRECTED ANY ERRORS.

1 ?0,1,199
2 ?1,6,112
3 ?6,11,436
4 ?11,-1,96

DO YOU WISH TO CHECK OR CORRECT ANY POINT ?YES
WHAT POINT ?2
THIS POINT IS: 2 , 1 , 6 , 112
DO YOU WISH TO CHANGE IT ?YES
TYPE CORRECT VALUES: 2 ?1,6,113

DO YOU WISH TO CHECK OR CORRECT ANY POINT ?YES
WHAT POINT ?3
```

```
THIS POINT IS: 3 , 6 , 11 , 436
DO YOU WISH TO CHANGE IT ?YES
TYPE CORRECT VALUES: 3 ?6,11,592

DO YOU WISH TO CHECK OR CORRECT ANY POINT ?NO
WHAT IS THE MIDPOINT FOR THE LAST GROUP ?12.5

NOW CALCULATING TOTAL COHORT.

NOW CALCULATING SURVIVORSHIP, RATE OF MORTALITY, AND
AVERAGE SURVIVORS. DON'T GO AWAY!

NOW CALCULATING EXPECTATION OF FURTHER LIFE.


TOTAL COHORT = 1000


NUMBER OF AGE GROUPS = 4
```

X	1	L	L(%)	D	M	M/YR	E
0	1000	900	43.189	199	0.199	0.199	6.73
1	801	744	35.707	113	0.141	0.028	7.27
6	688	392	18.801	592	0.860	0.172	3.06
11	96	48	2.302	96	1.000	0.333	1.50

PROGRAM

```
100   REM*  PROGRAM:        LIFE TABLE
110   REM*  AUTHOR:         PHILIP M. BREWER
120   REM*  LANGUAGE:       BASIC
130   REM*  LAST REVISED:   3 SEPTEMBER 1980
140   REM
150   REM*  PURPOSE:
160   REM*     THIS PROGRAM CREATES A LIFE TABLE. IT REQUIRES
170   REM*  THE FOLLOWING INPUT:
180   REM*  ANY NUMBER OF LINES, EACH CONSISTING OF A LOWER AGE
190   REM*  BOUND, AN UPPER AGE BOUND, AND THE MORTALITY WITHIN
200   REM*  THAT AGE GROUP. THE OUTPUT CONSISTS OF THE TOTAL
210   REM*  COHORT (CALCULATED FROM THE MORTALITY INPUT), THE
220   REM*  NUMBER OF AGE GROUPS INPUT, AND A LIFE TABLE WITH:
230   REM*  AGE                                          (X)
240   REM*  NUMBER SURVIVING AT START OF PERIOD          (1)
250   REM*  NUMBER SURVIVING AT MIDPOINT OF PERIOD       (L)
260   REM*  PERCENT OF TOTAL SURVIVING AT MIDPOINT       (L(%))
270   REM*  NUMBER OF DEATHS DURING PERIOD               (D)
280   REM*  DEATH RATE DURING PERIOD                     (M)
290   REM*  ANNUALIZED DEATH RATE DURING PERIOD          (M/YR)
300   REM*  EXPECTATION OF FURTHER LIFE                  (E)
310   REM
312   REM
313   REM*  FOR M/YR TO BE CORRECT, AGE MUST BE
314   REM*  IN YEARS OR FRACTIONS THEREOF; EXCEPT FOR THIS
```

```
315     REM*        COLUMN, ANY UNITS CAN BE USED.
320     REM*        PART ONE — INTRO AND INPUT INSTRUCTIONS
340     PRINT       "AT EACH QUESTION MARK TYPE THE LOWER AGE BOUND,"
350     PRINT       "THE UPPER AGE BOUND, AND THE MORTALITY, SEPARATED"
360     PRINT       "BY COMMAS. FOR THE UPPER AGE BOUND IN THE LAST 'AND"
370     PRINT       "OVER' GROUP, TYPE A NEGATIVE NUMBER. THE PROGRAM WILL"
380     PRINT       "TAKE THAT AS THE LAST POINT, AND REQUEST THE MIDPOINT"
390     PRINT       "TO BE USED IN CALCULATIONS FOR THE LAST GROUP"
400     PRINT       "AFTER YOU HAVE CORRECTED ANY ERRORS."
1000    REM*        PART TWO — INITIALIZATION AND DATA INPUT
1010    REM
1020    K=1
1030    PRINT
1040    PRINT K;
1050    INPUT       F(K), H(K), D(K)
1060    IF H(K) < 0 THEN 2000
1070    K=K+1
1080    GOTO 1040
2000    REM*        PART THREE — CHECK AND CORRECT DATA
2020    PRINT
2030    PRINT       "DO YOU WISH TO CHECK OR CORRECT ANY POINT";
2040    INPUT       A$
2050    IF A$<>"YES" THEN 3000
2060    PRINT       "WHAT POINT";
2070    INPUT       N1
2080    PRINT       "THIS POINT IS:",N1; ","; F(N1); ","; H(N1); ","; D(N1)
2090    PRINT       "DO YOU WISH TO CHANGE IT";
2100    INPUT       A$
2110    IF A$ <>"YES" THEN 2000
2120    PRINT       "TYPE CORRECT VALUES:";N1;
2130    INPUT       F(N1),H(N1),D(N1)
2135    GOTO        2000
2140    REM
3000    REM*        MAIN CALCULATION SECTION
3010    REM
3020    REM*        NOTE THAT AT THIS POINT K IS THE TOTAL NUMBER OF POINTS
3030    REM*        READ IN. LET'S CALL IT T FOR TOTAL.
3040    REM
3050    T=K
3060    REM*        WHILE WE'RE HERE, LET'S GET THE MIDPOINT OF THE
3070    REM*        LAST SECTION, AND FIGURE OUT WHAT OUR UPPER BOUND
3080    REM*        WOULD HAVE BEEN.
3090    REM
3100    PRINT       "WHAT IS THE MIDPOINT FOR THE LAST GROUP";
3110    INPUT       M1
3120    LET         M1=M1-F(T)
3130    H(T)=F(T)+2*M1
3140    REM
3150    REM*        NOW WE WANT TO GET THE TOTAL COHORT. WE DO THIS BY
3160    REM*        SUMMING D FOR ALL GROUPS.
411     DIM         F(90), H(90), L(90), D(90), M(90), E(90)
412     DIM         L2(90), L3(90), M3(90)
```

```
3170    REM
3180    PRINT
3190    PRINT   "NOW CALCULATING TOTAL COHORT."
3200    REM
3210    FOR     I=1 TO T
3220                LET S=S+D(I)
3230    NEXT I
3240    REM
3250    REM*    S NOW EQUALS THE TOTAL COHORT. WE ARE READY TO
3260    REM*    CALCULATE THE LIFE TABLE. FIRST WE CALCULATE THE
3270    REM*    VALUES FOR SURVIVORSHIP AND MORTALITY. WE HAVE TO WAIT
3280    REM*    ON THE CALCULATION FOR EXPECTATION OF FURTHER LIFE
3290    REM*    BECAUSE WE NEED TO HAVE THE WHOLE TABLE FILLED IN
3300    REM*    FOR THAT.
3310    REM
3320    REM*    ALSO, AT THE LAST MINUTE WE'VE BEEN ASKED TO CALCULATE
3330    REM*    THE AVERAGE SURVIVORSHIP AS BOTH AN ABSOLUTE NUMBER
3340    REM*    AND AS A PERCENTAGE OF THE COHORT FOR EACH GROUP.
3350    REM*    LET'S JUST SLIP THAT IN HERE, WHERE NO ONE WILL EVER
3360    REM*    SUSPECT.
3370    REM
3375    PRINT
3380    PRINT   "NOW CALCULATING SURVIVORSHIP, RATE OF MORTALITY, AND"
3390    PRINT   "AVERAGE SURVIVORS. DON'T GO AWAY!"
3400    LET K=1
3410    LET L(K)=S
3420    LET M(K)=D(K)/L(K)
3430    FOR K=2  TO T
3440            LET L(K)=L(K-1)-D(K-1)
3450            LET M(K)=D(K)/L(K)
3460    NEXT K
3470    REM
3480    FOR K=1  TO T
3490            LET L2(K)=(L(K)+L(K+1))/2
3500            LET L2=L2+L2(K)
3510    NEXT K
3520    REM
3530    FOR K=1  TO T
3540            LET L3(K)=L2(K)/L2*100
3550    NEXT K
3560    REM
3570    REM*    AH, NOW WE'RE READY TO FIND THE EXPECTATION OF FURTHER
3580    REM*    LIFE. THE STEPS IN THIS ARE:
3590    REM*    CALCULATE THE MIDPOINT BETWEEN THE LOWER AGE OF THE
3600    REM*    CURRENT GROUP AND THE UPPER AGES OF BOTH THE CURRENT
3610    REM*    GROUP AND EACH GROUP AFTER IT. THEN MULTIPLY THE NUMBER
3620    REM*    ALIVE IN EACH GROUP TIMES THE MIDPOINT OF THAT GROUP,
3630    REM*    AND SUM THEM TOGETHER. WHEN WE'VE DONE THAT FROM THE
3640    REM*    CURRENT AGE TO THE END OF THE TABLE, WE DIVIDE THE SUM
3650    REM*    BY THE NUMBER ALIVE AT THE BEGINNING OF THE CURRENT
3660    REM*    PERIOD AND THAT'S E FOR THIS GROUP. SOMEHOW, I SUSPECT
3670    REM*    THAT THE CODE IS LESS OPAQUE THAN THIS DOCUMENTATION.
```

```
3680    REM
3690    PRINT
3700    PRINT    "NOW CALCULATING EXPECTATION OF FURTHER LIFE."
3710    REM
3720    FOR K=1  TO T
3730             FOR I=K TO T
3740                       LET M=F(I)+(H(I)-F(I))/2
3750                       LET M2=M-F(K)
3760                       LET G=D(I)*M2
3770                       LET G2=G2+G
3780             NEXT I
3790             LET E(K)=G2/L(K)
3800             LET G2=0
3810    NEXT K
3820    FOR K=1  TO T
03830            LET M3(K)=M(K)/(H(K)-F(K))
03840   NEXT K
03850   REM
03860   REM*     WELL, THAT'S IT EXCEPT FOR PRINTING OUT THE TABLE.
03870   REM
04000   PRINT
04010   PRINT
04020   PRINT
04030   PRINT    "TOTAL COHORT =''; S
04040   PRINT
04050   PRINT    "NUMBER OF AGE GROUPS =''; T
04060   PRINT
04070   PRINT        X        1         L      L(%)        D          M      M/YR       E"
04080   PRINT    " ---   ------   ---------------   ------   ----------------   ------"
                   ######   ######   ###.###   ######   ###.###   ###.###   ###.##
04090   : ###
04100   FOR K=1 TO T
04110   PRINTUSING 4090, F(K),L(K),L2(K),L3(K),D(K),M(K),M3(K),E(K)
04120   NEXT K
04130   END
```

GROW

The second program is designed to be used with Exercise 11, Population Growth. It can be used for some of the data analysis and also for simple modeling purposes. In data analysis it applies particularly to questions 4 (note that if you have no data for a given day, such as 21, you can enter any value and the program will still compute the value to be expected under exponential growth), 6 and 8 (in both cases for producing the values to be expected from the logistic equation, given K and r_m). Sample output is followed by the program.

SAMPLE OUTPUT

```
HOW MANY TIMES WERE THE DUCKWEEDS COUNTED? 7
AT EACH QUESTION MARK TYPE THE COUNT DAY BEGINNING WITH
DAY 0, A COMMA, THEN THE MEAN NUMBER THAT DAY.
? 0, 3
? 2, 4
```

```
? 4, 5.5
? 7, 7.2
? 9, 8.3
? 11, 8.5
? 13, 8.6
TO START, R HAS BEEN CALCULATED FROM THE NUMBER ON THE
1ST AND 3RD COUNT ASSUMING EXPONENTIAL GROWTH DURING
THAT TIME. HERE'S HOW EXPONENTIAL GROWTH WITH R = .151534
PER DAY COMPARES TO THE ACTUAL NUMBER:

DAY        ACTUAL NO.     EXP.,R= .151534
 0          3              3
 2          4              4.06202
 4          5.5            5.49999
 7          7.2            8.66549
 9          8.3            11.7331
 11         8.5            15.8867
 13         8.6            21.5107
NOW ASSUME THAT GROWTH IS LOGISTIC. FOR EACH ROUND YOU
MAY TRY 4 PAIRS OF R AND K VALUES. THE PROGRAM WILL PRINT
THE NUMBER OF DUCKWEEDS FOR ALTERNATE DAYS FOR 4 WEEKS.
FIRST PAIR: R=? .15
            K=? 8.6
SECOND PAIR: R=? .15
             K=? 30
THIRD PAIR: R=? .3
            K=? 8.6
LAST PAIR: R=? .3
           K=? 45

            R = .15      R = .15      R = .3       R = .3
DAY         K = 8.6      K = 30       K = 8.6      K = 45
 0          3            3            3            3
 2          3.60911      3.91269      4.24807      5.18233
 4          4.24807      5.0511       5.50495      8.62612
 6          4.88934      6.43898      6.57212      13.578
 8          5.50495      8.08463      7.35458      19.8234
 10         6.07126      9.97284      7.86872      26.5171
 12         6.57212      12.0594      8.18265      32.549
 14         6.99992      14.2714      8.36583      37.192
 16         7.35458      16.5156      8.46988      40.3509
 18         7.64139      18.6933      8.5281       42.3238
 20         7.86872      20.717       8.56039      43.4908
 22         8.04604      22.5234      8.57822      44.159
 24         8.18265      24.0787      8.58803      44.5345
 26         8.28688      25.3769      8.59343      44.7433
 28         8.36583      26.4326      8.59639      44.8588

WOULD YOU LIKE TO TRY 4 MORE VALUES FOR R AND K?
IF SO, TYPE YES. IF NOT, THE PROGRAM WILL END.
?NO
```

PROGRAM

```
00010 REM   PROGRAM TITLE: GROW
00020 REM   SOURCE LANGUAGE: BASIC
00030 REM   AUTHOR: M. MCCANN
00040 REM   PURPOSE: THIS IS AN INTERACTIVE
00050 REM   PROGRAM TO GO WITH EXERCISE 11, POPULATION
00060 REM   GROWTH. R IS ROUGHLY ESTIMATED FROM CLASS
00070 REM   DATA THEN VARIOUS VALUES OF R AND K CAN BE
00080 REM   TRIED FOR COMPARISON.
00090 DIM T(30), N(30), E(30)
00100 PRINT "HOW MANY TIMES WERE THE DUCKWEEDS COUNTED";
00110 INPUT J
00130 PRINT "AT EACH QUESTION MARK TYPE THE COUNT DAY BEGINNING WITH"
00140 PRINT "DAY 0, A COMMA, THEN THE MEAN NUMBER THAT DAY."
00150 FOR I=1 TO J
00160 INPUT T(I), N(I)
00170 NEXT I
00180 R=(LOG(N(3)/N(1)))/(T(3)-T(1))
00190 PRINT "TO START, R HAS BEEN CALCULATED FROM THE NUMBER ON THE"
00200 PRINT "1ST AND 3RD COUNT ASSUMING EXPONENTIAL GROWTH DURING"
00210 PRINT "THAT TIME. HERE'S HOW EXPONENTIAL GROWTH WITH R=";R
00220 PRINT "PER DAY COMPARES TO THE ACTUAL NUMBER:"
00225 PRINT
00230 PRINT "DAY","ACTUAL NO.","EXP.,R=";R
00240 FOR I=1 TO J
00250 E(I)=N(1)*EXP(R*T(I))
00260 PRINT T(I),N(I),E(I)
00270 NEXT I
00280 PRINT "NOW ASSUME THAT GROWTH IS LOGISTIC. FOR EACH ROUND YOU"
00290 PRINT "MAY TRY 4 PAIRS OF R AND K VALUES. THE PROGRAM WILL PRINT"
00300 PRINT "THE NUMBER OF DUCKWEEDS FOR ALTERNATE DAYS FOR 4 WEEKS."
00310 PRINT "FIRST PAIR: R=";
00320 INPUT R1
00330 PRINT "                K=";
00340 INPUT K1
00350 PRINT "SECOND PAIR: R=";
00360 INPUT R2
00370 PRINT "                 K=";
00380 INPUT K2
00390 PRINT "THIRD PAIR: R=";
00400 INPUT R3
00410 PRINT "                K=";
00420 INPUT K3
00430 PRINT "LAST PAIR: R=";
00440 INPUT R4
00450 PRINT "               K=";
00460 INPUT K4
00465 PRINT
00470 PRINT, "R=";R1,"R=";R2,"R=";R3,"R=";R4
00480 PRINT "DAY","K=";K1,"K=";K2,"K=";K3,"K=",K4
00490 PRINT
```

```
00500 FOR T=0 TO 28 STEP 2
00510 A1=LOG((K1-N(1))/N(1))
00520 L1=K1/(1+EXP(A1-(R1*T)))
00530 A2=LOG((K2-N(1))/N(1))
00540 L2=K2/(1+EXP(A2-(R2*T)))
00550 A3=LOG((K3-N(1))/N(1))
00560 L3=K3/(1+EXP(A3-(R3*T)))
00570 A4=LOG((K4-N(1))/N(1))
00580 L4=K4/(1+EXP(A4-(R3*T)))
00590 PRINT T,L1,L2,L3,L4
00600 NEXT T
00610 PRINT "WOULD YOU LIKE TO TRY 4 MORE VALUES FOR R AND K?"
00620 PRINT "IF SO, TYPE YES. IF NOT, THE PROGRAM WILL END."
00630 INPUT A$
00640 IF A$="YES" GOTO 310
00650 END
```

CHANGE

The third program is designed for use with Exercise 20, Forest Succession (it can also be readily used for Exercise 12, Use of the Leslie Matrix). The program and some sample output is shown below.

```
05 REM     PROGRAM TITLE: CHANGE
06 REM     SOURCE LANGUAGE: BASIC
07 REM     AUTHOR: M. MCCANN
10 DIM R(3,3),P(3),N1(3),N2(3)
20 MAT READ R,P
30 DATA .5, .1, 0, .1, .1, .75, .4, .8, .25
40 DATA 120, 240, 600
50 MAT N1=R*P
60 PRINT "N1 COMPOSITION"
70 MAT PRINT N1
80 MAT N2=R*N1
90 PRINT "N2 COMPOSITION"
100 MAT PRINT N2
110 END
```

```
N1 COMPOSITION

 84
 486
 390

N2 COMPOSITION

 90.6
 349.5
 519.9
```

It is difficult to make programs involving matrices interactive. What the user must do is this: with the program as shown stored and ready to run, type in new DIM (dimension) and DATA statements (lines 10, 30, and 40) that specify the particular matrices he is dealing with. In the dimension statement, R is the replacement matrix. Here it is a 3×3 matrix. P is the column vector representing current composition; it will always be dimensioned equal to one side of the square matrix. The last two entries specify the sizes of the results of the multiplications. The result of the first multiplication of the replacement matrix by the column vector (P) will be N1. The results of another multiplication (the square matrix by N1) will be N2. The results in each case will be a column vector with the same number of species as the original composition (P). If one wished to deal with a situation involving four species, he would, then, type in a new dimension line as follows:

```
10 DIM R(4,4), P(4), N1(4), N2(4)
```

For the data statement, type in the square matrix from left to right across each row and then type in the column vector. In the sample program the square matrix

$$\begin{bmatrix} .5 & .1 & 0 \\ .1 & .1 & .75 \\ .4 & .8 & .25 \end{bmatrix}$$

is given in line 30 and the column vector is given in line 40. It is, however, unnecessary to separate the two; the computer will take the first nine values as the square matrix and the next three as the column vector. If we alter the dimensions as above to a 4×4 matrix, the computer will interpret the first 16 figures as the square matrix and the next 4 as the column vector. If you have a square matrix too large to fit on one line, you can break it up. The sample program above could have data statements

```
30 DATA .5,.1,0,.1
31 DATA .1,.75,.4,.8
32 DATA .25
```

and still give the same values.

After you have typed in new DIM and DATA statements, the program will be ready to run and provide output as shown above but with the new compositions derived from the multiplication of the newly entered square matrix by the newly entered column vector. Below is sample output showing the operations just described.

```
OLD FILE NAME- -CHANGE

READY
10 DIM R(4,4),P(4),N1(4),N2(4)
30 DATA.5,.1,0,.2,.1,.1,.7,.3,.3,.7,.2,.4,.1,.1,.1,.1
40 DATA 80,480,380,20
RUN

CHANGE

N1 COMPOSITION

  92
  328
  444
  96
```

N2 COMPOSITION

98
381.6
384.4
94

To obtain compositions for time 3 and 4 (N3 and N4) without having to alter the program much, simply enter a new line 40 using as data (for P) the figures generated as N2 composition.

BIBLIOGRAPHY

Bledsoe, C. J., and D. A. Jameson. 1969. Model structure for a grassland ecosystem. In R. L. Dix and R. G. Beidleman (eds.). *The Grassland Ecosystem: A Preliminary Synthesis*. Colorado State Univ. Range Sci. Dept. Sci. Ser. 2.

Gottfried, B. S. 1975. *Theory and Problems of Programming with BASIC*. Schaum's Outline Series, McGraw-Hill, New York.

Hall, C. A. S., and J. W. Day, Jr. eds. 1977. *Ecosystem Modeling in Theory and Practice: An Introduction with Case Histories*. Wiley, New York.

Jeffers, J. N. R. 1978. *An Introduction to Systems Analysis: With Ecological Applications*. University Park Press, Baltimore.

Levins, R. 1966. The strategy of model building in population biology. *Amer. Scientist* 54: 421–431.

Meadows, D. H., D. L. Meadows, J. Randers, and W. W. Behrens III. 1972. *The Limits to Growth*. Universe Books, New York.

Orr, H., J. C. Marshall, T. L. Isenhour, and P. C. Jurs. 1973. *Introduction to Computer Programming for Biological Scientists*. Allyn and Bacon, Boston.

Patten, B. C., ed. 1971, 1972, 1975, and 1976. *System Analysis and Simulation in Ecology*. *Vols. 1-4*. Academic Press, New York.

Randall, J. E. 1980. *Microcomputers and Physiological Simulation*. Addison-Wesley, Reading, Massachusetts.

Searle, S. R. 1966. *Matrix Algebra for the Biological Sciences (Including Applications in Statistics)*. Wiley, New York.

Spain, J. D. 1977. *Basic Computer Models in Biology*. Privately published. (Available from Bookstore, Michigan Technological University, Houghton, Michigan 49931.)

Spencer, D. D. 1970. *A Guide to BASIC Programming*. Addison-Wesley, Reading, Massachusetts.

Walters, C. J. 1971. Systems ecology: the systems approach and mathematical models in ecology. Pp. 276–292 in Odum, E. P. *Fundamentals of Ecology*. 3rd ed. Saunders, Philadelphia.

Wiegert, R. G. 1975. Simulation models of ecosystems. *Annual Rev. Ecol. Syst.* 6: 311–338.

Appendix 6

EXERCISES PARTICULARLY SUITABLE FOR SPECIAL SITUATIONS

Courses with Minimal Mathematical Content (16)

Ex. 1. Microclimate.
Ex. 2. Soils.
Ex. 3. Ecological Modifications of Leaves.
Ex. 5. Sampling and Density Estimation. This will be an important exercise for most ecology labs. It can be brought within the range of mathematically unsophisticated students by omitting calculations of confidence intervals and sticking to graphical treatments where choices exist.
Ex. 7. Spatial Relations in Humans.
Ex. 8. Territoriality.
Ex. 9. Predation.
Ex. 11. Population Growth. Many important features of population growth can be developed by having classes simply follow the cultures and plot population size against time, with no mathematical treatment.
Ex. 14. Biomass in a Grassland. This requires a lot of arithmetic but no mathematical thought. Leading the whole class through the calculations or just doing them and handing out the final figures are two ways of avoiding even the arithmetic.
Ex. 16. Forest Composition. As in Exercise 14, the arithmetic is substantial in quantity but simple.
Ex. 17. Island Biogeography. By using only graphical methods, many of the computations can be avoided.
Ex. 18. Ecological Use of Remote Sensing.
Ex. 19. Succession on Abandoned Fields.
Ex. 21. Lake Sampling. Omit question 21.
Ex. 22. Comparison of Polluted and Unpolluted Areas of a Stream. By using only the sequential comparison index to diversity, most computations are avoided.
Ex. 24. Noise.

Plant Ecology Courses (14)

Ex. 1. Microclimate.
Ex. 2. Soils.
Ex. 3. Ecological Modifications of Leaves.
Ex. 5. Sampling and Density Estimation.
Ex. 6. Spatial Relations in Plants.
Ex. 11. Population Growth.
Ex. 14. Biomass in a Grassland.
Ex. 15. Productivity.

Ex. 16. Forest Composition.

Ex. 17. Island Biogeography. The same analytical procedures can be followed but the artificial substrates can be microscope slides on which diatoms are studied.

Ex. 18. Ecological Use of Remote Sensing.

Ex. 19. Succession on Abandoned Fields.

Ex. 20. Forest Succession.

Ex. 23. The Trophic Ecology of Humans. A great deal of this exercise has to do with production and yields from agricultural ecosystems.

Animal Ecology Courses (18)

Ex. 1. Microclimate.

Ex. 2. Soils.

Ex. 4. Time and Energy Budgets.

Ex. 5. Sampling and Density Estimation.

Ex. 6. Spatial Relations in Plants. The same techniques can be applied to barnacles, ant lion pits, singing male birds, people in the student center, etc.

Ex. 7. Spatial Relations in Humans.

Ex. 8. Territoriality.

Ex. 9. Predation.

Ex. 10. Life Tables.

Ex. 11. Population Growth. Duckweeds work so much better than most other organisms that switching to *Paramecium* or grain beetles just because they are animals is hard to justify.

Ex. 12. Use of the Leslie Matrix.

Ex. 13. Population Problems.

Ex. 14. Biomass in a Grassland.

Ex. 17. Island Biogeography.

Ex. 18. Ecological Use of Remote Sensing. Nothing directly to do with animals but a good tool.

Ex. 21. Lake Sampling.

Ex. 22. Comparison of Polluted and Unpolluted Areas of a Stream.

Ex. 23. The Trophic Ecology of Humans.

Courses Emphasizing Human and Human-Influenced Ecology (13)

Ex. 1. Microclimate. The discussion can emphasize human alterations of climate, appropriate technology in housing, etc.

Ex. 2. Soils. Emphasize soils in agriculture, gardening, and land use planning.

Ex. 5. Sampling and Density Estimation. Much of this material is applicable to wildlife management.

Ex. 6. Spatial Relations in Plants. Use the techniques to study the distribution of humans or human products or artifacts.

Ex. 7. Spatial Relations in Humans. The statistics of Ex. 6 could be combined with the behavioral observations here. For example, the variance:mean ratio could be calculated for numbers of persons at library tables and the behavioral observations used to interpret the results.

Ex. 10. Life Tables.

Ex. 12. Use of the Leslie Matrix. If the class gets into demography at a reasonably sophisticated level.

Ex. 13. Population Problems.

Ex. 18. Ecological Use of Remote Sensing.

Ex. 19. Succession on Abandoned Fields.

Ex. 22. Comparison of Polluted and Unpolluted Areas of a Stream.
Ex. 23. The Trophic Ecology of Humans.
Ex. 24. Noise.

Courses Where There Is Little Access to Natural Habitats (19)

Ex. 1. Microclimate. The forest reactions section may have to be omitted.
Ex. 3. Ecological Modifications of Leaves.
Ex. 4. Time and Energy Budgets. Ducks on a pond in a park can be used.
Ex. 5. Sampling and Density Estimation. Can be done entirely in the laboratory if necessary.
Ex. 6. Spatial Relations in Plants. Use dandelions in the lawn or sowbugs in the courtyard.
Ex. 7. Spatial Relations in Humans.
Ex. 9. Predation. If the pellets can be obtained, the exercise itself is done in the lab.
Ex. 10. Life Tables. If it is not feasible to visit a cemetery, many genealogical societies publish books with transcriptions of gravestones.
Ex. 11. Population Growth.
Ex. 12. Use of the Leslie Matrix.
Ex. 13. Population Problems.
Ex. 14. Biomass in a Grassland. If even an overgrown vacant lot is available, most of this can be done.
Ex. 15. Productivity. Construction sites are ideal for the single harvest method.
Ex. 16. Forest Composition. The essentials of the point-quarter method can be gained using the artificial plant population (in Ex. 5), with the different symbols considered as different tree species.
Ex. 18. Ecological Use of Remote Sensing.
Ex. 21. Lake Sampling. Virtually any fair-sized pond is suitable.
Ex. 22. Comparison of Polluted and Unpolluted Areas of a Stream. Finding an unpolluted station may be a problem but comparisons can be made with the literature.
Ex. 23. The Trophic Ecology of Humans.
Ex. 24. Noise.

If there is an opportunity for an all-day or weekend trip to visit areas away from the city, a course taught in an urban area can cover most of the same topics, though with different emphasis, as courses with better access to the countryside.

Courses Taught During the Winter (20)

Ex. 1. Microclimate. A few modifications will be necessary; for example, you may need to use conifer forests or plantations. Avoid periods when the whole day is below freezing.
Ex. 2. Soils. Any time the soil is not frozen.
Ex. 3. Ecological Modifications of Leaves.
Ex. 4. Time and Energy Budgets.
Ex. 5. Sampling and Density Estimation.
Ex. 6. Spatial Relations in Plants. A fair number of herbs are winter- or evergreen; or photographs, such as Figure 3–14 in *Principles of Ecology* by Brewer, can be used.
Ex. 7. Spatial Relations in Humans.
Ex. 9. Predation.
Ex. 10. Life Tables. Use compilations of gravestone inscriptions published by genealogical societies, if necessary.
Ex. 11. Population Growth. You may wish to collect the duckweed in the fall.
Ex. 12. Use of the Leslie Matrix.

Ex. 13. Population Problems.
Ex. 16. Forest Composition. You may have to omit the smaller size classes.
Ex. 17. Island Biogeography.
Ex. 18. Ecological Use of Remote Sensing.
Ex. 20. Forest Succession.
Ex. 21. Lake Sampling.
Ex. 22. Comparison of Polluted and Unpolluted Areas of a Stream.
Ex. 23. The Trophic Ecology of Humans.
Ex. 24. Noise.

The aquatic exercises (17, 21, 22) are especially good because many students are not aware of the activity of lake and stream organisms in cold weather. A special winter field trip or two focusing on tracks and other animal signs, the wintergreen plants, and other winter ecological observations is worthwhile.

Courses Emphasizing Field Ecology (16)

Ex. 1. Microclimate.
Ex. 2. Soils.
Ex. 4. Time and Energy Budgets.
Ex. 5. Sampling and Density Estimation. Apply the methods to natural populations.
Ex. 6. Spatial Relations in Plants.
Ex. 8. Territoriality.
Ex. 9. Predation. Alter to include collection of pellets by class and live trapping of mammals to assess prey densities.
Ex. 14. Biomass in a Grassland.
Ex. 15. Productivity.
Ex. 16. Forest Composition.
Ex. 17. Island Biogeography.
Ex. 18. Ecological Use of Remote Sensing. Alter to emphasize aerial photographs of sites visited on field trips. Include ground-truthing.
Ex. 19. Succession on Abandoned Fields.
Ex. 20. Forest Succession.
Ex. 21. Lake Sampling.
Ex. 22. Comparison of Polluted and Unpolluted Areas of a Stream.

Large Classes (Divided into Lab Sections of 20–35) (19)

Nearly all the exercises can be done with multiple sections as long as sufficient material (such as owl pellets) or equipment (such as Hester-Dendy samplers) can be provided. Large numbers of students should not be taken to the more fragile natural areas and, of course, care should be exercised on field trips to any natural ecosystems. The following are readily done with classes of 40 to 120 students.

Ex. 1. Microclimate.
Ex. 2. Soils.
Ex. 3. Ecological Modifications of Leaves.
Ex. 4. Time and Energy Budgets. Have students work in pairs and observe for only a half-hour; or have half the class observe one species or sex and the other half another.
Ex. 5. Sampling and Density Estimation.
Ex. 6. Spatial Relations in Plants.
Ex. 8. Territoriality. Combine data from sections to assess effects of time of day on level of aggressive behavior.

Ex. 10. Life Tables.

Ex. 11. Population Growth. Obtain many replicates or, better, set up several experimental treatments with one section using a larger surface area, another replenishing water, etc.

Ex. 12. Use of the Leslie Matrix.

Ex. 13. Population Problems.

Ex. 14. Biomass in a Grassland. If done on an abandoned hayfield or an early stage of old-field succession, the impact of a large class may not be important. Different sections could be responsible for different aspects of the sampling.

Ex. 15. Productivity. If done on a cropfield, a hayfield, a construction site, or some similar area, the impact of a large class may not be important.

Ex. 16. Forest Composition. Avoid significant natural areas; otherwise, the increased sample size resulting from multiple sections is all to the good.

Ex. 18. Ecological Use of Remote Sensing.

Ex. 19. Succession on Abandoned Fields. Different sections can sample fields abandoned at different times and pool the data.

Ex. 20. Forest Succession. Avoid significant natural areas.

Ex. 21. Lake Samplings. Multiple sections allow such modifications as comparing two different lakes or different zones in the same lake.

Ex. 23. The Trophic Ecology of Humans.

Index

Note: Methods are listed in a separate index on the back cover.